ANATOMY & PHYSIOLOGY COLORING WORKBOOK

A Complete Study Guide

ELEVENTH EDITION

Elaine N. Marieb, R.N., Ph.D.

Holyoke Community College

PEARSON

Boston Columbus Indianapolis New York San Francisco Upper Saddle River
Amsterdam Cape Town Dubai London Madrid Milan Munich Paris Montréal Toronto
Delhi Mexico City São Paulo Sydney Hong Kong Seoul Singapore Taipei Tokyo

Editor-in-Chief: Serina Beauparlant

Senior Acquisitions Editor: Brooke Suchomel

Director of Development: Barbara Yien

Assistant Editor: Ashley Williams

Senior Managing Editor: Michael Early

Project Manager: Michael Penne

Compositor: Cenveo® Publisher Services

Senior Manufacturing Buyer: Stacey Weinberger

Senior Marketing Manager: Allison Rona

Design Management: Side by Side Studios

Cover Designer: Tandem Creative, Inc.

Interior & Cover Printer: Edwards Brothers Malloy

www.pearsonhighered.com

ISBN 10: 0-321-96077-7 (Student edition)
ISBN 13: 978-0-321-96077-1 (Student edition)
2 3 4 5 6 7 8 9 10—EBM—17 16 15 14

PREFACE

Although never a simple task, the study of the human body is always fascinating. Over the years, thousands of students have benefited in their studies and enjoyed the process of working through this book. Whether you are taking a 1- or 2-semester course, you will find this book invaluable to the study of anatomy and physiology.

What's New to This Edition?

The eleventh edition of the *Anatomy & Physiology Coloring Workbook* continues to serve as a review and reinforcement tool to help health professional and life-science students master the basic concepts of human anatomy and physiology. We have helped students by making the following revisions:

- **New *Finale: Multiple Choice* questions** have been added throughout.
- **New *At the Clinic* application questions** appear throughout the book.
- **Updated terminology** has been added throughout the book.
- **New figure illustrating the major tissue types** has been added.

Scope

Although this book reviews the human body from microscopic to macroscopic levels (that is, topics range from simple chemistry and cells to body organ systems), it is not intended to be encyclopedic. In fact, to facilitate learning, this workbook covers only the most important and useful aspects of human anatomy and physiology. Pathophysiology is briefly introduced with each system so that students can apply their learning. Where relevant, clinical aspects (for example, muscles used for injection sites, the role of ciliated cells in protection of the respiratory tract, and reasons for skin ulcer formation) are covered. To encourage a view of the human body as a dynamic and continually changing organism, developmental aspects of youth, adulthood, and old age are included.

Learning Aids

As in previous editions, multiple pedagogical devices are used throughout the book to test comprehension of key concepts. The integration of a traditional study guide approach with visualization and coloring exercises is unique. The variety of exercises demands learning on several levels, avoids rote memorization, and helps maintain a high level of interest.

The exercises include completion from a selection of key choices, matching terms or descriptions, and labeling diagrams. Elimination questions require the student to discover the similarities or dissimilarities among a number of structures or objects and to select the one that is not appropriate. Correctable true/false questions add a new dimension to the more traditional form of this exercise. Also, students are asked to provide important definitions. In the completion sections, the answer lines are long enough so that the student can write in either the key letter or the appropriate term. Both responses are provided in the answer section.

Coloring exercises are a proven motivating, effective approach to learning. Each illustration has been carefully prepared to show sufficient detail for learning without students becoming bored with coloring. There are more than 120 coloring exercises distributed throughout the text that should prove valuable to all students. Students who are visually oriented will find these exercises particularly beneficial. When completed, the color diagrams provide an ideal reference and review tool.

Visualization exercises are a truly unique feature of this book. With the exception of the introductory chapter on terminology, each chapter contains an "Incredible Journey." Students are asked to imagine themselves in miniature, traveling within the body through various organs and systems. These visualization exercises are optional, but they often summarize chapter content, allowing students to assimilate what they have learned in unusual and amusing ways.

Thought-provoking "At the Clinic" questions challenge students to apply their newly acquired knowledge to clinical situations. Additionally, the eleventh edition features a finale to each chapter with challenging multiple-choice questions.

Acknowledgments

To those educators, colleagues, and students who have provided feedback and suggestions during the preparation of all eleven editions of this workbook, I am sincerely grateful. In particular, I want to thank the following reviewers for their valuable comments and suggestions: LuAnne Clark, Lansing Community College; Catherine Elliott; Judy Garrett, University of Arkansas Community College; Judy Megaw, Indian River State College; Hal Nauman; Lyn Rivers, Henry Ford Community College; Tinna Ross, North Hennepin Community College; and Mary Weis, Collin College–Spring Creek Campus.

The staff at Pearson Education has continuously supported my efforts to turn out a study tool that will be well received and beneficial to both educator and student audiences. For this edition, Brooke Suchomel, Senior Acquisitions Editor, Ashley Williams, Assistant Editor, and Michael Penne, Project Manager, deserve special mention.

INSTRUCTIONS FOR THE STUDENT—
HOW TO USE THIS BOOK

Dear Student,

The *Anatomy & Physiology Coloring Workbook* has been created particularly for you. It is the outcome of years of personal attempts to find and create exercises helpful to my own students when they study and review for a lecture test or laboratory quiz.

I never cease to be amazed at how remarkable the human body is, but I would never try to convince you that studying it is easy. The study of human anatomy and physiology has its own special terminology. It requires that you become familiar with the basic concepts of chemistry to understand physiology, and often (sadly) it requires rote memorization of facts. It is my hope that this workbook will help simplify your task. To make the most of the exercises, read these instructions carefully before starting work.

Labeling and Coloring. Some of these questions ask you only to label a diagram, but most also ask that you do some coloring of the figure. You can usually choose whichever colors you prefer. Soft colored pencils are recommended so that the underlying diagram shows through. Most figures have several parts to color, so you will need a variety of colors—18 should be sufficient. In the coloring exercises, you are asked to choose a particular color for each structure to be colored. That color is then used to fill in both a color-coding circle found next to the name of the structure or organ, and the structure or organ on the figure. This allows you to identify the colored structure quickly and by name in cases where the diagram is not labeled. In a few cases you are given specific coloring instructions to follow.

Matching. Here you are asked to match a key term denoting a structure or physiological process with a descriptive phrase or sentence. Because you must write the chosen term in the appropriate answer blank, the learning is more enduring.

Completion. You select the correct term to answer a specific question, or you fill in blanks to complete a sentence. In many exercises, some terms are used more than once and others are not used at all.

Definitions. You are asked to provide a brief definition of a particular structure or process.

True or False. One word or phrase is underlined in a sentence. You decide if the sentence is true as it is written. If not, you correct the underlined word or phrase.

Elimination. Here you are asked to find the term that does not "belong" in a particular grouping of related terms. In this type of exercise, you must analyze how the various terms are similar to or different from the others.

Visualization. The "Incredible Journey" is a special type of completion exercise, found in every chapter except the first one. For this exercise, you are asked to imagine that you have been miniaturized and injected into the body of a human being (your host). Anatomical landmarks and physiological events are described from your miniaturized viewpoint, and you are then asked to identify your observations. Although this exercise is optional, my students have found them fun to complete and I hope you will too.

At the Clinic. "At the Clinic" sections ask you to apply your newly acquired knowledge to clinical situations.

The Finale: Multiple Choice. The multiple-choice questions test you from several vantage points and 1, 2, 3, or all of the answers may be correct—an approach that really tests your understanding of what you have studied.

Each exercise has complete instructions, which you should read carefully before beginning the exercise. When there are multiple instructions, complete them in the order given.

At times it may appear that information is duplicated in the different types of exercises. Although there is some overlap, the understandings being tested are different in the different exercises. Remember, when you understand a concept from several different perspectives, you have mastered that concept.

I sincerely hope that the *Anatomy & Physiology Coloring Workbook* challenges you to increase your knowledge, comprehension, retention, and appreciation of the structure and function of the human body.

Good luck!

Elaine Marieb

Elaine Marieb
Pearson Education
1301 Sansome Street
San Francisco, CA 94111

CONTENTS

1 THE HUMAN BODY: AN ORIENTATION

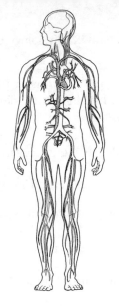

Most of us have a natural curiosity about our bodies, and a study of anatomy and physiology elaborates on this interest. Anatomists have developed a universally acceptable set of reference terms that allows body structures to be located and identified with a high degree of clarity. Initially, students might have difficulties with the language used to describe anatomy and physiology, but without such a special vocabulary, confusion is bound to occur.

The topics in this chapter enable students to test their mastery of terminology commonly used to describe the body and its various parts, and concepts concerning functions vital for life and homeostasis. Body organization from simple to complex levels and an introduction to the organ systems forming the body as a whole are also covered.

AN OVERVIEW OF ANATOMY AND PHYSIOLOGY

1. Match the terms in Column B to the appropriate descriptions provided in Column A. Enter the correct letter or its corresponding term in the answer blanks.

Column A	Column B
_____ 1. The branch of biological science that studies and describes how body parts work or function	A. Anatomy
_____ 2. The study of the shape and structure of body parts	B. Homeostasis
_____ 3. The tendency of the body's systems to maintain a relatively constant or balanced internal environment	C. Metabolism
_____ 4. The term that indicates *all* chemical reactions occurring in the body	D. Physiology

2. Circle all the terms or phrases that correctly relate to the study of *physiology*. Use a highlighter to identify those terms or phrases that pertain to the study of *anatomy*.

A. Measuring an organ's size, shape, and weight.   H. Dynamic

B. Can be studied in dead specimens   I. Dissection

C. Often studied in living subjects   J. Experimentation

D. Chemistry principles   K. Observation

E. Measuring the acid content of the stomach   L. Directional terms

F. Principles of physics   M. Static

G. Observing a heart in action

LEVELS OF STRUCTURAL ORGANIZATION

3. The structures of the body are organized into successively larger and more complex structures. Fill in the answer blanks with the correct terms for these increasingly larger structures.

Chemicals ⟶ _____ ⟶ _____ ⟶

_____ ⟶ _____ ⟶ Organism

4. Circle the term that does not belong in each of the following groupings.

1. Electron   Cell   Tissue   Alive   Organ

2. Brain   Stomach   Heart   Liver   Epithelium

3. Epithelium   Heart   Muscle tissue   Nervous tissue   Connective tissue

4. Human   Digestive system   Horse   Pine tree   Amoeba

5. Using the key choices, identify the organ systems to which the following organs or functions belong. Insert the correct letter or term in the answer blanks.

Key Choices

A. Cardiovascular   D. Integumentary   G. Nervous   J. Skeletal

B. Digestive   E. Lymphatic/Immune   H. Reproductive   K. Urinary

C. Endocrine   F. Muscular   I. Respiratory

_____ 1. Rids the body of nitrogen-containing wastes

_____ 2. Is affected by the removal of the thyroid gland

_____ 3. Provides support and levers on which the muscular system can act

_____ 4. Includes the heart

_____ 5. Protects underlying organs from drying out and mechanical damage

_____ 6. Protects the body; destroys bacteria and tumor cells

_____ 7. Breaks down foodstuffs into small particles that can be absorbed

_____ 8. Removes carbon dioxide from the blood

_____ 9. Delivers oxygen and nutrients to the body tissues

_____ 10. Moves the limbs; allows facial expression

_____ 11. Conserves body water or eliminates excesses

_____ 12. Provides for conception and childbearing

_____ 13. Controls the body with chemicals called hormones

_____ 14. Is damaged when you cut your finger or get a severe sunburn

6. Using the key choices from Exercise 5, choose the organ system to which each of the following sets of organs belongs. Enter the correct letter or term in the answer blanks.

_____ 1. Blood vessels, heart

_____ 2. Pancreas, pituitary, adrenal glands

_____ 3. Kidneys, bladder, ureters

_____ 4. Testis, vas deferens, urethra

_____ 5. Esophagus, large intestine, rectum

_____ 6. Breast bone, vertebral column, skull

_____ 7. Brain, nerves, sensory receptors

7. Figures 1–1 to 1–6, on pages 4–6, represent the various body organ systems. First identify and name each organ system by labeling the organ system under each illustration. Then select a different color for each organ and use it to color the coding circles and corresponding structures in the illustrations.

◯ Blood vessels   ◯ Nasal cavity

◯ Heart   ◯ Lungs

◯ Trachea

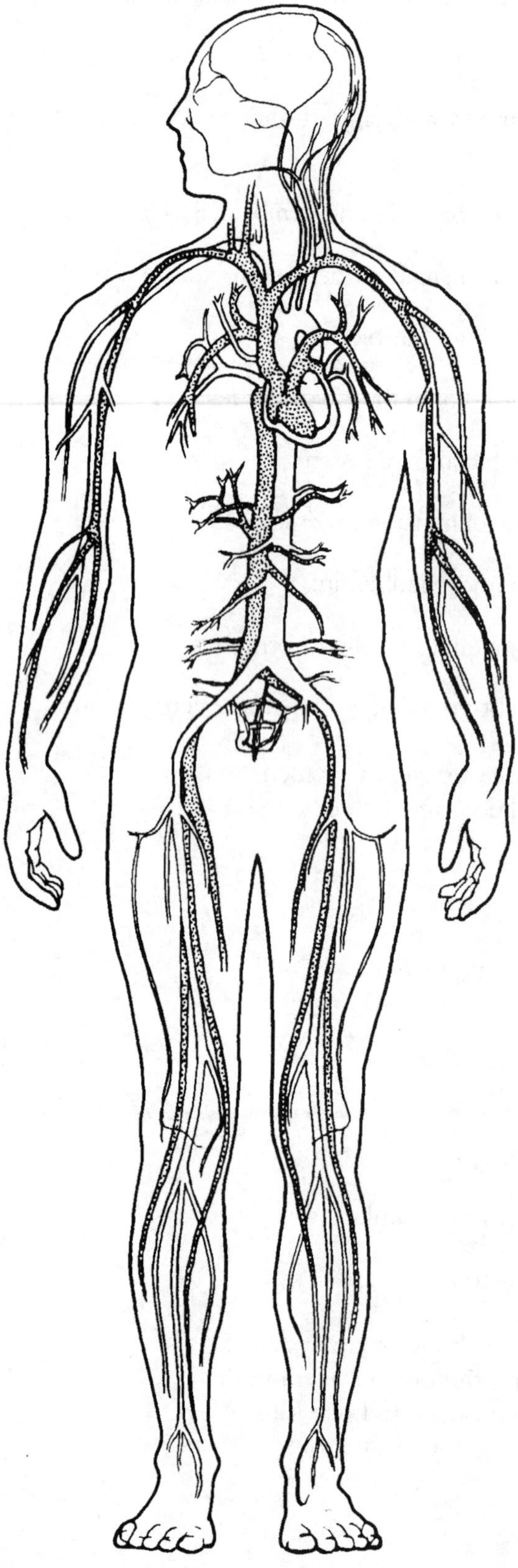

Figure 1–1 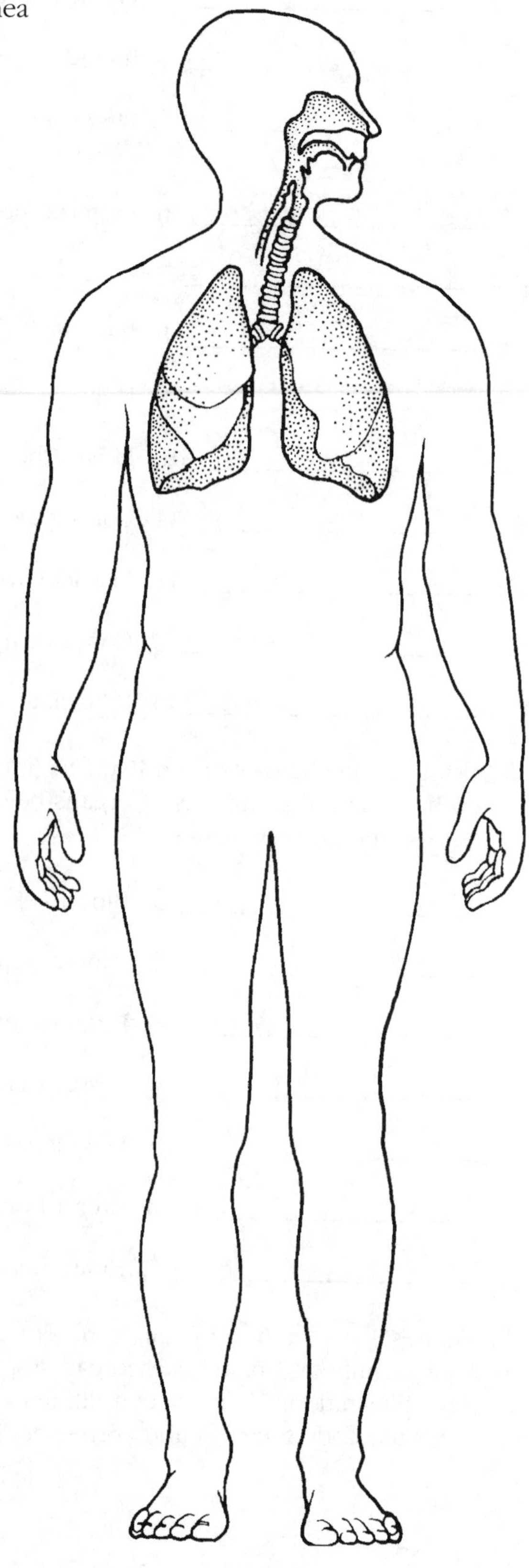   **Figure 1–2**

◯ Brain

◯ Spinal cord

◯ Nerves

◯ Kidneys

◯ Ureters

◯ Bladder

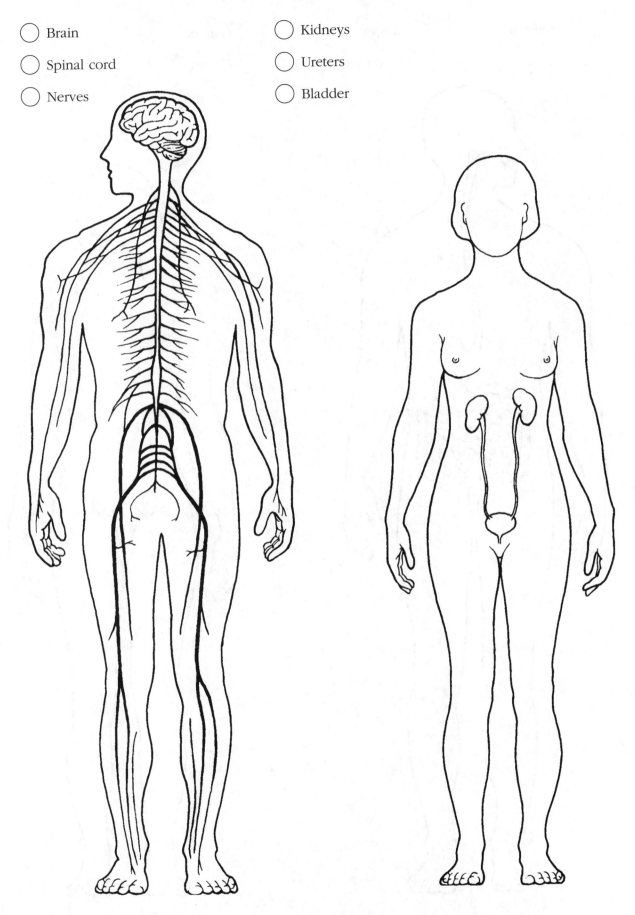

Figure 1–3   **Figure 1–4**

○ Stomach ○ Esophagus ○ Ovaries

○ Intestines ○ Oral cavity ○ Uterus

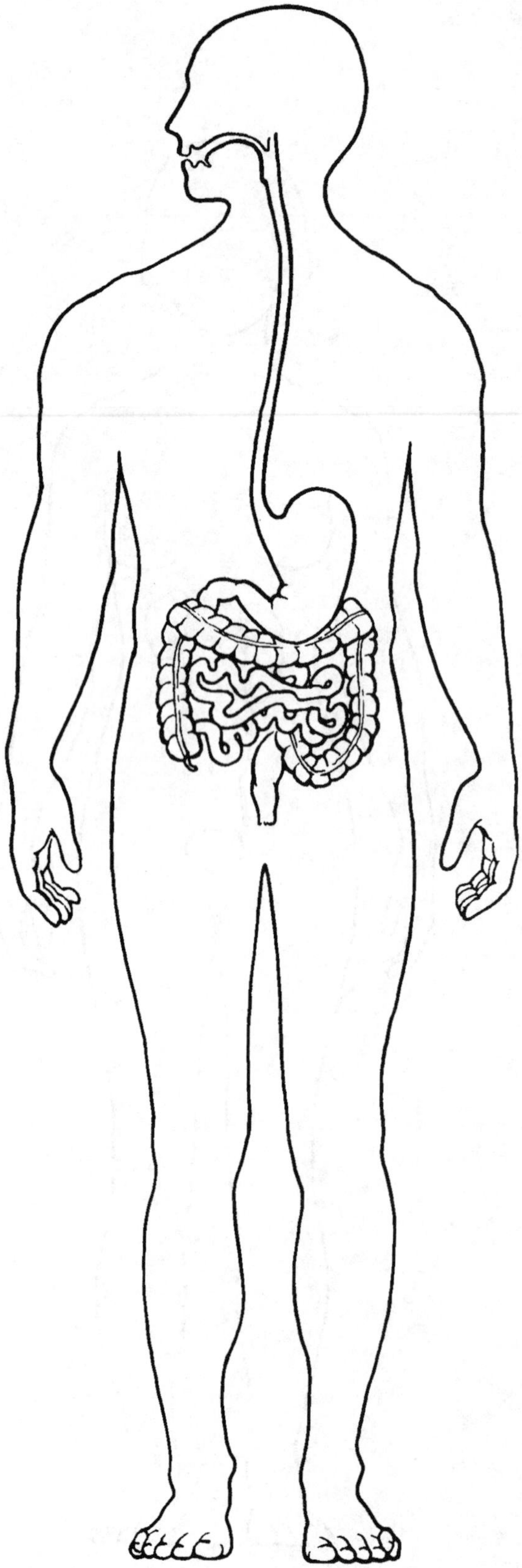

Figure 1–5

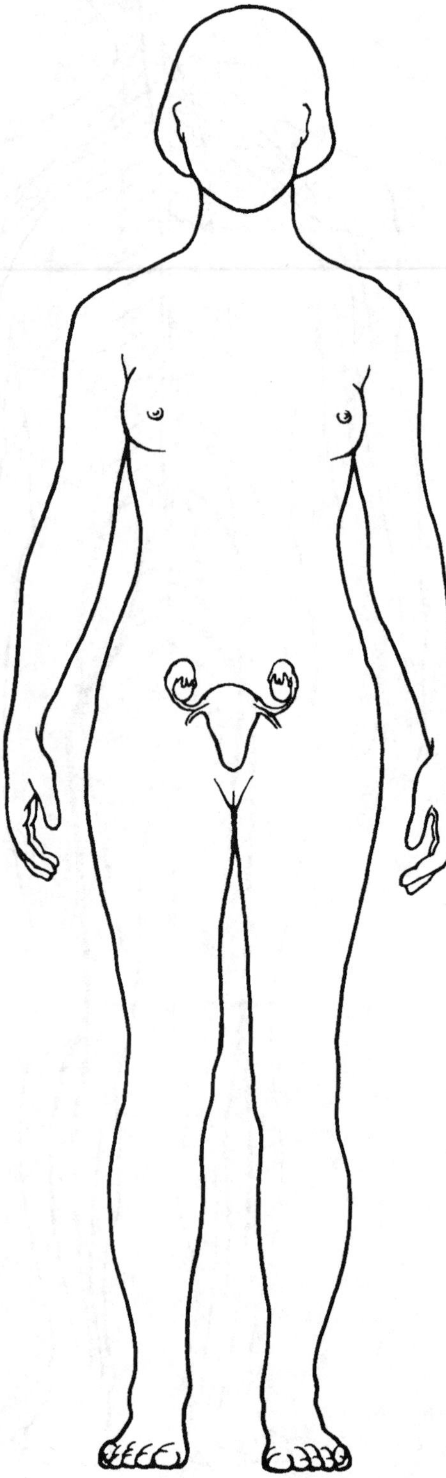

Figure 1–6

MAINTAINING LIFE

8. Match the terms pertaining to functional characteristics of organisms in Column B with the appropriate descriptions in Column A. Fill in the answer blanks with the appropriate letter or term.

Column A	**Column B**
_____ 1. Keeps the body's internal environment distinct from the external environment	A. Digestion
_____ 2. Provides new cells for growth and repair	B. Excretion
_____ 3. Occurs when constructive activities occur at a faster rate than destructive activities	C. Growth
_____ 4. The tuna sandwich you have just eaten is broken down to its chemical building blocks	D. Maintenance of boundaries
_____ 5. Elimination of carbon dioxide by the lungs and elimination of nitrogenous wastes by the kidneys	E. Metabolism
_____ 6. Ability to react to stimuli; a major role of the nervous system	F. Movement
_____ 7. Walking, throwing a ball, riding a bicycle	G. Responsiveness
_____ 8. All chemical reactions occurring in the body	H. Reproduction
_____ 9. At the cellular level, membranes; for the whole organism, the skin	

9. Using the key choices, correctly identify the survival needs that correspond to the following descriptions. Insert the correct letter or term in the answer blanks.

Key Choices

A. Appropriate body temperature   C. Nutrients   E. Water

B. Atmospheric pressure   D. Oxygen

_____ 1. Includes carbohydrates, proteins, fats, and minerals

_____ 2. Essential for normal operation of the respiratory system and breathing

_____ 3. Single substance accounting for more than 60% of body weight

_____ 4. Required for the release of energy from foodstuffs

_____ 5. Provides the basis for body fluids of all types

_____ 6. When too high or too low, physiological activities cease, primarily because molecules are destroyed or become nonfunctional

HOMEOSTASIS

10. The following statements refer to homeostatic control systems. Complete each statement by inserting your answers in the answer blanks.

_____ 1.

_____ 2.

_____ 3.

_____ 4.

_____ 5.

_____ 6.

_____ 7.

_____ 8.

_____ 9.

There are three essential components of all homeostatic control mechanisms: control center, receptor, and effector. The __(1)__ senses changes in the environment and responds by sending information (input) to the __(2)__ along the __(3)__ pathway. The __(4)__ analyzes the input, determines the appropriate response, and activates the __(5)__ by sending information along the __(6)__ pathway. When the response causes the initial stimulus to decline, the homeostatic mechanism is referred to as a __(7)__ feedback mechanism. When the response enhances the initial stimulus, the mechanism is called a __(8)__ feedback mechanism. __(9)__ feedback mechanisms are much more common in the body.

THE LANGUAGE OF ANATOMY

11. Complete the following statements by filling in the answer blanks with the correct term.

_____ 1.

_____ 2.

_____ 3.

The abdominopelvic and thoracic cavities are subdivisions of the __(1)__ body cavity; the cranial and spinal cavities are parts of the __(2)__ body cavity. The __(3)__ body cavity is totally surrounded by bone and provides very good protection to the structures it contains.

12. Circle the term or phrase that does not belong in each of the following groupings.

1. Transverse   Distal   Frontal   Sagittal

2. Lumbar   Thoracic   Antecubital   Abdominal

3. Calf   Brachial   Femoral   Popliteal

4. Epigastric   Hypogastric   Right iliac   Left upper quadrant

5. Orbital cavity   Nasal cavity   Ventral cavity   Oral cavity

13. Select different colors for the *dorsal* and *ventral* body cavities. Color the coding circles below and the corresponding cavities in part A of Figure 1–7. Complete the figure by labeling those body cavity subdivisions that have a leader line. Complete part B by labeling each of the abdominal regions indicated by a leader line.

◯ Dorsal body cavity   ◯ Ventral body cavity

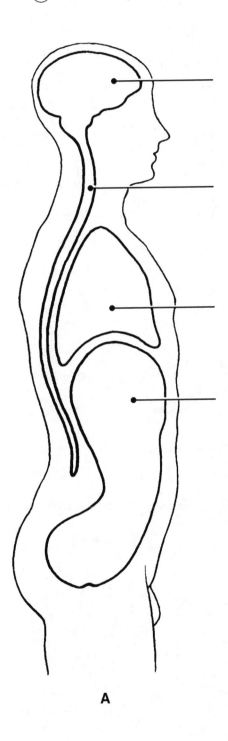

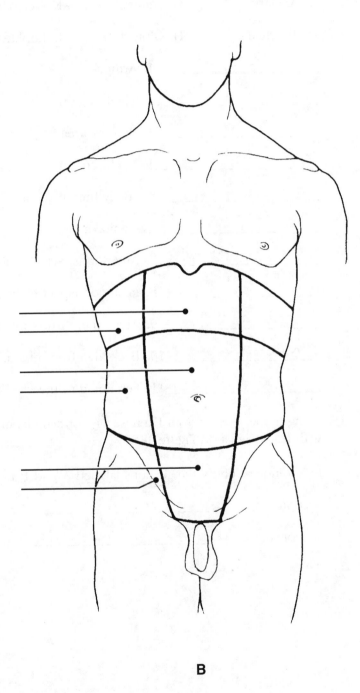

A   B

Figure 1–7

14. Select the key choices that identify the following body parts or areas. Enter the appropriate letter or corresponding term in the answer blanks.

Key Choices

A. Abdominal	E. Buccal	I. Inguinal	M. Pubic
B. Antecubital	F. Cervical	J. Lumbar	N. Scapular
C. Axillary	G. Femoral	K. Occipital	O. Sural
D. Brachial	H. Gluteal	L. Popliteal	P. Umbilical

_____ 1. Armpit

_____ 2. Thigh region

_____ 3. Buttock area

_____ 4. Neck region

_____ 5. "Belly button" area

_____ 6. Genital area

_____ 7. Anterior aspect of elbow

_____ 8. Posterior aspect of head

_____ 9. Area where trunk meets thigh

_____10. Back area from ribs to hips

_____11. Pertaining to the cheek

15. Using the key terms from Exercise 14, correctly label all body areas indicated with leader lines on Figure 1–8.

In addition, identify the sections labeled A and B in the figure.

Section A: _____

Section B: _____

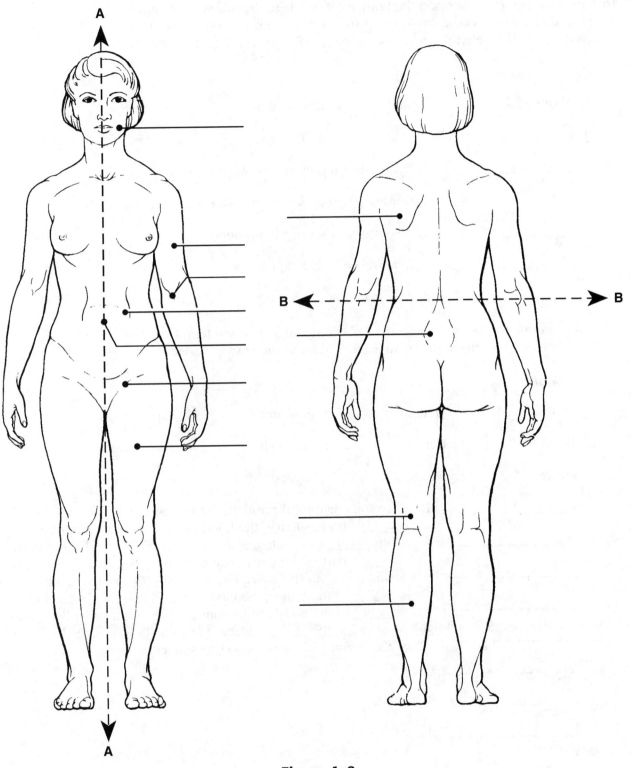

Figure 1–8

16. From the key choices, select the body cavities where the following surgical procedures would occur. Insert the correct letter or term in the answer blanks. Be precise. Also select the name of the cavity subdivision if appropriate.

Key Choices

A. Abdominal   C. Dorsal   E. Spinal   G. Ventral

B. Cranial   D. Pelvic   F. Thoracic

_____ 1. Removal of the uterus, or womb

_____ 2. Coronary bypass surgery (heart surgery)

_____ 3. Removal of a serious brain tumor

_____ 4. Removal of a "hot" appendix

_____ 5. A stomach ulcer operation

17. Complete the following statements by choosing an anatomical term from the key choices. Enter the appropriate letter or term in the answer blanks.

Key Choices

A. Anterior   D. Inferior   G. Posterior   J. Superior

B. Distal   E. Lateral   H. Proximal   K. Transverse

C. Frontal   F. Medial   I. Sagittal

_____ 1.

_____ 2.

_____ 3.

_____ 4.

_____ 5.

_____ 6.

_____ 7.

_____ 8.

_____ 9.

_____ 10.

_____ 11.

In the anatomical position, the face and palms are on the __(1)__ body surface, the buttocks and shoulder blades are on the __(2)__ body surface, and the top of the head is the most __(3)__ part of the body. The ears are __(4)__ to the shoulders and __(5)__ to the nose. The heart is __(6)__ to the spine and __(7)__ to the lungs. The elbow is __(8)__ to the fingers but __(9)__ to the shoulder. In humans, the dorsal surface can also be called the __(10)__ surface; however, in four-legged animals, the dorsal surface is the __(11)__ surface.

_____12.

_____13.

_____14.

_____15.

If an incision cuts the heart into right and left parts, the section is a __(12)__ section, but if the heart is cut so that anterior and posterior parts result, the section is a __(13)__ section. You are told to cut an animal along two planes so that the paired kidneys are observable in both sections. The two sections that meet this requirement are the __(14)__ and __(15)__ sections.

18. Using the key choices, identify the body cavities where the following body organs are located. Enter the appropriate letter or term in the answer blanks.

Key Choices

A. Abdominopelvic   B. Cranial   C. Spinal   D. Thoracic

_____ 1. Stomach   _____ 7. Bladder

_____ 2. Small intestine   _____ 8. Trachea

_____ 3. Large intestine   _____ 9. Lungs

_____ 4. Spleen   _____ 10. Pituitary gland

_____ 5. Liver   _____ 11. Rectum

_____ 6. Spinal cord   _____ 12. Ovaries

19. Number the following structures, from darkest (black) to lightest (white), as they would appear on an X-ray. Number the darkest one 1, the next darkest 2, etc.

_____ A. Soft tissue

_____ B. Femur (bone of the thigh)

_____ C. Air in lungs

_____ D. Gold (metal) filling in a tooth

 AT THE CLINIC

20. A jogger has stepped in a pothole and sprained his ankle. What organ systems have suffered damage?

21. A newborn baby is unable to hold down any milk. Examination reveals a developmental disorder in which the esophagus fails to connect to the stomach. What survival needs are most immediately threatened?

22. The Chan family was traveling in their van and had a minor accident. The children in the backseat were wearing lap belts, but they still sustained bruises around the abdomen and had some internal organ injuries. Why is this area more vulnerable to damage than others?

23. John, a patient at Jones City Hospital, is in tough shape. He has a hernia in his inguinal region, pain from an infected kidney in his lumbar region, and severe bruises and swelling in his pubic region. Explain where each of these regions is located.

24. The hormone thyroxine is released in response to a pituitary hormone called TSH. As thyroxine levels increase in the blood, they exert negative feedback on the release of TSH by the pituitary gland. What effect will this have on the release of TSH?

25. In congestive heart failure, the weakened heart is unable to pump with sufficient strength to empty its own chambers. As a result, blood backs up in the veins, blood pressure rises, and circulation is impaired. Describe what will happen as this situation worsens owing to positive feedback. Then, predict how a heart-strengthening medication will reverse the positive feedback.

26. The following advanced imaging techniques are discussed in the text: CT, DSA, PET, ultrasound, and MRI. Which of these techniques uses X-ray? Which uses radio waves and magnetic fields? Which uses radioisotopes? Which displays body regions in sections? (You may have more than one answer for each question.)

27. A patient reports stabbing pains in the right hypochondriac region. The medical staff suspects gallstones. What region of the body will be examined?

28. Mr. Harvey, a computer programmer, has been complaining of numbness and pain in his right hand. His nurse practitioner diagnoses his problem as carpal tunnel syndrome and prescribes use of a splint. Where will Mr. Harvey apply the splint?

✓ THE FINALE: MULTIPLE CHOICE

29. Select the best answer or answers from the choices given.

1. Which of the following activities would *not* represent an anatomical study?

  A. Making a section through the heart to observe its interior

  B. Drawing blood from recently fed laboratory animals at timed intervals to determine their blood sugar levels

  C. Examining the surface of a bone

  D. Viewing muscle tissue through a microscope

2. The process that increases the size of the body or its number of cells is:

  A. metabolism   C. growth

  B. responsiveness   D. digestion

3. Which of the following is (are) involved in maintaining homeostasis?

  A. Effector   D. Feedback

  B. Control center   E. Lack of change

  C. Receptor

4. When a capillary is damaged, a platelet plug is formed. The process involves platelets sticking to each other. The more platelets that stick together, the more the plug attracts additional platelets. This is an example of:

   A. negative feedback

   B. positive feedback

5. A coronal plane through the head:

   A. could pass through both the nose and the occiput

   B. could pass through both ears

   C. must pass through the mouth

   D. could lie in a horizontal plane

6. Which of the following statements is correct?

   A. The brachium is proximal to the antebrachium.

   B. The femoral region is superior to the tarsal region.

   C. The orbital region is inferior to the buccal region.

   D. The axillary region is lateral to the sternal region.

   E. The crural region is posterior to the sural region.

7. Which of the following body regions is (are) found on the torso?

   A. Gluteal   D. Acromial

   B. Inguinal   E. Olecranal

   C. Popliteal

8. A neurosurgeon orders a spinal tap for a patient. Into what body cavity will the needle be inserted?

   A. Ventral   D. Cranial

   B. Thoracic   E. Pelvic

   C. Dorsal

9. An accident victim has a collapsed lung. Which cavity has been entered?

   A. Mediastinal   D. Vertebral

   B. Pericardial   E. Ventral

   C. Pleural

10. Which body system would be affected by degenerative cartilage?

    A. Muscular   D. Skeletal

    B. Nervous   E. Lymphatic

    C. Cardiovascular

11. The position of the heart relative to the structures around it would be described accurately as:

    A. deep to the sternum (breast bone)

    B. lateral to the lungs

    C. superior to the diaphragm

    D. inferior to the ribs

    E. anterior to the vertebral column

12. What term(s) could be used to describe the position of the nose?

    A. Intermediate to the eyes

    B. Inferior to the brain

    C. Superior to the mouth

    D. Medial to the ears

    E. Anterior to the ears

13. The radiographic technique used to provide information about blood flow is:

    A. DSR   D. ultrasonography

    B. CT   E. any X-ray technique

    C. PET

14. A patient complains of pain in the lower right quadrant. Which system is most likely to be involved?

    A. Respiratory   D. Skeletal

    B. Digestive   E. Muscular

    C. Urinary

15. Harry was sweating profusely as he ran in the 10-K race. The sweat glands producing the sweat would be considered which part of a feedback system?

    A. Stimulus   C. Control center

    B. Effectors   D. Receptors

2 BASIC CHEMISTRY

Everything in the universe is composed of one or more elements, the unique building blocks of all matter. Although over 100 elemental substances exist, only four of these (carbon, hydrogen, oxygen, and nitrogen) make up more than 96% of all living material.

The student activities in this chapter consider basic concepts of both inorganic and organic chemistry. Chemistry is the science that studies the composition of matter. Inorganic chemistry studies the chemical composition of nonliving substances that (generally) do not contain carbon. Organic chemistry studies the carbon-based chemistry (or biochemistry) of living organisms, whether they are maple trees, fish, or humans.

Understanding of atomic structure, bonding behavior of elements, and the structure and activities of the most abundant biological molecules (proteins, fats, carbohydrates, and nucleic acids) is tested in various ways. Mastering these concepts is necessary to understand how the body functions.

CONCEPTS OF MATTER AND ENERGY

1. Select *all* phrases that apply to each of the following statements and insert the letters in the answer blanks.

_____ 1. The energy located in the bonds of food molecules:
    A. is called thermal energy
    B. is a form of potential energy
    C. causes molecular movement
    D. can be transformed to the bonds of ATP (adenosine triphosphate)

_____ 2. Heat is:
    A. thermal energy
    B. infrared radiation
    C. kinetic energy
    D. molecular movement

_____ 3. Whenever energy is transformed:
    A. the amount of useful energy decreases
    B. some energy is lost as heat
    C. some energy is created
    D. some energy is destroyed

2. Use choices from the key to identify the energy *form* in use in each of the following examples.

Key Choices

A. Chemical   B. Electrical   C. Mechanical   D. Radiant

_____ 1. Chewing food

_____ 2. Vision (two types, please—think!)

_____ 3. Bending your fingers to make a fist

_____ 4. Breaking the bonds of ATP molecules to energize your muscle cells to make that fist

_____ 5. Lying under a sunlamp

COMPOSITION OF MATTER

3. Complete the following table by inserting the missing words.

Particle	Location	Electrical charge	Mass
		+ 1	
Neutron			
	Orbitals		

4. Insert the *chemical symbol* (the chemist's shorthand) in the answer blank for each of the following elements.

_____ 1. Oxygen   _____ 4. Iodine   _____ 7. Calcium   _____ 10. Magnesium

_____ 2. Carbon   _____ 5. Hydrogen   _____ 8. Sodium   _____ 11. Chlorine

_____ 3. Potassium   _____ 6. Nitrogen   _____ 9. Phosphorus   _____ 12. Iron

5. Using the key choices, select the correct responses to the following descriptive statements. Insert the appropriate answers in the answer blanks.

Key Choices

A. Atom   C. Element   E. Ion   G. Molecule   I. Protons

B. Electrons   D. Energy   F. Matter   H. Neutrons   J. Valence

_____ 1. An electrically charged atom or group of atoms

_____ 2. Anything that takes up space and has mass (weight)

_____ 3. A unique substance composed of atoms having the same atomic number

_____ 4. Negatively charged particles, forming part of an atom

_____ 5. Subatomic particles that determine an atom's chemical behavior, or bonding ability

_____ 6. The ability to do work

_____ 7. The smallest particle of an element that retains the properties of the element

_____ 8. The smallest particle of a compound, formed when atoms combine chemically

_____ 9. Positively charged particles forming part of an atom

_____ 10. Name given to the electron shell that contains the most reactive electrons

_____ 11. _____ 12. Subatomic particles responsible for most of an atom's mass

6. For each of the following statements that is true, insert *T* in the answer blank. If any of the statements are false, correct the <u>underlined</u> term by inserting your correction in the answer blank.

_____ 1. Na$^+$ and K$^+$ are <u>needed</u> for nerve cells to conduct electrical impulses.

_____ 2. The atomic number of oxygen is 8. Therefore, oxygen atoms always contain 8 <u>neutrons</u>.

_____ 3. The greater the distance of an electron from the nucleus, the <u>less</u> energy it has.

_____ 4. Electrons are located in more or less designated areas of space around the nucleus called <u>orbitals</u>.

_____ 5. An unstable atom that decomposes and emits energy is called <u>retroactive</u>.

_____ 6. <u>Iron</u> is necessary for oxygen transport in red blood cells.

_____ 7. The most abundant negative ion in extracellular fluid is <u>calcium</u>.

_____ 8. The element essential for the production of thyroid hormones is <u>magnesium</u>.

_____ 9. <u>Calcium</u> is found as a salt in bones and teeth.

MOLECULES, CHEMICAL BONDS, AND CHEMICAL REACTIONS

7. Match the terms in Column B to the chemical equations listed in Column A. Enter the correct letter or term in the answer blanks.

Column A	Column B
_____ 1. $A + B \rightarrow AB$	A. Decomposition
_____ 2. $AB + CD \rightarrow AD + CB$	B. Exchange
_____ 3. $XY \rightarrow X + Y$	C. Synthesis

8. Figure 2–1 is a diagram of an atom. Select two different colors and use them to color the coding circles and corresponding structures on the figure. Complete this exercise by responding to the questions that follow, referring to the atom in this figure. Insert your answers in the answer blanks provided.

◯ Nucleus

◯ Electrons

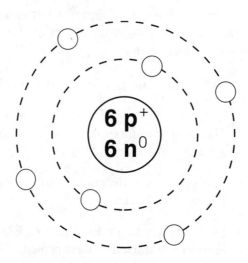

Figure 2–1

1. What is the atomic number of this atom? _____

2. What is its atomic mass? _____

3. What atom is this? _____

4. If this atom had one additional neutron but the other subatomic particles remained the same as shown, this slightly different atom (of the same element) would be called a(n) _____

5. Is this atom chemically active or inert? _____

6. How many electrons would be needed to fill its outer (valence) shell? _____

7. Would this atom most likely take part in forming ionic or

    covalent bonds? _____ Why? _____

    _____

9. Both H_2O_2 and $2OH^-$ are chemical species with two hydrogen atoms and two oxygen atoms. Briefly explain how these species are different:

    _____

    _____

10. Two types of chemical bonding are shown in Figure 2–2. In the figure, identify each type as a(n) *ionic* or *covalent* bond. In the case of the ionic bond, indicate which atom has lost an electron by adding a colored arrow to show the direction of electron transfer. For the covalent bond, indicate the shared electrons.

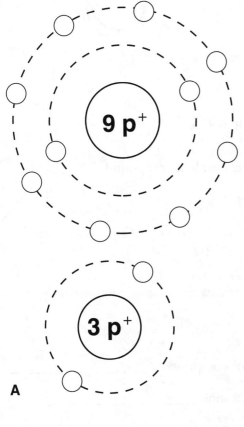

A

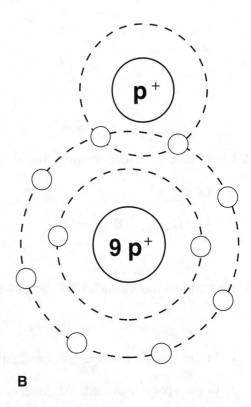

B

Type of bond: _____

Type of bond: _____

Figure 2–2

11. Figure 2–3 illustrates five water molecules held together by hydrogen bonds. First, correctly identify the oxygen and hydrogen atoms both by color and by inserting their atomic symbols on the appropriate circles (atoms). Then label the following structures in the figure:

○ Oxygen

○ Hydrogen

○ Positive pole (end)

○ Negative pole (end)

○ Hydrogen bonds

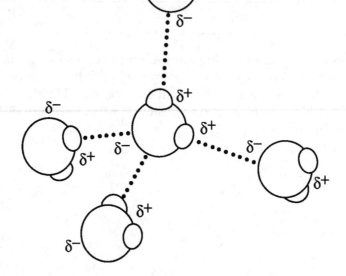

Figure 2–3

12. Circle each structural formula that is *likely* to be a polar covalent compound.

$$\textbf{A} \quad Cl-\underset{\underset{Cl}{|}}{\overset{\overset{Cl}{|}}{C}}-Cl \qquad \textbf{B} \quad H-Cl \qquad \textbf{C} \quad \overset{H}{\diagdown}\underset{\diagup}{N}\overset{\diagup}{\diagdown}\underset{H}{H} \qquad \textbf{D} \quad Cl-Cl \qquad \textbf{E} \quad \overset{H}{\diagdown}\underset{\diagup}{O}\underset{H}{}$$

13. Respond to the instructions following the equation:

$$H_2CO_3 \longrightarrow H^+ + HCO_3^-$$

1. In the space provided, list the chemical formula(s) of compounds. _____

2. In the space provided, list the chemical formula(s) of ions. _____

3. Circle the product(s) of the reaction.

4. Modify the equation by adding a colored arrow in the proper place to indicate that the reaction is reversible.

BIOCHEMISTRY: THE COMPOSITION OF LIVING MATTER

14. Use the key choices to identify the substances described in the following statements. Insert the appropriate letter(s) or corresponding term(s) in the answer blanks.

Key Choices

A. Acid(s)   B. Base(s)   C. Buffer   D. Salt(s)

_____ 1. _____ 2. _____ 3. Substances that ionize in water; good electrolytes

_____ 4. Proton (H⁺) acceptor

_____ 5. Ionize in water to release hydrogen ions and a negative ion other than hydroxide (OH⁻)

_____ 6. Ionize in water to release ions other than H⁺ and OH⁻

_____ 7. Formed when an acid and a base are combined

_____ 8. Substances such as lemon juice and vinegar

_____ 9. Prevents rapid/large swings in pH

15. Complete the following statements concerning the properties and biological importance of water.

_____ 1.

_____ 2.

_____ 3.

_____ 4.

_____ 5.

_____ 6.

_____ 7.

_____ 8.

The ability of water to maintain a relatively constant temperature and thus prevent sudden changes is because of its high __(1)__. Biochemical reactions in the body must occur in __(2)__. About __(3)__ % of the volume of a living cell is water. Water molecules are bonded to other water molecules because of the presence of __(4)__ bonds. Water, as H⁺ and OH⁻ ions, is essential in biochemical reactions such as __(5)__ and __(6)__ reactions. Because of its __(7)__, water is an excellent solvent and forms the basis of mucus and other body __(8)__.

16. Use an *X* to designate which of the following are inorganic compounds or substances.

_____ Carbon dioxide   _____ Fats   _____ Proteins   _____ H_2O

_____ Oxygen   _____ KCl   _____ Glucose   _____ DNA

17. Using the key choices, fully characterize weak and strong acids.

Key Choices

A. Ionize completely in water   E. Ionize at high pH

B. Ionize incompletely in water   F. Ionize at low pH

C. Act as part of a buffer system   G. Ionize at pH 7

D. When placed in water, always act to change the pH

Weak acid: _____   Strong acid: _____

18. Match the terms in Column B to the descriptions provided in Column A.
Enter the correct letter(s) or term(s) in the answer blanks.

Column A	Column B
_____ 1. Building blocks of carbohydrates	A. Amino acids
_____ 2. Building blocks of fat	B. Carbohydrates
_____ 3. Building blocks of protein	C. Lipids (fats)
_____ 4. Building blocks of nucleic acids	D. Fatty acids
_____ 5. Cellular cytoplasm is primarily composed of this substance	E. Glycerol
_____ 6. The single most important fuel source for body cells	F. Nucleotides
_____ 7. Not soluble in water	G. Monosaccharides
_____ 8. Contain C, H, and O in the ratio CH_2O	H. Proteins
_____ 9. Contain C, H, and O, but have relatively small amounts of oxygen	
_____10. _____11. These building blocks contain N in addition to C, H, and O	
_____12. Contain P in addition to C, H, O, and N	
_____13. Used to insulate the body and found in all cell membranes	
_____14. Primary components of meat and cheese	
_____15. Primary components of bread and lollipops	
_____16. Primary components of egg yolk and peanut oil	
_____17. Include collagen and hemoglobin	
_____18. Class that usually includes cholesterol	
_____19. The alpha helix and beta pleated sheet are both examples of the secondary structure of these molecules.	

19. Using the key choices, correctly select *all* terms that correspond to the following descriptions. Insert the correct letter(s) or their corresponding term(s) in the answer blanks.

Key Choices

A. Cholesterol	D. Enzyme	G. Hormones	J. Maltose
B. Collagen	E. Glycogen	H. Keratin	K. RNA
C. DNA	F. Hemoglobin	I. Lactose	L. Starch

_____ 1. Example(s) of fibrous (structural) proteins

_____ 2. Example(s) of globular (functional) proteins

_____ 3. Biological catalyst

_____ 4. Plant storage carbohydrate

_____ 5. Animal storage carbohydrate

_____ 6. The "stuff" of the genes

_____ 7. A steroid

_____ 8. Double sugars, or disaccharides

20. Five simplified diagrams of biological molecules are depicted in Figure 2–4. First, identify the molecules and insert the correct names in the answer blanks on the figure. Then select a different color for each molecule listed below and use them to color the coding circles and the corresponding molecules on the illustration.

◯ Fat   ◯ Nucleotide   ◯ Monosaccharide

◯ Globular protein   ◯ Polysaccharide

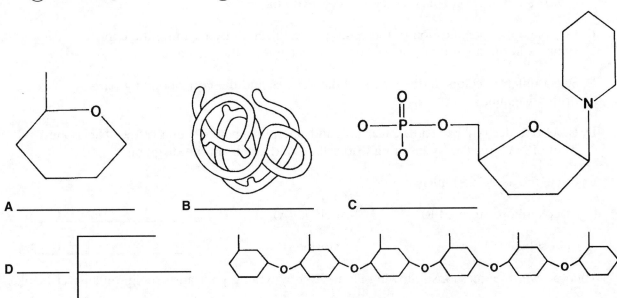

Figure 2–4

21. Circle the term that does not belong in each of the following groupings.

1. Adenine   Guanine   Glucose   Thymine

2. DNA   Ribose   Phosphate   Deoxyribose

3. Galactose   Glycogen   Fructose   Glucose

4. Amino acid   Polypeptide   Glycerol   Protein

5. Glucose   Sucrose   Lactose   Maltose

22. For each true statement, insert T in the answer blank. If any are false, correct the underlined term and insert your correction in the answer blank.

_____ 1. Phospholipids are underlined polarized molecules.

_____ 2. Steroids are the major form in which body fat is stored.

_____ 3. Water is the most abundant compound in the body.

_____ 4. Nonpolar molecules are generally soluble in water.

_____ 5. The bases of RNA are A, G, C, and U.

_____ 6. The universal energy currency of living cells is RNA.

_____ 7. RNA is single stranded.

_____ 8. The four elements that make up more than 90% of living matter are C, H, N, and Na.

23. Figure 2–5 shows the molecular structure of DNA, a nucleic acid.

A. First, identify the two unnamed nitrogen (N) bases and insert their names and symbols in the two blanks beside the color-coding circles.

B. Complete the identification of the bases on the diagram by inserting the correct symbols in the appropriate spaces on the right side of the diagram.

C. Select different colors and color the coding circles and the corresponding parts of the diagram.

D. Label one deoxyribose (d-R) sugar unit and one phosphate (P) unit of the "backbones" of the DNA structure by inserting leader lines and labels on the diagram.

E. Circle the associated nucleotide.

◯ Deoxyribose sugar (d-R)   ◯ Adenine (A)   ◯ _____ ( )

◯ Phosphate (P)   ◯ Cytosine (C)   ◯ _____ ( )

Then answer the questions following Figure 2–5 by writing your answers in the answer blanks.

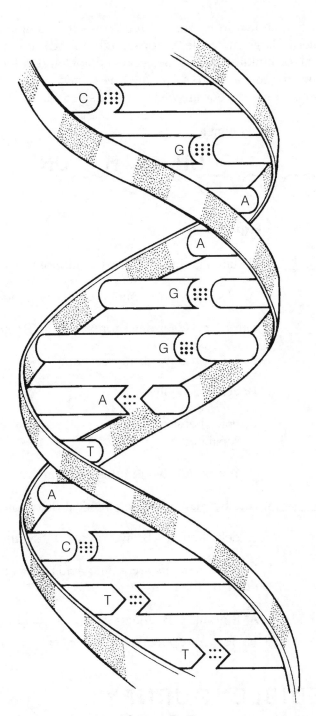

Figure 2–5

1. Name the bonds that help to hold the two DNA strands together. _____

2. Name the three-dimensional shape of the DNA molecule. _____

3. How many base pairs are present in this segment of a DNA model? _____

4. What is the term that means "base pairing"? _____

24. The biochemical reaction shown in Figure 2–6 represents the complete digestion of a polymer (a large molecule as consumed in food) down to its constituent monomers, or building blocks. Select two colors and color the coding circles and the structures. Then, select the one correct answer for each statement below and insert your answer in the answer blank.

○ Monomer   ○ Polymer

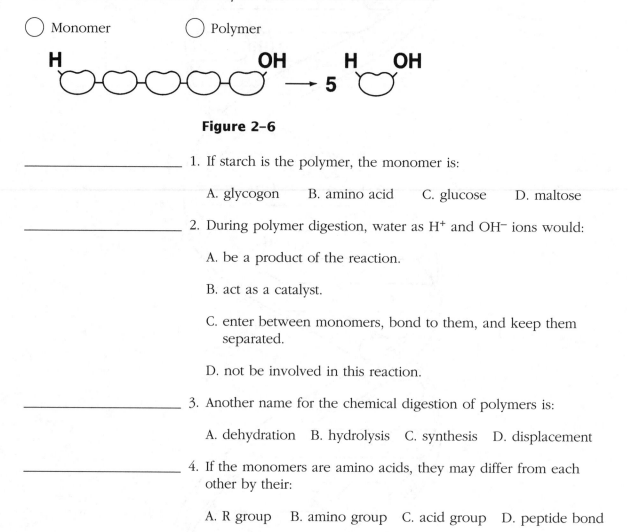

Figure 2–6

_____ 1. If starch is the polymer, the monomer is:

         A. glycogon   B. amino acid   C. glucose   D. maltose

_____ 2. During polymer digestion, water as H^+ and OH^- ions would:

         A. be a product of the reaction.

         B. act as a catalyst.

         C. enter between monomers, bond to them, and keep them separated.

         D. not be involved in this reaction.

_____ 3. Another name for the chemical digestion of polymers is:

         A. dehydration   B. hydrolysis   C. synthesis   D. displacement

_____ 4. If the monomers are amino acids, they may differ from each other by their:

         A. R group   B. amino group   C. acid group   D. peptide bond

 INCREDIBLE JOURNEY

A Visualization Exercise for Biochemistry

... you are suddenly upended and are carried along in a sea of water molecules at almost unbelievable speed.

25. Complete the narrative by inserting the missing words in the answer blanks.

For this journey, you are miniaturized to the size of a very small molecule by colleagues who will remain in contact with you by radio. Your instructions are to play the role of a water molecule and to record any reactions that involve water molecules. Considering water molecules are polar

_____ 1.

_____ 2.

_____ 3.

_____ 4.

_____ 5.

_____ 6.

_____ 7.

_____ 8.

_____ 9.

_____10.

_____11.

_____12.

_____13.

_____14.

_____15.

molecules, you are outfitted with an insulated rubber wet suit with a __(1)__ charged helmet and two __(2)__ charges, one at the end of each leg.

As soon as you are injected into your host's bloodstream, you feel as though you are being pulled apart. Some large, attractive forces are pulling at your legs from different directions! You look about but can see only water molecules. After a moment's thought, you remember the polar nature of your wet suit. You record that these forces must be the __(3)__ that are easily formed and easily broken in water.

After this initial surprise, you are suddenly upended and carried along in a sea of water molecules at almost unbelievable speed. You have just begun to observe some huge, red, disk-shaped structures (probably __(4)__) taking up O_2 molecules when you are swept into a very turbulent environment. Your colleagues radio that you are in the small intestine. With difficulty, because of numerous collisions with other molecules, you begin to record the various types of molecules you see.

In particular, you notice a very long helical molecule made of units with distinctive R-groups. You identify and record this type of molecule as a __(5)__, made of units called __(6)__ that are joined together by __(7)__ bonds. As you move too close to the helix during your observations, you are nearly pulled apart to form two ions, __(8)__, but you breathe a sigh of relief as two ions of another water molecule take your place. You watch as these two ions move between two units of the long helical molecule. Then, in a fraction of a second, the bond between the two units is broken. As you record the occurrence of this chemical reaction, called __(9)__, you are jolted into another direction by an enormous globular protein, the very same __(10)__ that controls and speeds up this chemical reaction.

Once again you find yourself in the bloodstream, heading into an organ identified by your colleagues as the liver. Inside a liver cell, you observe many small monomers, made up only of C, H, and O atoms. You identify these units as __(11)__ molecules because the liver cells are bonding them together to form very long, branched polymers called __(12)__. You record that this type of chemical reaction is called __(13)__, and you happily note that this reaction also produces __(14)__ molecules like you!

After another speedy journey through the bloodstream, you reach the skin. You move deep into the skin and finally gain access to a sweat gland. In the sweat gland, you collide with millions of water molecules and some ionized salt molecules that are continually attracted to your positive and negative charges. Suddenly, the internal temperature rises, and molecular collisions __(15)__ at an alarming rate, propelling you through the pore of the sweat gland onto the surface of the skin. So that you will be saved from the fate of evaporating into thin air, you contact your colleagues and are speedily rescued.

AT THE CLINIC

26. It is determined that a patient is in acidosis. What does this mean, and would you treat the condition with a chemical that would *raise* or *lower* the pH?

27. A newborn is diagnosed with sickle cell anemia, a genetic disease in which substitution of one amino acid results in abnormal hemoglobin. Explain to the parents how the substitution can have such a drastic effect on the structure of the protein.

28. Johnny's body temperature is spiking upward. When it reaches 104°F, his mother puts in a call to the pediatrician. She is advised to give Johnny children's acetaminophen or ibuprofen and sponge his body with cool to tepid water to prevent a further rise in temperature. How might a fever (excessively high body temperature) be detrimental to Johnny's welfare?

29. Mrs. Gallo's physician suspects that she is showing the initial signs of multiple sclerosis, a disease characterized by the formation of hardened plaques in the insulating sheaths surrounding nerve fibers. What medical imaging technique will the physician probably order to determine if such plaques are present?

30. Stanley has indigestion and is doubled over with pain. How could an antacid reduce his stomach discomfort?

31. Explain why the formation of ATP from ADP (adenosine diphosphate) and P_i requires more energy than the amount released for cellular use when ATP is broken down.

✓ THE FINALE: MULTIPLE CHOICE

32. Select the best answer or answers from the choices given.

1. Which of the following is (are) true concerning the atomic nucleus?

   A. Contains the mass of the atom

   B. The negatively charged particles are here

   C. Particles can be ejected

   D. Contains particles that determine atomic number

   E. Contains particles that interact with other atoms

2. Organic compounds include:

   A. water   D. carbonic acid

   B. carbon dioxide   E. glycerol

   C. oxygen

3. Important functions of water include:

   A. cushioning

   B. transport medium

   C. participation in chemical reactions

   D. solvent for sugars, salts, and other solutes

   E. reducing temperature fluctuations

4. Which of the elements listed is the most abundant extracellular anion?

   A. Phosphorus   D. Chloride

   B. Sulfur   E. Calcium

   C. Potassium

5. The element essential for normal thyroid function is:

   A. iodine

   B. iron

   C. copper

   D. selenium

   E. zinc

6. Alkaline substances include:

   A. gastric juice   D. orange juice

   B. water   E. ammonia

   C. blood

7. Which of the following is (are) not a monosaccharide?

   A. Glucose   D. Glycogen

   B. Fructose   E. Deoxyribose

   C. Sucrose

8. Which is a building block of neutral fats?

   A. Ribose   D. Glycine

   B. Guanine   E. Glucose

   C. Glycerol

9. Which of the following is primarily responsible for the helical structure of a polypeptide chain?

   A. Hydrogen bonding

   B. Tertiary folding

   C. Peptide bonding

   D. Quaternary associations

   E. Complementary base pairing

10. Which of the following is (are) not true of RNA?

    A. Double stranded

    B. Contains cytosine

    C. Directs protein synthesis

    D. Found primarily in the nucleus

    E. Can act as an enzyme

11. DNA:

    A. contains uracil   C. is the "genes"

    B. is a helix   D. contains ribose

12. Glucose is to starch as:

    A. a steroid is to a lipid

    B. a nucleotide is to nucleic acid

    C. an amino acid is to a protein

    D. a polypeptide is to an amino acid

13. An organic sample is analyzed and shown to have C, H, O, N, and P as its constituents. The orgranic molecule is identified as a:

    A. carbohydrate   C. lipid

    B. protein   D. nucleic acid

14. Which of the following forms of energy is the stimulus for vision?

    A. Mechanical   C. Electrical

    B. Light   D. Chemical

15. Which of the following describe energy?

    A. Has mass

    B. Massless

    C. Occupies space

    D. Puts matter into motion

16. Which of the following is (are) a synthetic reaction?

    A. Glucose to glycogen

    B. Glucose and fructose to sucrose

    C. Starch to glucose

    D. Amino acids to dipeptide

3 CELLS AND TISSUES

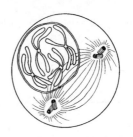

The basic unit of structure and function in the human body is the cell. Each of a cell's parts, or organelles, as well as the entire cell, is organized to perform a specific function. Cells have the ability to metabolize, grow and reproduce, move, and respond to stimuli. The cells of the body differ in shape, size, and in specific roles in the body. Cells that are similar in structure and function form tissues, which, in turn, construct the various body organs.

Student activities in this chapter include questions relating to the structure and function of the generalized animal cell and to the general arrangement of tissues and their contribution to the activities of the various body organs.

CELLS
Overview

1. Answer the following questions by inserting your responses in the answer blanks.

_____ 1.

_____ 2.

_____ 3.

_____ 4.

_____ 5.

_____ 6.

_____ 7.

_____ 8.

_____ 9.

_____ 10.

1–4. Name the four elements that make up the bulk of living matter.

5. Name the single most abundant material or substance in living matter.

6. Name the trace element most important for making bones hard.

7. Name the element, found in small amounts in the body, that is needed to make hemoglobin for oxygen transport.

8–12. Although there are many specific "jobs" that certain cells are able to do, name five functions common to all cells.

_____ 11.

_____ 12. →

_____13.

_____14.

_____15.

_____16.

_____17.

13–15. List three different cell shapes.

16. Name the fluid, similar to seawater, that surrounds and bathes all body cells.

17. Name the flattened cells, important in protection against damage, that fit together like tiles. (This is just one example of the generalization that a cell's structure is very closely related to its function in the body.)

Anatomy of a Generalized Cell

2. Using the list of terms on the following page, correctly label all cell parts indicated by leader lines in Figure 3–1. Then select different colors for each structure and use them to color the coding circles and the corresponding structures in the illustration.

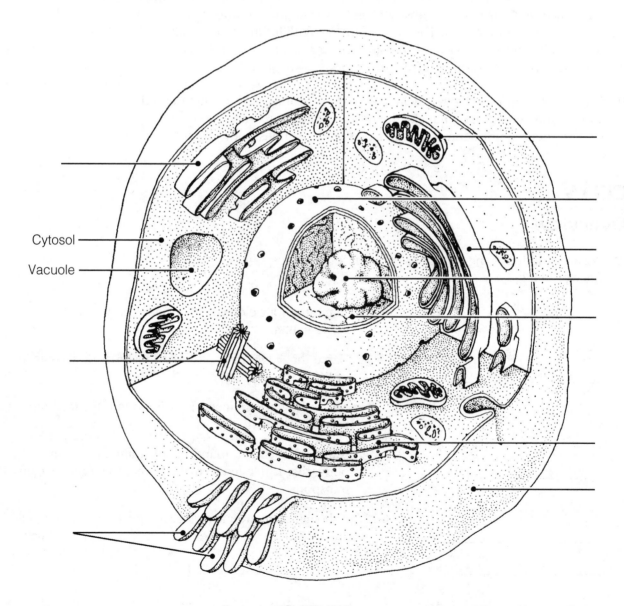

Cytosol

Vacuole

Figure 3–1

⭕ Plasma membrane   ⭕ Mitochondrion

⭕ Centriole(s)   ⭕ Nuclear membrane

⭕ Chromatin thread(s)   ⭕ Nucleolus

⭕ Golgi apparatus   ⭕ Rough endoplasmic reticulum (ER)

⭕ Microvilli   ⭕ Smooth endoplasmic reticulum (ER)

3. Figure 3–2 is a diagram of a portion of a plasma membrane. Select three different colors and color the coding circles and the corresponding structures in the diagram. Then respond to the questions that follow, referring to Figure 3–2, and insert your answers in the answer blanks.

⭕ Phospholipid molecules   ⭕ Carbohydrate molecules   ⭕ Protein molecules

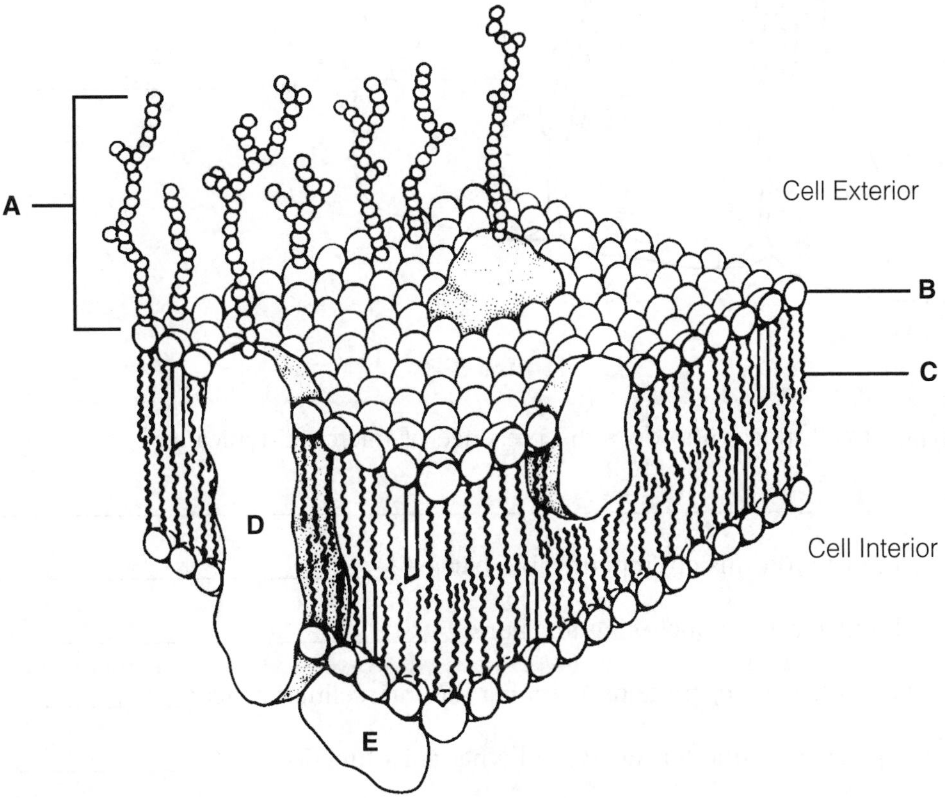

Figure 3–2

1. Name the carbohydrate-rich area at the cell surface (indicated by bracket A). _____

2. Which label, B or C, indicates the nonpolar region of a phospholipid molecule? _____

3. Does nonpolar mean hydrophobic or hydrophilic? _____

4. What are two roles of the membrane proteins? _____

   and _____

4. Label the specializations of the plasma membrane, shown in Figure 3–3, and color the diagram as you wish. Then, answer the questions provided below that refer to this figure.

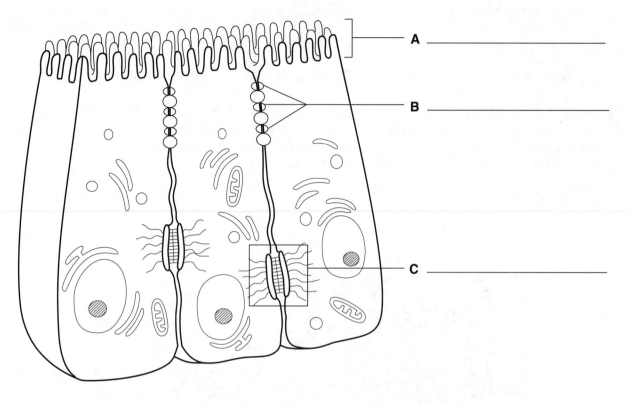

A _____

B _____

C _____

Figure 3–3

1. What type of cell function(s) does the presence of microvilli typically

   indicate? _____

2. Which cell junction forms an impermeable barrier? _____

3. Which cell junction is an anchoring junction? _____

4. Which junction has linker proteins spanning the intercellular space? _____

5. Which cell junction is not illustrated, and what is its function? _____

   _____

6. What two types of membrane junctions would you expect to find between cells of the heart?

   _____ and _____

5. Relative to cellular organelles, circle the term or phrase that does not belong in each of the following groupings.

1. Peroxisomes   Enzymatic breakdown   Centrioles   Lysosomes

2. Microtubules   Intermediate filaments   Cytoskeleton   Cilia

3. Ribosomes   Smooth ER   Rough ER   Protein synthesis

4. Mitochondrion   Cristae   ATP production   Vitamin A storage

5. Centrioles   Mitochondria   Cilia   Flagella

6. ER   Nuclear pores   Ribosomes   Transport vesicles   Golgi apparatus

7. Nucleus   DNA   Lysosomes   Chromatin   Nucleolus

6. Name the cytoskeletal element (microtubules, microfilaments, or intermediate filaments) described by each of the following phrases.

_____ 1. Give the cell its shape

_____ 2. Resist tension placed on a cell

_____ 3. Radiate from the cell center

_____ 4. Involved in moving intracellular structures

_____ 5. Are the most stable

_____ 6. Have the thickest diameter

7. Different organelles are abundant in different cell types. Match the cell types with their abundant organelles by selecting a letter from the key choices.

Key Choices

A. Mitochondria   C. Rough ER   E. Microfilaments   G. Intermediate filaments

B. Smooth ER   D. Peroxisomes   F. Lysosomes   H. Golgi apparatus

_____ 1. Cell lining the small intestine (assembles fats)

_____ 2. White blood cell; a phagocyte

_____ 3. Liver cell that detoxifies carcinogens

_____ 4. Muscle cell (contractile cell)

_____ 5. Mucus-secreting cell (secretes a protein product)

_____ 6. Cell at external skin surface (withstands friction and tension)

_____ 7. Kidney tubule cell (makes and uses large amounts of ATP)

Cell Physiology

Membrane Transport

8. Figure 3–4 shows a semipermeable sac, containing 4% NaCl, 9% glucose, and 10% albumin, suspended in a solution with the following composition: 10% NaCl, 10% glucose, and 40% albumin. Assume the sac is permeable to all substances *except* albumin. Using the key choices, insert the letter indicating the correct event in the answer blanks.

Key Choices

A. Moves into the sac   B. Moves out of the sac   C. Does not move

_____ 1. Glucose   _____ 3. Albumin

_____ 2. Water   _____ 4. NaCl

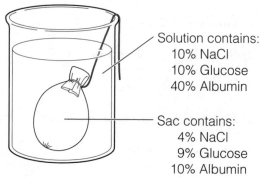

Solution contains:
10% NaCl
10% Glucose
40% Albumin

Sac contains:
4% NaCl
9% Glucose
10% Albumin

Figure 3–4

9. Figure 3–5 shows three microscopic fields (A–C) containing red blood cells. Arrows indicate the direction of net osmosis. Respond to the following questions, referring to Figure 3–5, by inserting your responses in the spaces provided.

1. Which microscopic field contains a *hypertonic* solution? _____

   The cells in this field are said to be _____

2. Which microscopic field contains an isotonic bathing solution? _____

   What does *isotonic* mean? _____

   _____

3. Which microscopic field contains a *hypotonic* solution? _____

   What is happening to the cells in this field and why? _____

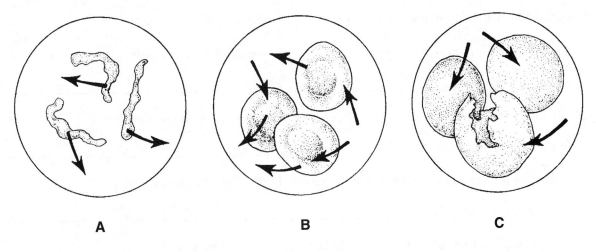

Figure 3–5

10. Figure 3–6 is a simplified diagram of the plasma membrane. Structure A represents channel proteins constructing a pore, structure B represents an ATP-energized solute pump, and structure C is a transport protein that does not depend on energy from ATP. Identify these structures and the membrane phospholipids by color before continuing.

◯ Channel   ◯ Solute pump   ◯ Passive transport protein carrier   ◯ Phospholipids

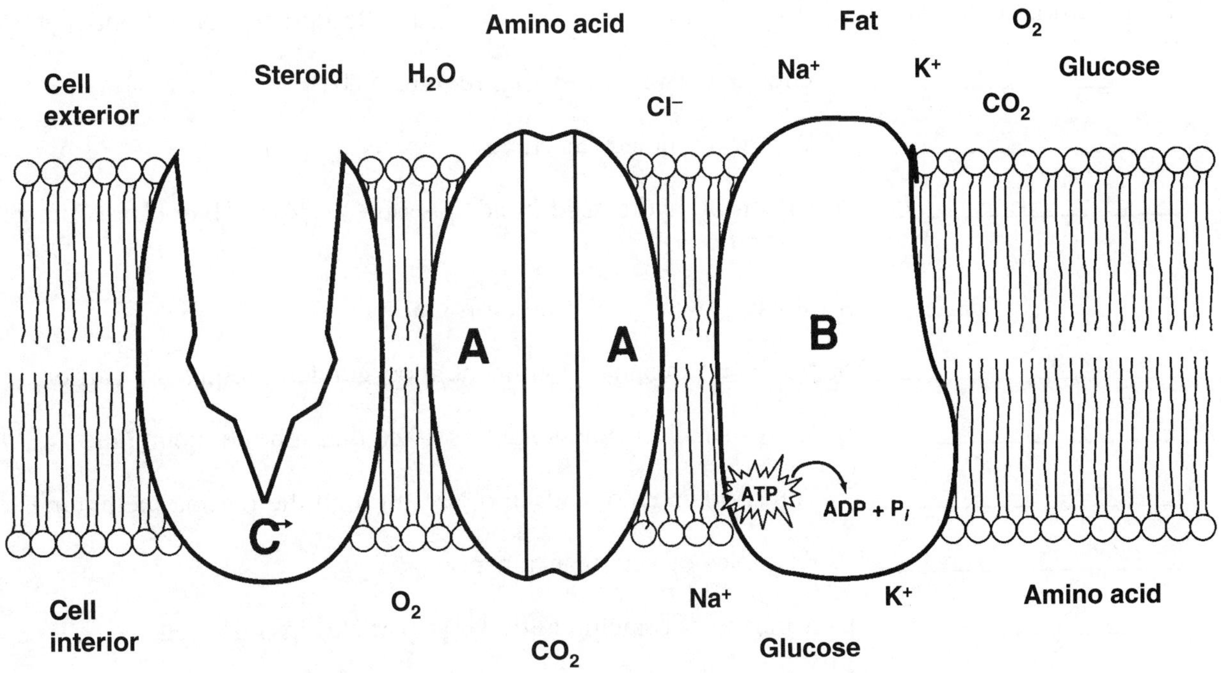

Figure 3–6

Now add arrows to Figure 3–6 as instructed next: For each substance that moves through the plasma membrane, draw an arrow indicating its (most likely) direction of movement (into or out of the cell). If it is moved actively, use a red arrow; if it is moved passively, use a blue arrow.

→

Finally, answer the following questions referring to Figure 3–6:

1. Which of the substances shown moves passively *through the lipid* part
   of the membrane? _____

2. Which of the substances shown enters the cell by attachment to a passive
   transport protein carrier? _____

3. Which of the substances shown moves passively through the membrane
   by moving through its pores? _____

4. Which of the substances shown would have to use a solute pump to be
   transported through the membrane? _____

11. Select the key choices that characterize each of the following statements.
Insert the appropriate answers in the answer blanks.

Key Choices

A. Active transport   D. Exocytosis   G. Phagocytosis

B. Diffusion, simple   E. Facilitated diffusion   H. Pinocytosis

C. Diffusion, osmosis   F. Filtration   I. Receptor-mediated endocytosis

_____ 1. Engulfment processes that require ATP

_____ 2. Driven by molecular energy

_____ 3. Driven by hydrostatic (fluid) pressure (typically blood pressure
                    in the body)

_____ 4. Moves down a concentration gradient

_____ 5. Moves up (against) a concentration gradient; requires a carrier

_____ 6. Moves small or lipid-soluble solutes through the membrane

_____ 7. Transports amino acids and Na^+ through the plasma membrane

_____ 8. Examples of vesicular transport

_____ 9. A means of bringing fairly large particles into the cell

_____10. Used to eject wastes and to secrete cell products

_____11. Membrane transport using channels or carrier proteins that does
                    not require ATP

Cell Division

12. The following statements provide an overview of the structure of DNA (genetic material) and its role in the body. Choose responses from the key choices that complete the statements. Insert the appropriate answers in the answer blanks.

Key Choices

A. Adenine	G. Enzymes	M. Nucleotides	S. Ribosome
B. Amino acids	H. Genes	N. Old	T. Sugar (deoxyribose)
C. Bases	I. Growth	O. Phosphate	U. Template, or model
D. Codons	J. Guanine	P. Proteins	V. Thymine
E. Complementary	K. Helix	Q. Replication	W. Transcription
F. Cytosine	L. New	R. Repair	X. Uracil

_____ 1.

_____ 2.

_____ 3.

_____ 4.

_____ 5.

_____ 6.

_____ 7.

_____ 8.

_____ 9.

_____ 10.

_____ 11.

_____ 12.

_____ 13.

_____ 14.

_____ 15.

_____ 16.

_____ 17.

_____ 18.

DNA molecules contain information for building specific __(1)__. In a three-dimensional view, a DNA molecule looks like a spiral staircase; this is correctly called a __(2)__. The constant parts of DNA molecules are the __(3)__ and __(4)__ molecules, forming the DNA-ladder uprights, or backbones. The information of DNA is actually coded in the sequence of nitrogen-containing __(5)__, which are bound together to form the "rungs" of the DNA ladder. When the four DNA bases are combined in different three-base sequences, called triplets, different __(6)__ of the protein are called for. It is said that the N-containing bases of DNA are __(7)__, which means that only certain bases can fit or interact together. Specifically, this means that __(8)__ can bind with guanine, and adenine binds with __(9)__.

The production of proteins involves the cooperation of DNA and RNA. RNA is another type of nucleic acid that serves as a "molecular slave" to DNA. That is, it leaves the nucleus and carries out the instructions of the DNA for the building of a protein on a cytoplasmic structure called a __(10)__. When a cell is preparing to divide, in order for its daughter cells to have all its information, it must oversee the __(11)__ of its DNA so that a "double dose" of genes is present for a brief period. For DNA synthesis to occur, the DNA must uncoil, and the bonds between the N bases must be broken. Then the two single strands of __(12)__ each act as a __(13)__ for the building of a whole DNA molecule. When completed, each DNA molecule formed is half __(14)__ and half __(15)__. The fact that DNA replicates before a cell divides ensures that each daughter cell has a complete set of __(16)__. Cell division, which then follows, provides new cells so that __(17)__ and __(18)__ can occur.

13. Identify the phases of mitosis depicted in Figure 3–7 by inserting the correct name in the blank under the appropriate diagram. Then select different colors to represent the structures listed below and use them to color in the coding circles and the corresponding structures in the illustration.

◯ Nuclear membrane(s), if present

◯ Nucleoli, if present

◯ Chromosomes

◯ Centrioles

◯ Spindle fibers

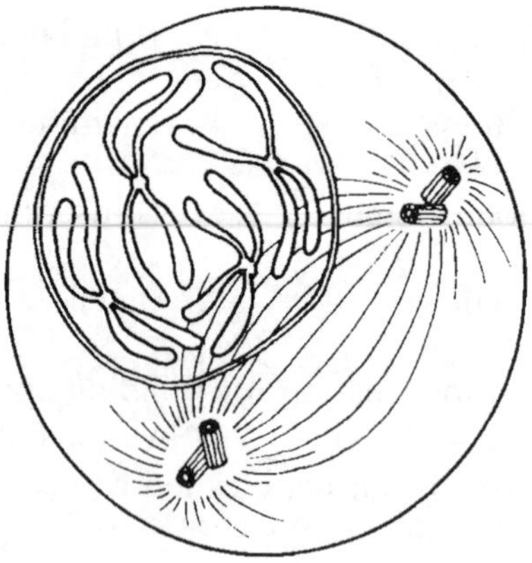

A _____

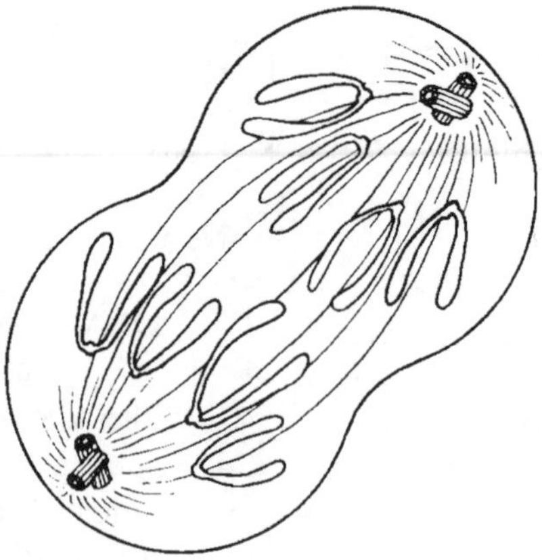

B _____

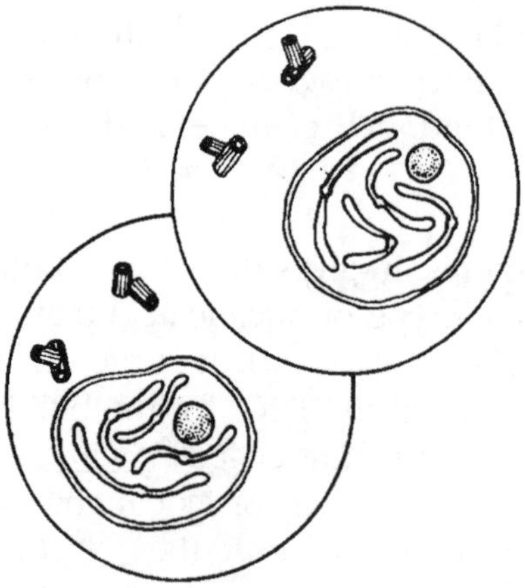

C _____

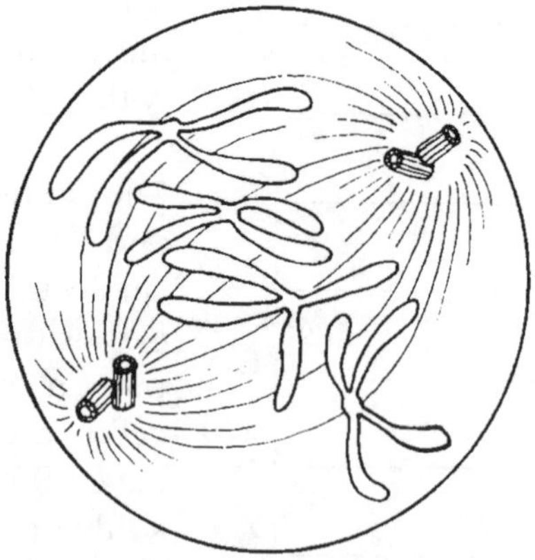

D _____

Figure 3–7

14. The following statements describe events that occur during the different
phases of mitosis. Identify the phase by choosing the correct response(s) from
the key choices and inserting the letter(s) or term(s) in the answer blanks.

Key Choices

A. Anaphase   C. Prophase   E. None of these

B. Metaphase   D. Telophase

_____ 1. Chromatin coils and condenses to form deeply staining bodies.

_____ 2. Centromeres break, and chromosomes begin migration toward opposite poles of the cell.

_____ 3. The nuclear membrane and nucleoli reappear.

_____ 4. When chromosomes cease their poleward movement, this phase begins.

_____ 5. Chromosomes align on the equator of the spindle.

_____ 6. The nucleoli and nuclear membrane disappear.

_____ 7. The spindle forms through the migration of the centrioles.

_____ 8. Chromosomal material replicates.

_____ 9. Chromosomes first appear to be duplex structures.

_____10. Chromosomes attach to the spindle fibers.

_____11. A cleavage furrow forms during this phase.

_____12. The nuclear membrane is absent during the entire phase.

_____13. A cell carries out its _usual_ metabolic activities.

15. Complete the following statements. Insert your answers in the answer blanks.

_____ 1. Division of the __(1)__ is referred to as mitosis. Cytokinesis is the division of the __(2)__. The major structural difference

_____ 2. between chromatin and chromosomes is that the latter are __(3)__. Chromosomes attach to the spindle fibers by undivided

_____ 3. structures called __(4)__. If a cell undergoes nuclear division but not cytoplasmic division, the product is a __(5)__. The structure

_____ 4. that acts as a scaffolding for chromosomal attachment and movement is called the __(6)__. __(7)__ is the period of cell life

_____ 5. when the cell is not involved in division.

_____ 6.

_____ 7.

Protein Synthesis

16. Figure 3–8 is a diagram illustrating protein synthesis. Select four different colors, and use them to color the coding circles and the corresponding structures in the diagram. Next, using the letters of the genetic code, label the nitrogen bases on strand 2 of the DNA double helix, on the mRNA strands, and on the tRNA molecules. Then, answer the questions that follow referring to Figure 3–8, inserting your answers in the answer blanks.

○ Backbones of the DNA double helix

○ Backbone of the mRNA strands

○ tRNA molecules

○ Amino acid molecules

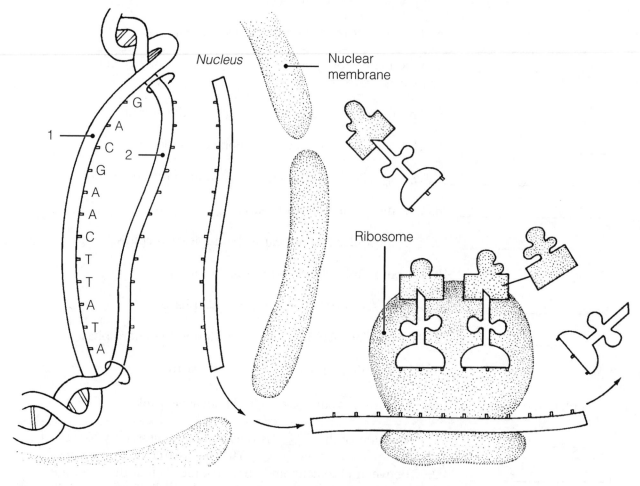

Figure 3–8

1. Transfer of the genetic message from DNA to mRNA is called _____.

2. Assembly of amino acids according to the genetic information carried by mRNA is called

_____.

3. The set of three nitrogen bases on tRNA that is complementary to an mRNA codon is called

a _____. The complementary three-base sequence on DNA is called a

_____.

BODY TISSUES

17. The 4 major tissue types are named in Figure 3–9. For each tissue type, provide its major function(s) after the tissue name. Then, select 2 leader lines under each tissue type and identify the body structure at the end of the leader line.

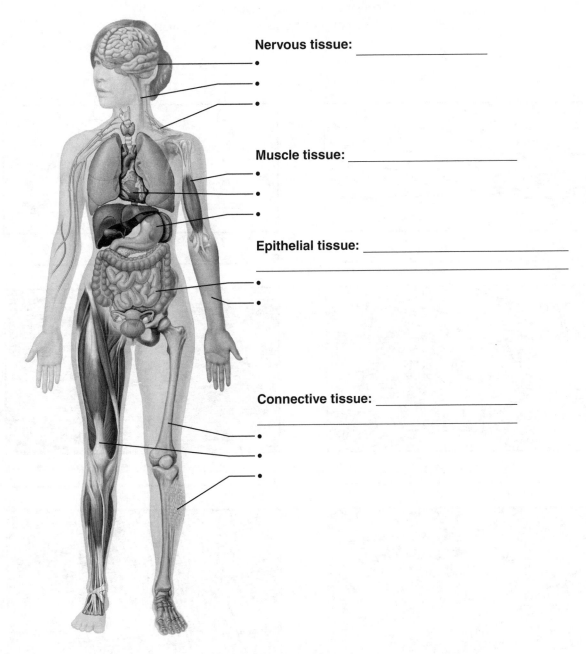

Nervous tissue: _____

Muscle tissue: _____

Epithelial tissue: _____
_____

Connective tissue: _____
_____

Figure 3–9

18. Twelve tissue types are diagrammed in Figure 3–10. Identify each tissue type by inserting the correct name in the blank below it on the diagram. Select different colors for the following structures and use them to color the coding circles and corresponding structures in the diagrams.

◯ Epithelial cells

◯ Muscle cells

◯ Nerve cells

◯ Matrix (Where found, matrix should be colored differently from the living cells of that tissue type. Be careful; this may not be as easy as it seems!)

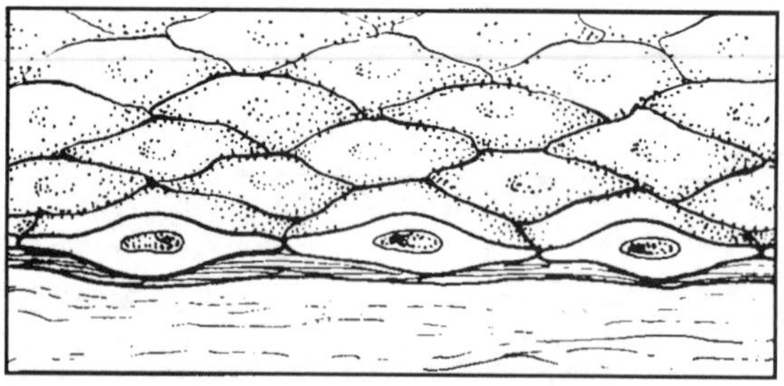

A _____

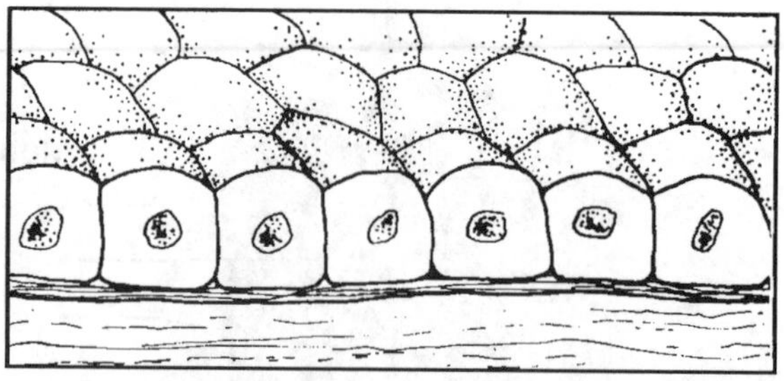

B _____

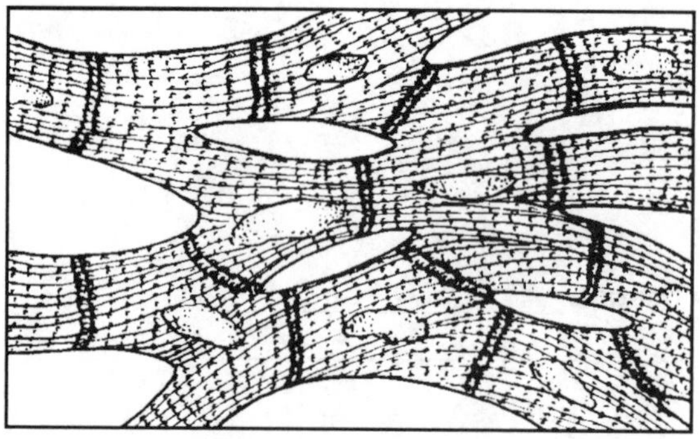

C _____

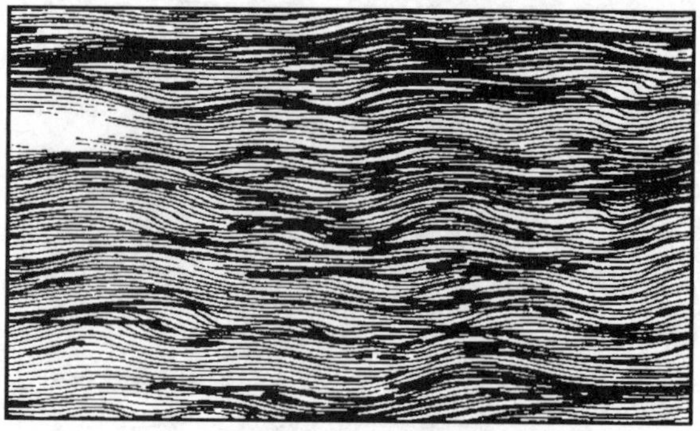

D _____

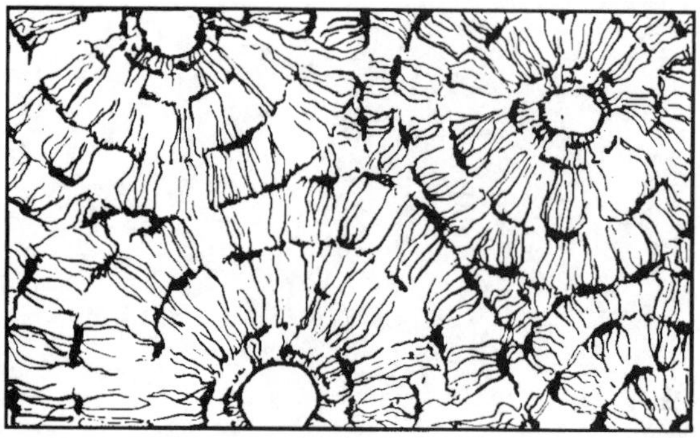

E _____

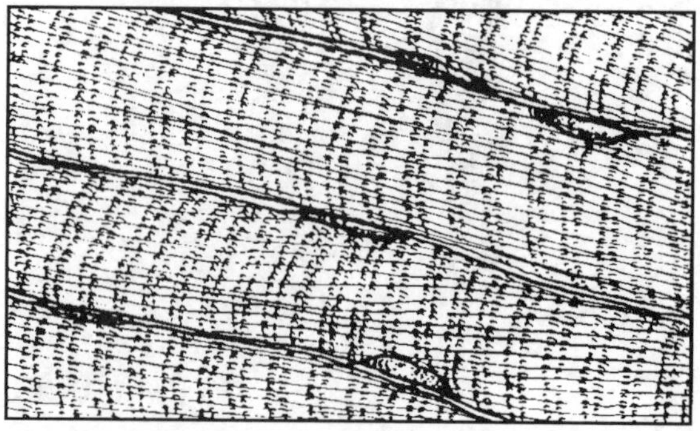

F _____

Figure 3–10, A–F

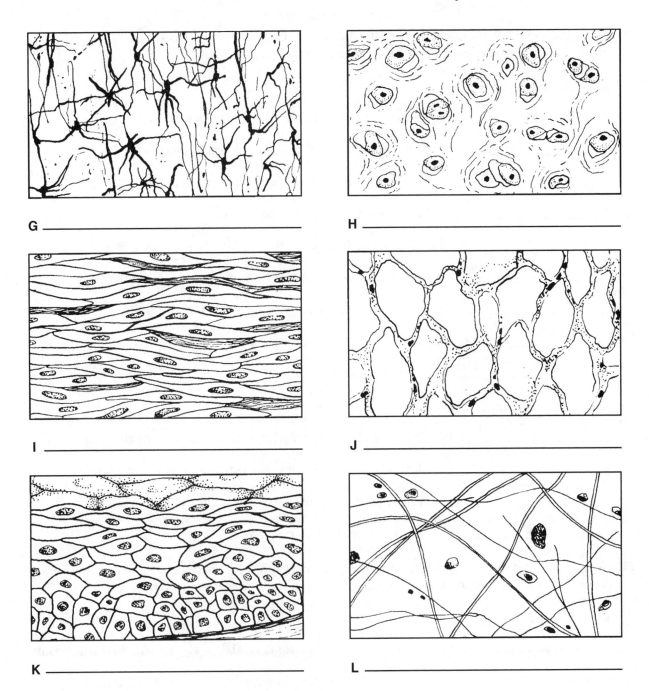

G _____

H _____

I _____

J _____

K _____

L _____

Figure 3–10, G–L

19. Describe briefly how the particular structure of a neuron relates to its function

in the body. _____

_____

_____

20. Using the key choices, correctly identify the *major* tissue types described. Enter the appropriate letter or tissue type term in the answer blanks.

Key Choices

A. Connective   B. Epithelium   C. Muscle   D. Nervous

_____ 1. Forms mucous, serous, and epidermal membranes

_____ 2. Allows for organ movements within the body

_____ 3. Transmits electrochemical impulses

_____ 4. Supports body organs

_____ 5. Cells of this tissue may absorb and/or secrete substances

_____ 6. Basis of the major controlling system of the body

_____ 7. The cells of this tissue shorten to exert force

_____ 8. Forms hormones

_____ 9. Packages and protects body organs

_____10. Characterized by having large amounts of nonliving matrix

_____11. Allows you to smile, grasp, swim, ski, and shoot an arrow

_____12. Most widely distributed tissue type in the body

_____13. Forms the brain and spinal cord

21. Using the key choices, identify the following specific type(s) of epithelial tissue. Enter the appropriate letter or classification term in the answer blanks.

Key Choices

A. Pseudostratified columnar (ciliated)   C. Simple cuboidal   E. Stratified squamous

B. Simple columnar   D. Simple squamous   F. Transitional

_____ 1. Lines the esophagus and forms the skin epidermis

_____ 2. Forms the lining of the stomach and small intestine

_____ 3. Best suited for areas subjected to friction

_____ 4. Lines much of the respiratory tract

_____ 5. Propels substances (e.g., mucus) across its surface

_____ 6. Found in the bladder lining; peculiar cells that slide over one another

_____ 7. Forms thin serous membranes; a single layer of flattened cells

22. The three types of muscle tissue exhibit certain similarities and differences. Check (✓) the appropriate spaces in the following table to indicate which muscle types exhibit each characteristic.

Characteristic	Skeletal	Cardiac	Smooth
1. Voluntarily controlled			
2. Involuntarily controlled			
3. Banded appearance			
4. Single nucleus in each cell			
5. Multinucleate			
6. Found attached to bones			
7. Allows you to direct your eyeballs			
8. Found in the walls of stomach, uterus, and arteries			
9. Contains spindle-shaped cells			
10. Contains cylindrical cells with branching ends			
11. Contains long, nonbranching cylindrical cells			
12. Displays intercalated discs			
13. Concerned with locomotion of the body as a whole			
14. Changes the internal volume of an organ as it contracts			
15. Tissue of the circulatory pump			

23. Circle the term that does not belong in each of the following groupings.

1. Collagen   Cell   Matrix   Cell product

2. Cilia   Flagellum   Microvilli   Elastic fibers

3. Glands   Bones   Epidermis   Mucosae

4. Adipose   Hyaline   Osseous   Nervous

5. Blood   Smooth   Cardiac   Skeletal

24. Using the key choices, identify the following connective tissue types. Insert the appropriate letter or corresponding term in the answer blanks.

Key Choices

A. Adipose connective tissue   C. Dense fibrous connective tissue   E. Reticular connective tissue

B. Areolar connective tissue   D. Osseous tissue   F. Hyaline cartilage

_____ 1. Provides great strength through parallel bundles of collagenic fibers; found in tendons

_____ 2. Acts as a storage depot for fat

_____ 3. Composes the dermis of the skin

_____ 4. Forms the bony skeleton

_____ 5. Composes the basement membrane and packages organs; includes a gel-like matrix with all categories of fibers and many cell types

_____ 6. Forms the embryonic skeleton and the surfaces of bones at the joints; reinforces the trachea

_____ 7. Provides insulation for the body

_____ 8. Structurally amorphous matrix, heavily invaded with fibers; appears glassy and smooth

_____ 9. Contains cells arranged concentrically around a nutrient canal; matrix is hard due to calcium salts

_____ 10. Forms the stroma or internal "skeleton" of lymph nodes, the spleen, and other lymphoid organs

Tissue Repair

25. For each of the following statements about tissue repair that is true, enter *T* in the answer blank. For each false statement, correct the underlined words by writing the correct words in the answer blank.

_____ 1. The nonspecific response of the body to injury is called underlined{regeneration}.

_____ 2. Intact capillaries near an injury dilate, leaking plasma, blood cells, and underlined{antibodies}, which cause the blood to clot. The clot at the surface dries to form a scab.

_____ 3. During the first phase of tissue repair, capillary buds invade the clot, forming a delicate pink tissue called underlined{endodermal} tissue.

_____ 4. When damage is not too severe, the surface epithelium migrates beneath the dry scab and across the surface of the granulation tissue. This repair process is called underlined{proliferation}.

_____ 5. If tissue damage is very severe, tissue repair is more likely to occur by <u>fibrosis</u>, or scarring.

_____ 6. During fibrosis, fibroblasts in the granulation tissue lay down <u>keratin</u> fibers, which form a strong, compact, but inflexible mass.

_____ 7. The repair of cardiac muscle and nervous tissue occurs mainly by <u>fibrosis</u>.

DEVELOPMENTAL ASPECTS OF CELLS AND TISSUES

26. Correctly complete each statement by inserting your responses in the answer blanks.

_____ 1.

_____ 2.

_____ 3.

_____ 4.

_____ 5.

_____ 6.

_____ 7.

_____ 8.

_____ 9.

_____ 10.

_____ 11.

_____ 12.

_____ 13.

_____ 14.

_____ 15.

_____ 16.

_____ 17.

_____ 18.

_____ 19.

_____ 20.

During embryonic development, cells specialize to form __(1)__. Mitotic cell division is very important for overall body __(2)__. All tissues except __(3)__ tissue continue to undergo cell division until the end of adolescence. After this time, __(4)__ tissue also becomes amitotic. When amitotic tissues are damaged, they are replaced by __(5)__ tissue, which does not function in the same way as the original tissue. This is a serious problem when heart cells are damaged.

Aging begins almost as soon as we are born. Three explanations of the aging process have been offered. One states that __(6)__ insults, such as the presence of toxic substances in the blood, are important. Another theory states that external __(7)__ factors, such as X-rays, help to cause aging. A third theory suggests that aging is programmed in our __(8)__. Three examples of aging processes seen in all people are __(9)__, __(10)__, and __(11)__.

Neoplasms occur when cells "go wild" and the normal controls of cell __(12)__ are lost. The two types of neoplasms are __(13)__ and __(14)__. The __(15)__ type tends to stay localized and have a capsule. The __(16)__ type is likely to invade other body tissues and spread to other (distant) parts of the body. To correctly diagnose the type of neoplasm, a microscopic examination of the tissue called a __(17)__ is usually done. Whenever possible, __(18)__ is the treatment of choice for neoplasms.

An overgrowth of tissue that is not considered to be a neoplasm is referred to as __(19)__. Conversely, a decrease in the size of an organ or tissue, resulting from loss of normal stimulation, is called __(20)__.

 INCREDIBLE JOURNEY

A Visualization Exercise for the Cell

A long, meandering membrane with dark globules clinging to its outer surface now comes into sight.

27. Where necessary, complete statements by inserting the missing words in the answer blanks.

_____ 1.

_____ 2.

_____ 3.

_____ 4.

_____ 5.

_____ 6.

_____ 7.

_____ 8.

_____ 9.

_____ 10.

For your second journey, you will be miniaturized to the size of a small protein molecule and will travel in a microsubmarine, specially designed to enable you to pass easily through living membranes. You are injected into the intercellular space between two epithelial cells, and you are instructed to observe one of these cells firsthand and to identify as many of its structures as possible.

You struggle briefly with the controls and then maneuver your microsub into one of these cells. Once inside the cell, you find yourself in a kind of "sea." This salty fluid that surrounds you is the __(1)__ of the cell.

Far below looms a large, dark, oval structure, much larger than anything else you can see. You conclude that it is the __(2)__. As you move downward, you pass a cigar-shaped structure with strange-looking folds on its inner surface. Although you have a pretty good idea that it must be a __(3)__, you decide to investigate more thoroughly. After passing through the external membrane of the structure, you are confronted with yet another membrane. Once past this membrane, you are inside the strange-looking structure. You activate the analyzer switch in your microsub for a readout indicating which molecules are in your immediate vicinity. As suspected, there is an abundance of energy-rich __(4)__ molecules. Having satisfied your curiosity, you leave this structure to continue the investigation.

A long, meandering membrane with dark globules clinging to its outer surface now comes into sight. You maneuver closer and sit back to watch the activity. As you watch, amino acids are joined together, and a long, threadlike protein molecule is built. The globules must be __(5)__, and the membrane, therefore, is the __(6)__. Once again, you head toward the large dark structure seen and tentatively identified earlier. On approach, you observe that this huge structure has very large openings in its outer wall; these openings must be the __(7)__. Passing through one of these openings, you discover that from the inside, the color of this structure is a result of dark, coiled, intertwined masses of __(8)__, which your analyzer confirms contain genetic material, or __(9)__ molecules. Making your way through this tangled mass, you pass two round, dense structures that appear to be full of the same type of globules you saw outside. These two round structures are __(10)__. All this information confirms your earlier identification of this cellular structure, so now you move to its exterior to continue observations.

_____ 11.
_____ 12.

Just ahead, you see what appears to be a mountain of flattened sacs with hundreds of small saclike vesicles at its edges. The vesicles seem to be migrating away from this area and heading toward the outer edges of the cell. The mountain of sacs must be the __(11)__. Eventually you come upon a rather simple-looking membrane-bound sac. Although it doesn't look too exciting and has few distinguishing marks, it does not resemble anything else you have seen so far. Deciding to obtain a chemical analysis before entering this sac, you activate the analyzer and on the screen you see "Enzymes — Enzymes — Hydrolases — Hydrolases — Danger — Danger." There is little doubt that this innocent-appearing structure is actually a __(12)__.

Completing your journey, you count the number of organelles identified so far. Satisfied that you have observed most of them, you request retrieval from the intercellular space.

AT THE CLINIC

28. Johnny lacerated his arm and rushed home to Mom so she could "fix it." His mother poured hydrogen peroxide over the area, and it bubbled vigorously where it came in contact with the wound. Because you can expect that cells were ruptured in the injured area, what do you *think* was happening here?

29. The epidermis (epithelium of the cutaneous membrane or skin) is a keratinized stratified squamous epithelium. Explain why that epithelium is much better suited for protecting the body's external surface than a mucosa consisting of a simple columnar epithelium would be.

30. Streptomycin (an antibiotic) binds to the small ribosomal subunit of bacteria (but not to the ribosomes of the host cells infected by bacteria). The result is the misreading of bacteria mRNA and the breakup of polysomes. What process is being affected, and how does this kill the bacterial cells?

31. Systemic lupus erythematosus (often simply called lupus) is a condition that primarily affects young women. It is a chronic (persistent) inflammation that affects all or most of the connective tissue proper in the body. Suzy is told by her doctor that she has lupus, and she asks if it will have widespread or merely localized effects within the body. What would the physician answer?

32. Mrs. Linsey sees her gynecologist because she is unable to become pregnant. The doctor discovers granulation tissue in her vaginal canal and explains that sperm are susceptible to some of the same chemicals as bacteria. What is inhibiting the sperm?

33. Sarah, a trainee of the electron microscopist at the local hospital, is reviewing some micrographs of muscle cells and macrophages (phagocytic cells). She notices that the muscle cells are loaded with mitochondria while the macrophages have abundant lysosomes. Why is this so?

34. Bradley tripped and tore one of the tendons surrounding his ankle. In anguish with pain, he asked his doctor how quickly he could expect it to heal. What do you think the doctor's response was and why?

35. In normally circulating blood, the plasma proteins cannot leave the bloodstream easily and, thus, tend to remain in the blood. But if stasis (blood flow stoppage) occurs, the proteins will begin to leak out into the interstitial fluid. Explain why this leads to edema (water buildup in the tissues).

36. Phagocytes gather in the air sacs of the lungs, especially in the lungs of smokers. What is the connection?

✔ THE FINALE: MULTIPLE CHOICE

37. Select the best answer or answers from the choices given.

1. A cell's plasma membrane would not contain:

   A. phospholipid   D. cholesterol

   B. nucleic acid   E. glycolipid

   C. protein

2. Which of the following would you expect to find in or on cells whose main function is absorption?

   A. Microvilli   D. Gap junctions

   B. Cilia   E. Secretory vesicles

   C. Desmosomes

3. Which cytoskeletal element interacts with myosin to produce contractile force in muscle cells?

   A. Microtubules

   B. Microfilaments

   C. Intermediate filaments

   D. None of the above

4. If a 10% sucrose solution within a semipermeable sac causes the fluid volume in the sac to increase a given amount when the sac is immersed in water, what would be the effect of replacing the sac solution with a 20% sucrose solution?

   A. The sac would lose fluid.

   B. The sac would gain the same amount of fluid.

   C. The sac would gain more fluid.

   D. There would be no effect.

5. Which of the following are possible functions of the glycocalyx?

   A. Determination of blood groups

   B. Binding sites for toxins

   C. Aiding the binding of sperm to egg

   D. Guiding embryonic development

   E. Increasing the efficiency of absorption

6. A cell stimulated to increase steroid production will have:

   A. abundant ribosomes

   B. a rough ER

   C. a smooth ER

   D. a Golgi apparatus

   E. abundant secretory vesicles

7. A cell's ability to replenish its ATP stores has been diminished by a metabolic poison. What organelle is most likely to be affected?

   A. Nucleus   D. Microtubule

   B. Plasma membrane   E. Mitochondrion

   C. Centriole

8. The fundamental structure of the plasma membrane is determined almost exclusively by:

   A. phospholipid molecules

   B. peripheral proteins

   C. cholesterol molecules

   D. integral proteins

9.–11. Consider the following information for Questions 9–11:

A DNA segment has this nucleotide sequence:

A A G C T C T T A C G A A T A T T C

9. Which mRNA is complementary?

A. A A G C T C T T A C G A A T A T T C

B. T T C G A G A A T G C T T A T A A G

C. A A G C U C U U A C G A A U A U U C

D. U U C G A G A A U G C U U A U A A G

10. How many amino acids are coded in this segment?

    A. 18   C. 6

    B. 9   D. 3

11. What is the tRNA anticodon sequence for the fourth codon from the left?

    A. G   C. GCU

    B. GC   D. CGA

12. The organelle that consists of a stack of 3–10 membranous discs associated with vesicles is:

A. mitochondrion

B. smooth ER

C. Golgi apparatus

D. lysosome

13. An epithelium "built" to stretch is:

A. simple squamous

B. stratified squamous

C. simple cuboidal

D. pseudostratified

E. transitional

14. Which of the following fibrous elements give a connective tissue high tensile strength?

A. Reticular fibers

B. Elastic fibers

C. Collagen fibers

D. Myofilaments

15. Viewed through the microscope, most cells in this type of tissue have only a rim of cytoplasm.

A. Reticular connective

B. Adipose connective

C. Areolar connective

D. Osseous tissue

E. Hyaline cartilage

16. Which type of cartilage is most abundant throughout life?

A. Elastic cartilage

B. Fibrocartilage

C. Hyaline cartilage

17. Which of the following terms describe skeletal muscle?

A. Striated

B. Intercalated discs

C. Multinucleated

D. Voluntary

E. Branching

18. Events of tissue repair include:

A. regeneration

B. organization

C. granulation

D. fibrosis

E. inflammation

19. Which of the following does *not* describe nervous tissue?

A. Cells may have long extensions

B. When activated, shortens

C. Found in the brain and spinal cord

D. Involved in fast-acting body control

4 SKIN AND BODY MEMBRANES

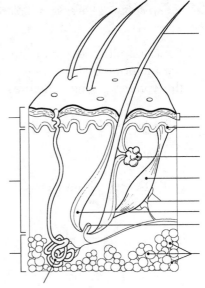

Body membranes, which cover body surfaces, line its cavities, and form protective sheets around organs, fall into two major categories. These are epithelial membranes (skin epidermis, mucosae, and serosae) and the connective tissue synovial membranes.

Topics for review in this chapter include a comparison of structure and function of various membranes, anatomical characteristics of the skin (composed of the connective tissue dermis and the epidermis) and its derivatives, and the manner in which the skin responds to both internal and external stimuli to protect the body.

CLASSIFICATION OF BODY MEMBRANES

1. Complete the following table relating to body membranes. Enter your responses in the areas left blank.

Membrane	Tissue type (epithelial/connective)	Common locations	Functions
Mucous	Epithelial sheet with underlying connective tissue (lamina propria)		
Serous		Lines internal ventral body cavities and covers their organs	
Cutaneous			Protection from external insults and water loss
Synovial		Lines cavities of synovial joints	

2. Four simplified diagrams are shown in Figure 4–1. Select different colors for the membranes listed below, and use them to color the coding circles and the corresponding structures.

◯ Cutaneous membrane   ◯ Parietal pleura (serosa)   ◯ Synovial membrane

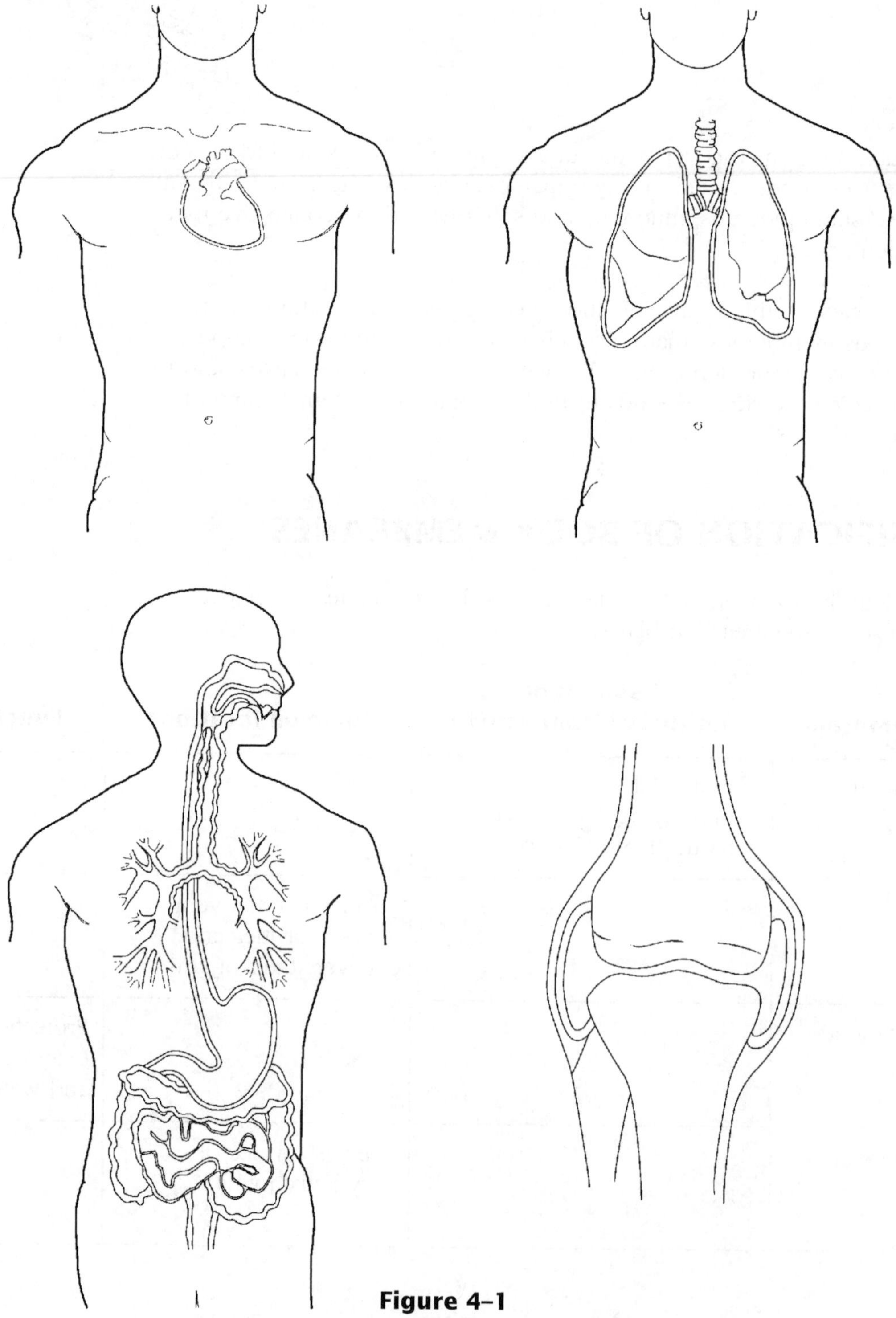

◯ Mucosae   ◯ Visceral pericardium (serosa)

◯ Visceral pleura (serosa)   ◯ Parietal pericardium (serosa)

Figure 4–1

INTEGUMENTARY SYSTEM (SKIN)
Basic Functions of the Skin

3. The skin protects the body by providing three types of barriers. Classify each of the protective factors listed below as an example of a chemical barrier (*C*), a biological barrier (*B*), or a mechanical (physical) barrier (*M*).

_____ 1. Epidermal dendritic cells and macrophages

_____ 2. Intact epidermis

_____ 3. Bactericidal secretions

_____ 4. Keratin

_____ 5. Melanin

_____ 6. Acid mantle

4. In what way does a sunburn impair the body's ability to defend itself?

(Assume the sunburn is mild.) _____

_____

5. Explain the role of sweat glands in maintaining body temperature homeostasis.

In your explanation, indicate how their activity is regulated. _____

_____

_____

_____

6. Complete the following statements. Insert your responses in the answer blanks.

_____ 1.

_____ 2.

_____ 3.

_____ 4.

_____ 5.

_____ 6.

_____ 7.

_____ 8.

The cutaneous sensory receptors that reside in the skin are actually part of the __(1)__ system. Four types of stimuli that can be detected by certain of the cutaneous receptors are __(2)__, __(3)__, __(4)__, and __(5)__.

Vitamin D is synthesized when modified __(6)__ molecules in the skin are irradiated by __(7)__ light. Vitamin D is important in the absorption and metabolism of __(8)__ ions.

Basic Structure of the Skin

7. Figure 4–2 depicts a longitudinal section of the skin. Label the skin structures and areas indicated by leader lines and brackets on the figure. Select different colors for the structures below and color the coding circles and the corresponding structures on the figure.

○ Arrector pili muscle

○ Adipose tissue

○ Hair follicle

○ Nerve fibers

○ Sweat (sudoriferous) gland

○ Sebaceous gland

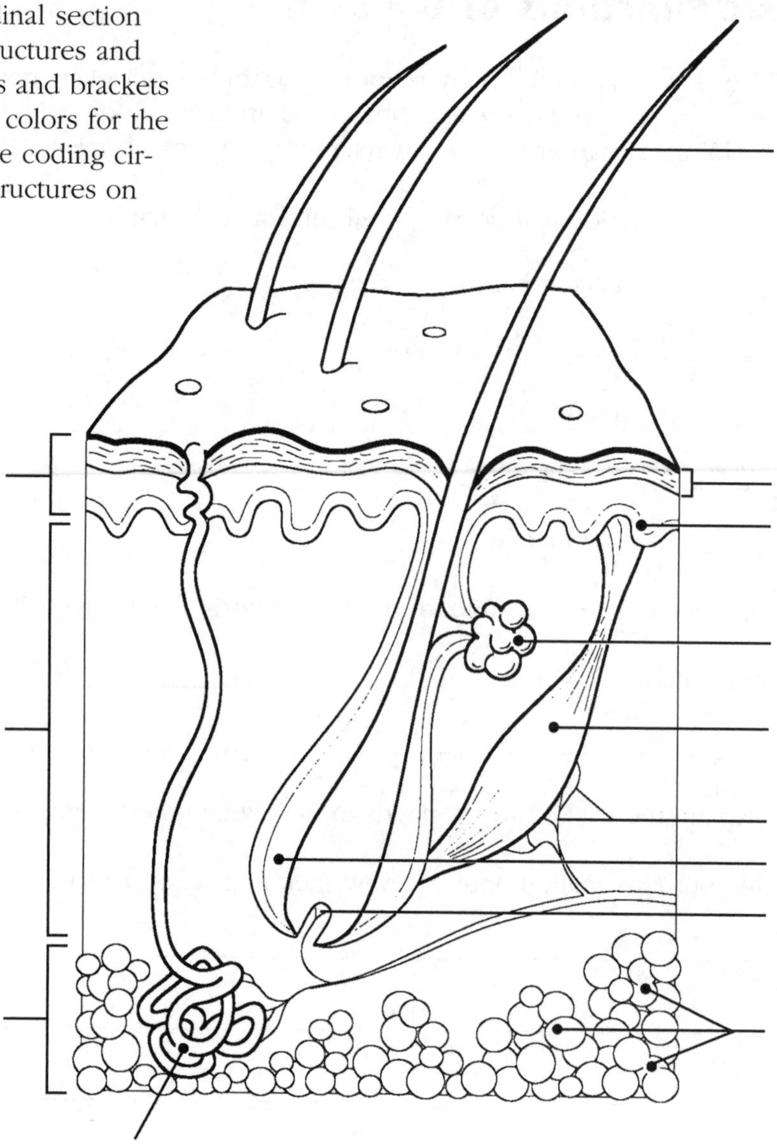

Figure 4–2

8. The more superficial cells of the epidermis become less viable and ultimately die. What two factors account for this natural demise of the epidermal cells?

1. _____

2. _____

_____

9. Using the key choices, choose all responses that apply to the following descriptions. Enter the appropriate letter(s) or term(s) in the answer blanks.

Key Choices

A. Stratum basale   D. Stratum lucidum   G. Reticular layer

B. Stratum corneum   E. Stratum spinosum   H. Epidermis as a whole

C. Stratum granulosum   F. Papillary layer   I. Dermis as a whole

_____ 1. Translucent cells, containing keratin

_____ 2. Strata containing all or mostly dead cells

_____ 3. Dermis layer responsible for fingerprints

_____ 4. Vascular region

_____ 5. Epidermal region involved in rapid cell division; most inferior epidermal layer

_____ 6. Scalelike cells full of keratin that constantly flake off

_____ 7. Site of elastic and collagen fibers

_____ 8. Site of melanin formation

_____ 9. Major skin area from which the derivatives (hair, nails) arise

_____ 10. Epidermal layer containing the oldest cells

_____ 11. When tanned becomes leather

10. Circle the term that does not belong in each of the following groupings.

1. Reticular layer   Keratin   Dermal papillae   Meissner's corpuscles

2. Melanin   Freckle   Wart   Malignant melanoma

3. Prickle cells   Stratum basale   Stratum spinosum   Cell shrinkage

4. Meissner's corpuscles   Lamellar corpuscles   Merkel's cells   Arrector pili

11. This exercise examines the relative importance of three pigments in determining skin color. Indicate which pigment is identified by the following descriptions by inserting the appropriate answer from the key choices in the answer blanks.

Key Choices

A. Carotene   B. Hemoglobin   C. Melanin

_____ 1. Most responsible for the skin color of dark-skinned people

_____ 2. Provides an orange cast to the skin

_____ 3. Provides a natural sunscreen

_____ 4. Most responsible for the skin color of light-skinned (Caucasian) people

_____ 5. Phagocytized by keratinocytes

_____ 6. Found predominantly in the stratum corneum

_____ 7. Found within red blood cells in the blood vessels

12. Complete the following statements in the blanks provided.

_____ 1. Radiation from the skin surface and evaporation of sweat are two ways in which the skin helps to get rid of body __(1)__.

_____ 2. Fat in the __(2)__ tissue layer beneath the dermis helps to insulate the body.

_____ 3. A vitamin that is manufactured in the skin is __(3)__.

_____ 4. Wrinkling of the skin is caused by loss of the __(4)__ of the skin.

_____ 5. A decubitus ulcer results when skin cells are deprived of __(5)__.

_____ 6. __(6)__ is a bluish cast of the skin resulting from inadequate oxygenation of the blood.

Appendages of the Skin

13. For each true statement, write *T*. For each false statement, correct the underlined word(s) and insert your correction in the answer blank.

_____ 1. A saltwater solution is secreted by sebaceous glands.

_____ 2. The most abundant protein in dead epidermal structures such as hair and nails is melanin.

_____ 3. Sebum is an oily mixture of lipids, cholesterol, and cell fragments.

_____ 4. The externally observable part of a hair is called the root.

_____ 5. The epidermis provides mechanical strength to the skin.

14. Figure 4–3 is a diagram of a cross-sectional view of a hair in its follicle. Complete this figure by following the directions in steps 1–3.

1. Identify the two portions of the follicle wall by placing the correct name of the sheath at the end of the appropriate leader line.

2. Use different colors to color these regions.

3. Label, color-code, and color the three following regions of the hair.

◯ Cortex   ◯ Cuticle   ◯ Medulla

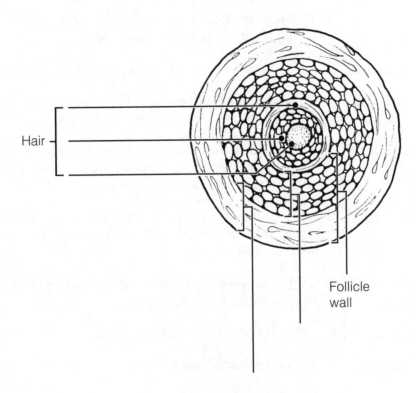

Hair

Follicle
wall

Figure 4–3

15. Circle the term that does not belong in each of the following groupings.

1. Luxuriant hair growth   Testosterone   Poor nutrition   Good blood supply

2. Vitamin D   Cholesterol   UV radiation   Keratin

3. Stratum corneum   Nail matrix   Hair bulb   Stratum basale

4. Scent glands   Eccrine glands   Apocrine glands   Axilla

5. Terminal hair   Vellus hair   Dark, coarse hair   Eyebrow hair

16. What is the scientific term for baldness? _____

17. Using the key choices, complete the following statements. Insert the appropriate letter(s) or term(s) in the answer blanks.

Key Choices

A. Arrector pili   C. Hair   E. Sebaceous glands   G. Sweat gland (eccrine)

B. Cutaneous receptors   D. Hair follicle(s)   F. Sweat gland (apocrine)

_____ 1. A blackhead is an accumulation of oily material produced by __(1)__.

_____ 2. Tiny muscles attached to hair follicles that pull the hair upright during fright or cold are called __(2)__.

_____ 3. The most numerous variety of perspiration gland is the __(3)__.

_____ 4. A sheath formed of both epithelial and connective tissues is the __(4)__.

_____ 5. A less numerous variety of perspiration gland is the __(5)__. Its secretion (often milky in appearance) contains proteins and other substances that favor bacterial growth.

_____ 6. __(6)__ is found everywhere on the body except the palms of the hands, soles of the feet, and lips, and it primarily consists of dead keratinized cells.

_____ 7. __(7)__ are specialized nerve endings that respond to temperature and touch, for example.

_____ 8. __(8)__ become more active at puberty.

_____ 9. Part of the heat-liberating apparatus of the body is the __(9)__.

_____ 10. Secretin contains bacteria-killing substances.

18. Circle the term that does not belong in each of the following groupings.

1. Sebaceous gland   Hair   Arrector pili   Epidermis

2. Radiation   Absorption   Conduction   Evaporation

3. Cortex   Medulla   Cuticle   Epithelial sheath

4. Scent glands   Eccrine glands   Apocrine glands   Axilla

5. Cyanosis   Erythema   Wrinkles   Pallor

19. Relative to nails:

1. What is the common name for the eponychium? _____

_____

2. Why does the lunula appear whiter than the rest of the nail? _____

_____

_____

Homeostatic Imbalances of the Skin

20. Overwhelming infection is one of the most important causes of death in burn patients. What is the other major problem they face, and what are its possible consequences?

_____

_____

21. This section reviews the severity of burns. Using the key choices, select the correct burn type for each of the following descriptions. Enter your answers in the answer blanks.

Key Choices

A. First-degree burn   B. Second-degree burn   C. Third-degree burn

_____ 1. Full-thickness burn; epidermal and dermal layers destroyed; skin is blanched

_____ 2. Blisters form

_____ 3. Epidermal damage, redness, and some pain (usually brief)

_____ 4. Epidermal and some dermal damage; pain; regeneration is possible

_____ 5. Regeneration impossible; requires grafting

_____ 6. Pain is absent because nerve endings in the area are destroyed

22. What is the importance of the "rule of nines" in treatment of burn patients?

_____

23. Fill in the type of skin cancer that matches each of the following descriptions:

_____ 1. Epithelial cells, not in contact with the basement membrane, develop lesions; metastasize

_____ 2. Cells of the lowest level of the epidermis invade the dermis and hypodermis; exposed areas develop ulcer; slow to metastasize

_____ 3. Rare but often deadly cancer of pigment-producing cells

24. What does ABCD mean in reference to examination of pigmented areas? _____

_____

_____

DEVELOPMENTAL ASPECTS OF THE SKIN AND BODY MEMBRANES

25. Match the choices (letters or terms) in Column B with the appropriate descriptions in Column A.

Column A

_____ 1. Skin inflammations that increase in frequency with age

_____ 2. Cause of graying hair

_____ 3. Small white bumps on the skin of newborn babies, resulting from accumulations of sebaceous gland material

_____ 4. Reflects the loss of insulating subcutaneous tissue with age

_____ 5. A common consequence of accelerated sebaceous gland activity during adolescence

_____ 6. Oily substance produced by the fetus's sebaceous glands

_____ 7. The hairy "cloak" of the fetus

Column B

A. Acne

B. Cold intolerance

C. Dermatitis

D. Delayed-action gene

E. Lanugo

F. Milia

G. Vernix caseosa

INCREDIBLE JOURNEY

A Visualization Exercise for the Skin

Your immediate surroundings resemble huge grotesquely twisted vines . . . you begin to climb upward.

26. Where necessary, complete statements by inserting the missing words in the answer blanks.

For this trip, you are miniaturized for injection into your host's skin. Your journey begins when you are deposited in a soft gel-like substance. Your immediate surroundings resemble huge grotesquely twisted vines. But when you peer carefully at the closest "vine," you realize you are actually seeing

_____ 1.

_____ 2.

_____ 3.

_____ 4.

_____ 5.

_____ 6.

_____ 7.

_____ 8.

_____ 9.

_____10.

connective tissue fibers. Although tangled together, most of the fibers are fairly straight and look like strong cables. You identify these as the __(1)__ fibers. Here and there are fibers that resemble coiled springs. These must be the __(2)__ fibers that help give skin its springiness. At this point, there is little question that you are in the __(3)__ region of the skin, particularly considering that you can also see blood vessels and nerve fibers around you.

Carefully, using the fibers as steps, you begin to climb upward. After climbing for some time and finding that you still haven't reached the upper regions of the skin, you stop for a rest. As you sit, a strange-looking cell approaches, moving slowly with parts alternately flowing forward and then receding. Suddenly you realize that this must be a __(4)__ that is about to dispose of an intruder (you) unless you move in a hurry! You scramble to your feet and resume your upward climb. On your right is a large fibrous structure that looks like a tree trunk anchored in place by muscle fibers. By scurrying up this __(5)__ sheath, you are able to escape from the cell. Once safely out of harm's way, you again scan your surroundings. Directly overhead are tall cubelike cells, forming a continuous sheet. In your rush to escape you have reached the __(6)__ layer of the skin. As you watch the activity of the cells in this layer, you notice that many of the cells are pinching in two, and the daughter cells are being forced upward. Obviously, this is the layer that continually replaces cells that rub off the skin surface, and these cells are the __(7)__ cells.

Looking through the transparent cell membrane of one of the basal cells, you see a dark mass hanging over its nucleus. You wonder if this cell could have a tumor; but then, looking through the membranes of the neighboring cells, you find that they also have dark umbrella-like masses hanging over their nuclei. As you consider this matter, a black cell with long tentacles begins to pick its way carefully between the other cells. As you watch with interest, one of the transparent cells engulfs the tip of one of the black cell's tentacles. Within seconds a black substance appears above the transparent cell's nucleus. Suddenly, you remember that one of the skin's functions is to protect the deeper layers from sun damage; the black substance must be the protective pigment __(8)__.

Once again you begin your upward climb and notice that the cells are becoming shorter and harder and are full of a waxy-looking substance. This substance has to be __(9)__, which would account for the increasing hardness of the cells. Climbing still higher, the cells become flattened like huge shingles. The only material apparent in the cells is the waxy substance—there is no nucleus, and there appears to be no activity in these cells. Considering the clues—shingle-like cells, no nuclei, full of the waxy substance, no activity—these cells are obviously __(10)__ and therefore very close to the skin surface.

Suddenly, you feel a strong agitation in your immediate area. The pressure is tremendous. Looking upward through the transparent cell layers, you see your host's fingertips vigorously scratching the area directly overhead. You wonder if you are causing his skin to sting or tickle. Then, within seconds, the cells around you begin to separate and fall apart, and you are catapulted out into the sunlight. Since the scratching fingers might descend once again, you quickly advise your host of your whereabouts.

AT THE CLINIC

27. Mrs. Ibañez volunteered to help at a hospital for children with cancer. When she first entered the cancer ward, she was upset by the fact that most of the children had no hair. What is the explanation for their baldness?

28. Linda, new mother, brings her infant to the clinic, worried about a yellowish, scummy deposit that has built up on the baby's scalp. What is this condition called, and is it serious?

29. Patients in hospital beds are rotated every 2 hours to prevent bedsores. Exactly why is this effective?

30. Eric and his wife are of northern European descent. Eric is a proud new father who was in the delivery room during his daughter's birth. He tells you that when she was born, her skin was purple and covered with a cream cheese–like substance. Shortly after birth, her skin turned pink. Can you explain his observations?

31. Would you expect to find the highest rate of skin cancer among the blacks of tropical Africa, research scientists in the Arctic, Norwegians in the southern United States, or blacks in the United States? Explain your choice.

32. After studying the skin in anatomy class, Toby grabbed the large "love handles" at his waist and said, "I have too thick a hypodermis, but that's okay because this layer performs some valuable functions!" What are the functions of the hypodermis?

33. A man got his finger caught in a machine at the factory. The damage was less serious than expected, but nonetheless, the entire nail was torn from his right index finger. The parts lost were the body, root, bed, matrix, and cuticle of the nail. First, define each of these parts. Then, tell if this nail is likely to grow back.

34. In cases of a ruptured appendix, what serous membrane is likely to become infected? Why can this be life threatening?

35. Mrs. Gaucher received second-degree burns on her abdomen when she dropped a kettle of boiling water. She asked the clinic physician (worriedly) if she would have to have a skin graft. What do you think he told her?

36. What two factors in the treatment of critical third-degree burn patients are absolutely essential?

37. Both newborn and aged individuals have very little subcutaneous tissue. How does this affect their sensitivity to cold?

✓ THE FINALE: MULTIPLE CHOICE

38. Select the best answer or answers from the choices given.

1. Which is *not* part of the skin?

   A. Epidermis   C. Dermis

   B. Hypodermis   D. Superficial fascia

2. Which of the following is *not* a tissue type found in the skin?

   A. Stratified squamous epithelium

   B. Loose connective tissue

   C. Dense irregular connective tissue

   D. Ciliated columnar epithelium

   E. Vascular tissue

3. Epidermal cells that aid in the immune response include:

   A. Merkel's cells   C. melanocytes

   B. dendritic cells   D. spinosum cells

4. Which epidermal layer has a high concentration of Langerhans' cells and has numerous desmosomes and thick bundles of keratin filaments?

   A. Stratum corneum

   B. Stratum lucidum

   C. Stratum granulosum

   D. Stratum spinosum

5. Fingerprints are caused by:

   A. the genetically determined arrangement of dermal papillae

   B. the conspicuous epidermal ridges

   C. the sweat pores

   D. all of these

6. Some infants are born with a fuzzy skin; this is due to:

   A. vellus hairs   C. lanugo

   B. terminal hairs   D. hirsutism

7. What is the major factor accounting for the waterproof nature of the skin?

   A. Desmosomes in stratum corneum

   B. Glycolipid between stratum corneum cells

   C. The thick insulating fat of the hypodermis

   D. The leathery nature of the dermis

8. Which of the following is *true* concerning oil production in the skin?

   A. Oil is produced by sudoriferous glands.

   B. Secretion of oil is the job of the apocrine glands.

   C. The secretion is called sebum.

   D. Oil is usually secreted into hair follicles.

9. Contraction of the arrector pili would be "sensed" by:

   A. Merkel's discs

   B. tactile corpuscles

   C. hair follicle receptors

   D. lamellated corpuscles

10. A dermatologist examines a patient with lesions on the face. Some of the lesions appear as shiny, raised spots; others are ulcerated with beaded edges. What is the diagnosis?

    A. Melanoma

    B. Squamous cell carcinoma

    C. Basal cell carcinoma

    D. Either squamous or basal cell carcinoma

11. Components of sweat include:

    A. water   D. ammonia

    B. sodium chloride   E. sebum

    C. vitamin D

12. A burn patient reports that the burns on her hands and face are not painful, but she has blisters on her neck and forearms and the skin on her arms is very red. This burn would be classified as:

    A. first-degree only

    B. second-degree only

    C. third-degree only

    D. critical

13. The reticular layer of the dermis is most important in providing:

    A. strength and elasticity to the skin

    B. toughness to the skin

    C. insulation to prevent heat loss

    D. the dermal papilla, which produce fingerprints

14. Which of the following is *not* associated with sweat?

    A. Sweat glands

    B. Holocrine glands

    C. Eccrine glands

    D. Apocrine glands

5 THE SKELETAL SYSTEM

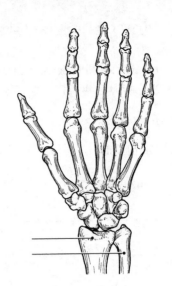

The skeleton is constructed of two of the most supportive tissues found in the human body—cartilage and bone. Besides supporting and protecting the body as an internal framework, the skeleton provides a system of levers that the skeletal muscles use to move the body. In addition, the bones provide a storage depot for substances such as lipids and calcium, and blood cell formation goes on within the red marrow cavities of bones.

The skeleton consists of bones connected at joints, or articulations, and is subdivided into two divisions. The axial skeleton includes those bones that lie around the body's center of gravity. The appendicular skeleton includes the bones of the limbs.

Topics for student review include structure and function of long bones, location and naming of specific bones in the skeleton, fracture types, and a classification of joint types in the body.

BONES—AN OVERVIEW

1. Classify each of the following terms as a projection (*P*) or a depression or opening (*D*). Enter the appropriate letter in the answer blanks.

   _____ 1. Condyle   _____ 4. Foramen   _____ 7. Ramus

   _____ 2. Crest   _____ 5. Head   _____ 8. Spine

   _____ 3. Fissure   _____ 6. Meatus   _____ 9. Tuberosity

2. Group each of the following bones into one of the four major bone categories. Use *L* for long bone, *S* for short bone, *F* for flat bone, and *I* for irregular bone. Enter the appropriate letter in the space provided.

   _____ 1. Calcaneus   _____ 4. Humerus   _____ 7. Radius

   _____ 2. Frontal   _____ 5. Mandible   _____ 8. Sternum

   _____ 3. Femur   _____ 6. Metacarpal   _____ 9. Vertebra

3. Using the key choices, characterize the following statements relating to long bones. Enter the appropriate term(s) or letter(s) in the answer blanks.

Key Choices

A. Diaphysis   C. Epiphysis   E. Yellow marrow cavity

B. Epiphyseal plate   D. Red marrow

_____ 1. Site of spongy bone in the adult

_____ 2. Site of compact bone in the adult

_____ 3. Site of hematopoiesis in the adult

_____ 4. Scientific name for bone shaft

_____ 5. Site of fat storage in the adult

_____ 6. Site of longitudinal growth in a child

4. Complete the following statements concerning bone formation and destruction, using the terms provided in the key choices. Insert the key letter or corresponding term in the answer blanks.

Key Choices

A. Atrophy   C. Gravity   E. Osteoclasts   G. Parathyroid hormone

B. Calcitonin   D. Osteoblasts   F. Osteocytes   H. Stress and/or tension

_____ 1. When blood calcium levels begin to drop below homeostatic levels, __(1)__ is released, causing calcium to be released from bones.

_____ 2. Mature bone cells, called __(2)__, maintain bone in a viable state.

_____ 3. Disuse such as that caused by paralysis or severe lack of exercise results in muscle and bone __(3)__.

_____ 4. Large tubercles and/or increased deposit of bony matrix occur at sites of __(4)__.

_____ 5. Immature, or matrix-depositing, bone cells are referred to as __(5)__.

_____ 6. __(6)__ causes blood calcium to be deposited in bones as calcium salts.

_____ 7. Bone cells that liquefy bone matrix and release calcium to the blood are called __(7)__.

_____ 8. Our astronauts must do isometric exercises when in space because bones atrophy under conditions of weightlessness or lack of __(8)__.

5. Five descriptions of bone structure are provided in Column A. First identify the structure by choosing the appropriate term from Column B and placing the corresponding answer in the answer blank. Then consider Figure 5–1A, a diagrammatic view of a cross section of bone, and Figure 5–1B, a higher magnification view of compact bone tissue. Select different colors for the structures and bone areas in Column B, and use them to color the coding circles and corresponding structures on the figure diagrams. Because the concentric lamellae would be difficult to color without confusing other elements, identify one lamella by using a bracket and label.

Column A

_____ 1. Layers of calcified matrix

_____ 2. "Residences" of osteocytes

_____ 3. Longitudinal canal, carrying blood vessels and nerves

_____ 4. Nonliving, structural part of bone

_____ 5. Tiny canals, connecting lacunae

Column B

A. Central (Haversian) canal ◯

B. Concentric lamellae

C. Lacunae ◯

D. Canaliculi ◯

E. Bone matrix ◯

F. Osteocyte ◯

A

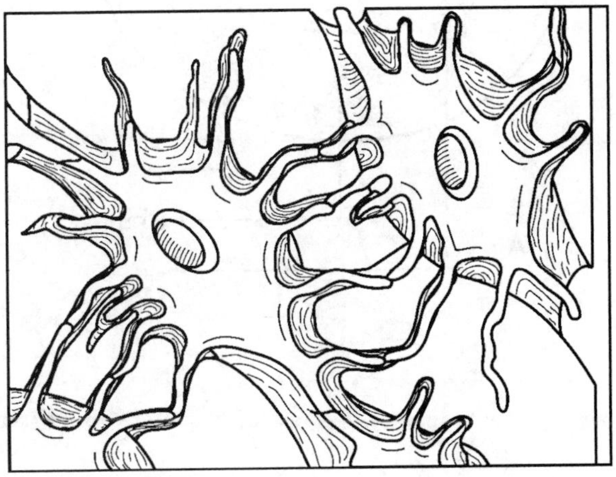

B

Figure 5–1

6. Circle the term that does not belong in each of the following groupings.

1. Hematopoiesis   Red marrow   Yellow marrow   Spongy bone

2. Lamellae   Canaliculi   Circulation   Osteoblasts

3. Osteon   Marrow cavity   Central canal   Canaliculi

4. Epiphysis surface   Articular cartilage   Periosteum   Hyaline cartilage

7. Figure 5–2A is a mid level, cross-sectional view of the diaphysis of the femur. Label the membrane that lines the cavity and the membrane that covers the outside surface.

Figure 5–2B is a drawing of a longitudinal section of the femur. Color the bone tissue gold. Do *not* color the articular cartilage; leave it white. Select different colors for the bone regions listed at the coding circles below. Color the coding circles and the corresponding regions on the drawing. Complete Figure 5–2B by labeling compact bone and spongy bone.

◯ Diaphysis

◯ Epiphyseal plate

◯ Area where red marrow is found

◯ Area where yellow marrow is found

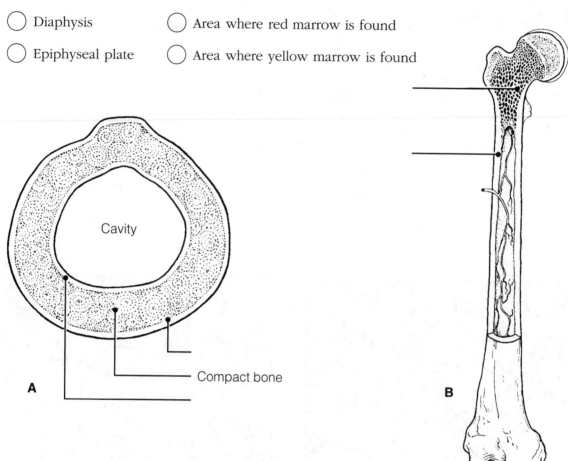

Figure 5–2

8. The following events apply to the endochondral ossification process as it occurs in the primary ossification center. Put these events in their proper order by assigning each a number (1–6).

_____ 1. Cavity formation occurs within the hyaline cartilage.

_____ 2. Collar of bone is laid down around the hyaline cartilage model just beneath the periosteum.

_____ 3. Periosteal bud invades the marrow cavity.

_____ 4. Perichondrium becomes vascularized to a greater degree and becomes a periosteum.

_____ 5. Osteoblasts lay down bone around the cartilage spicules in the bone's interior.

_____ 6. Osteoclasts remove the cancellous bone from the shaft interior, leaving a marrow cavity that then houses fat.

AXIAL SKELETON
Skull

9. Using the key choices, identify the bones indicated by the following descriptions. Enter the appropriate term or letter in the answer blanks.

_____ 1. Forehead bone

_____ 2. Cheek bone

_____ 3. Lower jaw

_____ 4. Bridge of nose

_____ 5. Posterior part of hard palate

_____ 6. Much of the lateral and superior cranium

_____ 7. Most posterior part of cranium

_____ 8. Single, irregular, bat-shaped bone, forming part of the cranial floor

_____ 9. Tiny bones, bearing tear ducts

_____ 10. Anterior part of hard palate

_____ 11. Superior and middle nasal conchae formed from its projections

_____ 12. Site of mastoid process

_____ 13. Site of sella turcica

_____ 14. Site of cribriform plate

_____ 15. Site of mental foramen

_____ 16. Site of styloid process

_____ 17. _____ 18. Four bones, containing paranasal sinuses

_____ 19. _____ 20.

_____ 21. Its condyles articulate with the atlas

_____ 22. Foramen magnum contained here

_____ 23. Middle ear found here

_____ 24. Nasal septum

_____ 25. Bears an upward protrusion, the "cock's comb," or crista galli

_____ 26. Site of external acoustic meatus

Key Choices

A. Ethmoid

B. Frontal

C. Hyoid

D. Lacrimals

E. Mandible

F. Maxillae

G. Nasals

H. Occipital

I. Palatines

J. Parietals

K. Sphenoid

L. Temporals

M. Vomer

N. Zygomatic

10. For each statement that is true, insert *T* in the answer blank. For false statements, correct the <u>underlined</u> words by inserting the correct words in the answer blanks.

_____ 1. When a bone forms from a fibrous membrane, the process is called <u>endochondral</u> ossification.

_____ 2. When trapped in lacunae, osteoblasts change into <u>osteocytes</u>.

_____ 3. Large numbers of <u>osteocytes</u> are found in the inner periosteum layer.

_____ 4. <u>Primary</u> ossification centers appear in the epiphyses of a long bone.

_____ 5. Epiphyseal plates are made of <u>spongy bone</u>.

_____ 6. In appositional growth, bone reabsorption occurs on the <u>periosteal</u> surface.

_____ 7. "Maturation" of newly formed (noncalcified) bone matrix takes about <u>10 days</u>.

11. Figure 5–3, A–C shows lateral, anterior, and inferior views of the skull. Select different colors for the bones listed below and color the coding circles and corresponding bones in the figure. Complete the figure by labeling the bone markings indicated by leader lines.

◯ Frontal

◯ Parietal

◯ Mandible

◯ Maxilla

◯ Sphenoid

◯ Ethmoid

◯ Temporal

◯ Zygomatic

◯ Palatine

◯ Occipital

◯ Nasal

◯ Lacrimal

◯ Vomer

Figure 5–3, A–C

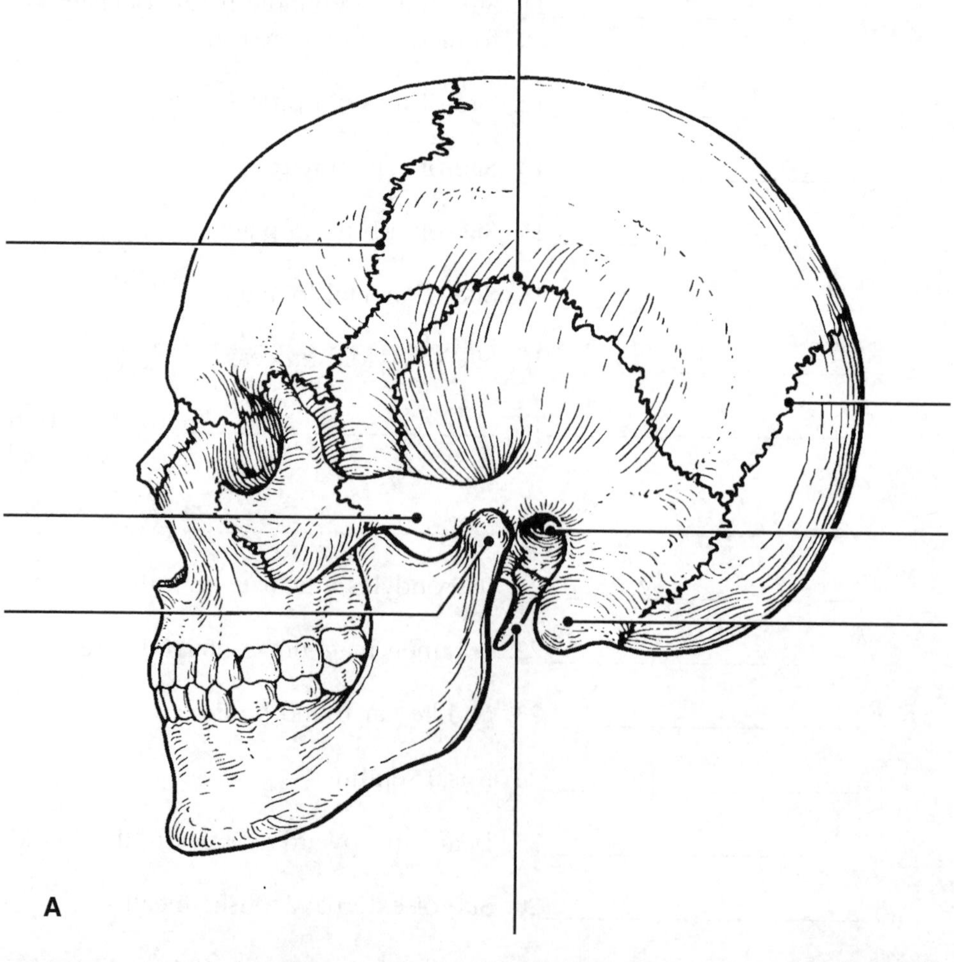

A

B

C

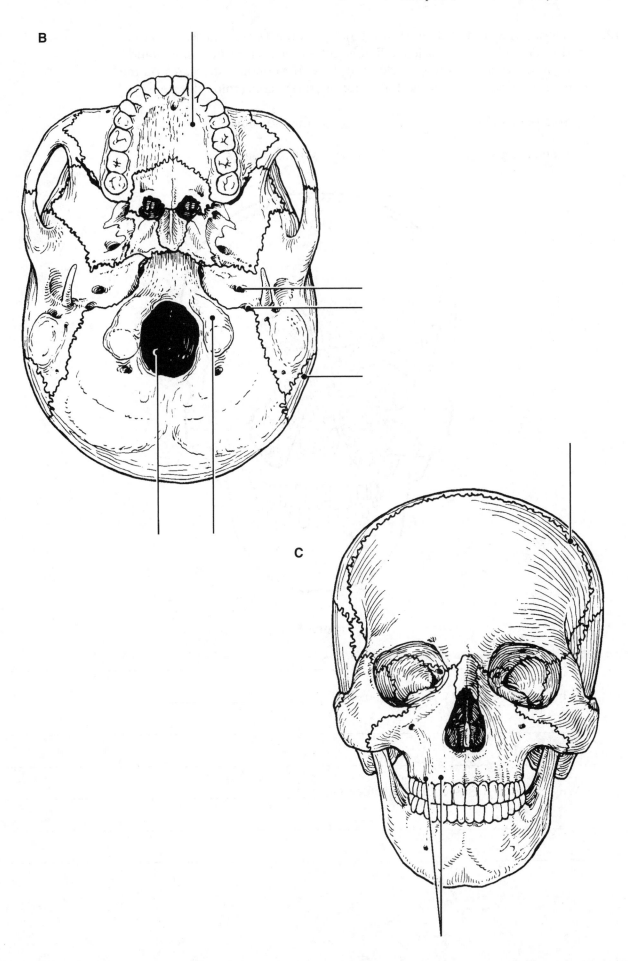

12. An anterior view of the skull, showing the positions of the sinuses, is provided in Figure 5–4. First select different colors for each of the sinuses and use them to color the coding circles and the corresponding structures on the figure. Then briefly answer the following questions concerning the sinuses.

◯ Sphenoid sinus   ◯ Ethmoid sinuses

◯ Frontal sinus   ◯ Maxillary sinus

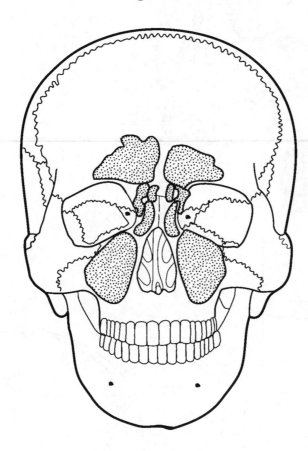

Figure 5–4

1. What *are* sinuses? _____

2. What purpose do they serve in the skull? _____

_____

_____

3. Why are they so susceptible to infection? _____

_____

_____

Vertebral Column

13. Using the key choices, correctly identify the vertebral parts/areas described as follows. Enter the appropriate term(s) or letter(s) in the spaces provided.

Key Choices

A. Body

B. Intervertebral foramina

C. Spinous process

D. Superior articular process

E. Transverse process

F. Vertebral arch

_____ 1. Structure that encloses the nerve cord

_____ 2. Weight-bearing part of the vertebra

_____ 3. Provide(s) levers for the muscles to pull against

_____ 4. Provide(s) an articulation point for the ribs

_____ 5. Openings allowing spinal nerves to pass

14. The following statements provide distinguishing characteristics of the vertebrae composing the vertebral column. Using the key choices, identify each described structure or region by inserting the appropriate term(s) or letter(s) in the spaces provided.

Key Choices

A. Atlas

B. Axis

C. Cervical vertebra—typical

D. Coccyx

E. Lumbar vertebra

F. Sacrum

G. Thoracic vertebra

_____ 1. Type of vertebra(e) containing foramina in the transverse processes, through which the vertebral arteries ascend to reach the brain

_____ 2. Its dens provides a pivot for rotation of the first cervical vertebra

_____ 3. Transverse processes have facets for articulation with ribs; spinous process points sharply downward

_____ 4. Composite bone; articulates with the hip bone laterally

_____ 5. Massive vertebrae; weight-sustaining

_____ 6. Tail bone; vestigal fused vertebrae

_____ 7. Supports the head; allows the rocking motion of the occipital condyles

_____ 8. Seven components; unfused

_____ 9. Twelve components; unfused

15. Complete the following statements by inserting your answers in the answer blanks.

_____ 1. In describing abnormal curvatures, it could be said that __(1)__ is an exaggerated thoracic curvature, and in __(2)__ the verte-

_____ 2. bral column is displaced laterally.

_____ 3. Invertebral discs are made of __(3)__ tissue. The discs provide __(4)__ to the spinal column.

_____ 4.

16. Figure 5–5, A–D shows superior views of four types of vertebrae. In the spaces provided below each vertebra, indicate in which region of the spinal column it would be found. In addition, specifically identify Figure 5–5A. Where indicated by leader lines, identify the vertebral body, spinous and transverse processes, superior articular processes, and vertebral foramen.

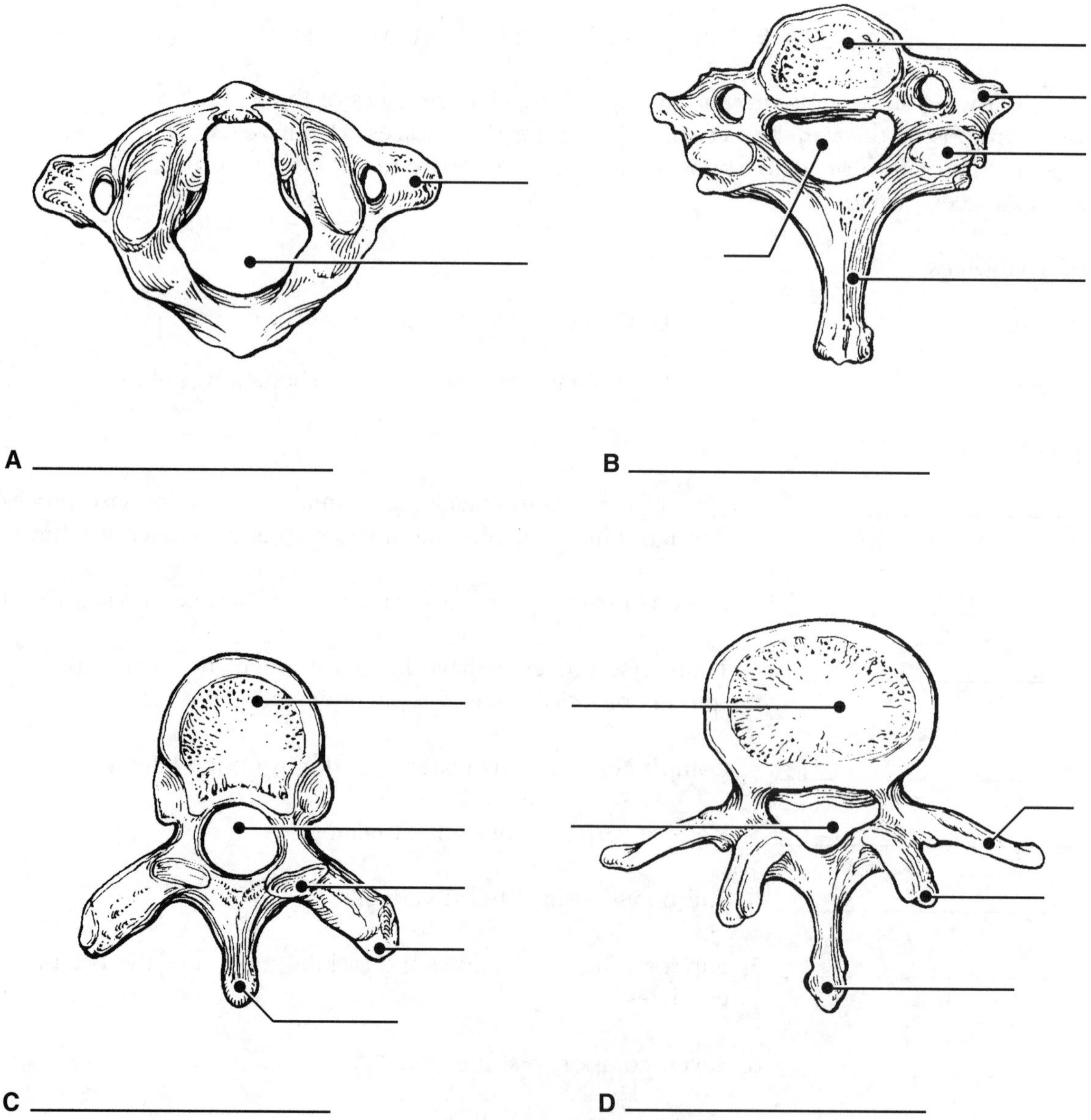

A _____   B _____

C _____   D _____

Figure 5–5

17. Figure 5–6 is a lateral view of the vertebral column. Identify each numbered region of the column by listing in the numbered answer blanks the region name first and then the specific vertebrae involved (for example, sacral region, S# to S#). Also identify the modified vertebrae indicated by numbers 6 and 7 in Figure 5–6. Select different colors for each vertebral region and use them to color the coding circles and the corresponding regions.

1. _____ ◯

2. _____ ◯

3. _____ ◯

4. _____ ◯

5. _____ ◯

6. _____ ◯

7. _____ ◯

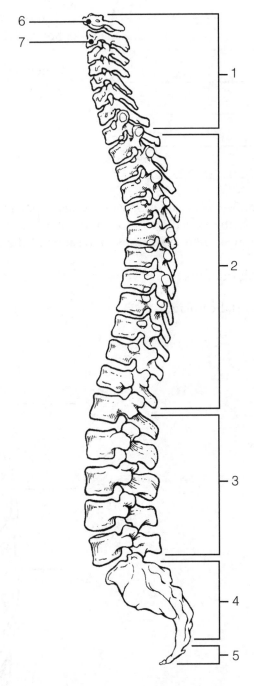

Figure 5–6

Thoracic Cage

18. Complete the following statements referring to the thoracic cage by inserting your responses in the answer blanks.

_____ 1.

_____ 2.

_____ 3.

_____ 4.

_____ 5.

_____ 6.

_____ 7.

_____ 8.

1. The organs protected by the thoracic cage include the __(1)__ and the __(2)__. Ribs 1 through 7 are called __(3)__ ribs, whereas ribs 8 through 12 are called __(4)__ ribs. Ribs 11 and 12 are also called __(5)__ ribs. All ribs articulate posteriorly with the __(6)__, and most connect anteriorly to the __(7)__, either directly or indirectly.

The general shape of the thoracic cage is __(8)__.

19. Figure 5–7 is an anterior view of the thoracic cage. Select different colors to identify the structures below and color the coding circles and corresponding structures. Then label the subdivisions of the sternum indicated by leader lines.

○ All true ribs   ○ All false ribs

○ Costal cartilages ○ Sternum

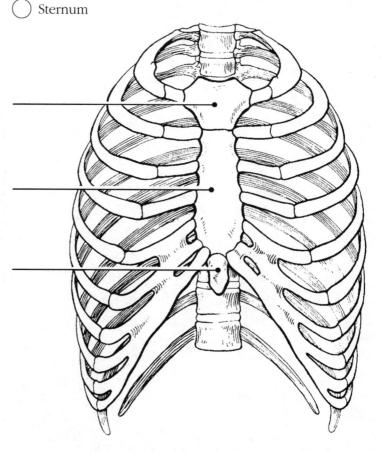

Figure 5–7

APPENDICULAR SKELETON

Several bones forming part of the upper limb and/or shoulder girdle are shown in Figures 5–8 to 5–11. Follow the specific directions for each figure.

20. Identify the bone in Figure 5–8. Insert your answer in the blank below the illustration. Select different colors for each structure listed below and use them to color the coding circles and the corresponding structures in the diagram. Then, label the angles indicated by leader lines.

○ Spine   ○ Glenoid cavity   ○ Coracoid process   ○ Acromion

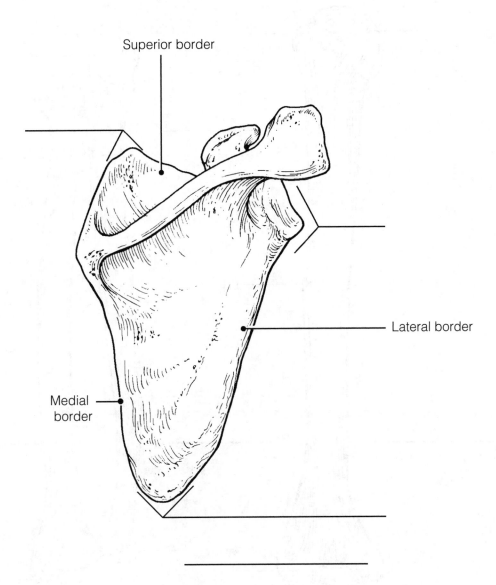

Superior border

Lateral border

Medial border

Figure 5–8

21. Identify the bones in Figure 5–9 by labeling the leader lines identified as A, B, and C. Color the bones different colors. Using the following terms, complete the illustration by labeling all bone markings provided with leader lines.

Trochlear notch   Capitulum   Coronoid process

Trochlea   Deltoid tuberosity   Olecranon process

Radial tuberosity   Head (three)   Greater tubercle

                         Styloid process   Lesser tubercle

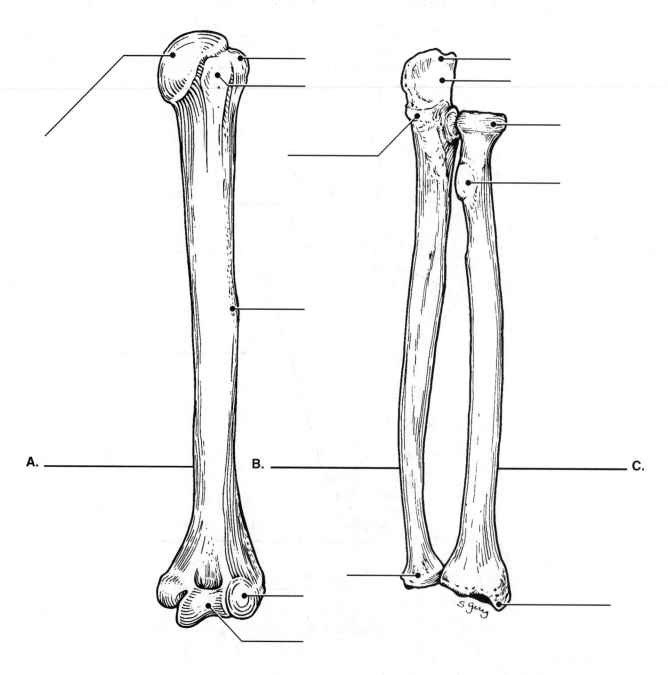

Figure 5–9

22. Figure 5–10 is a diagram of the hand. Select different colors for the following structures, and use them to color the coding circles and the corresponding structures in the diagram.

◯ Carpals   ◯ Metacarpals   ◯ Phalanges

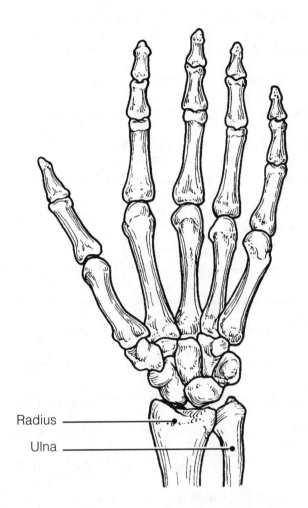

Radius

Ulna

Figure 5–10

23. Compare the pectoral and pelvic girdles by choosing descriptive terms from the key choices. Insert the appropriate key letters in the answer blanks.

Key Choices

A. Flexibility   D. Shallow socket for limb attachment

B. Massive   E. Deep, secure socket for limb attachment

C. Lightweight   F. Weight-bearing

Pectoral: _____, _____, _____   Pelvic: _____, _____, _____

24. Using the key choices, identify the bone names or markings according to the descriptions that follow. Insert the appropriate term or letter in the answer blanks.

Key Choices

A. Acromion	F. Coronoid fossa	K. Olecranon fossa	P. Scapula
B. Capitulum	G. Deltoid tuberosity	L. Olecranon process	Q. Sternum
C. Carpals	H. Glenoid cavity	M. Phalanges	R. Styloid process
D. Clavicle	I. Humerus	N. Radial tuberosity	S. Trochlea
E. Coracoid process	J. Metacarpals	O. Radius	T. Ulna

_____ 1. Raised area on lateral surface of humerus to which deltoid muscle attaches

_____ 2. Arm bone

_____ 3. _____ 4. Bones composing the shoulder girdle

_____ 5. _____ 6. Forearm bones

_____ 7. Point where scapula and clavicle connect

_____ 8. Shoulder girdle bone that has no attachment to the axial skeleton

_____ 9. Shoulder girdle bone that articulates anteriorly with the sternum

_____ 10. Socket in the scapula for the arm bone

_____ 11. Process above the glenoid cavity that permits muscle attachment

_____ 12. Commonly called the collar bone

_____ 13. Distal medial process of the humerus; joins the ulna

_____ 14. Medial bone of the forearm in anatomical position

_____ 15. Rounded knob on the humerus that articulates with the radius

_____ 16. Anterior depression; superior to the trochlea; receives part of the ulna when the forearm is flexed

_____ 17. Forearm bone involved in formation of elbow joint

_____ 18. _____ 19. Bones that articulate with the clavicle

_____ 20. Bones of the wrist

_____ 21. Bones of the fingers

_____ 22. Heads of these bones form the knuckles

25. Figure 5–11 is a diagram of the articulated pelvis. Identify the bones and bone markings indicated by leader lines on the figure. Select different colors for the structures listed below and use them to color the coding circles and the corresponding structures in the figure. Also, label the dashed line showing the dimensions of the true pelvis and that showing the diameter of the false pelvis. Complete the illustration by labeling the following bone markings: obturator foramen, iliac crest, anterior superior iliac spine, ischial spine, pubic ramus, and pelvic brim. Last, list three ways in which the female pelvis differs from the male pelvis and insert your answers in the answer blanks.

◯ Coxal bone (hip bone)   ◯ Pubic symphysis

◯ Sacrum   ◯ Acetabulum

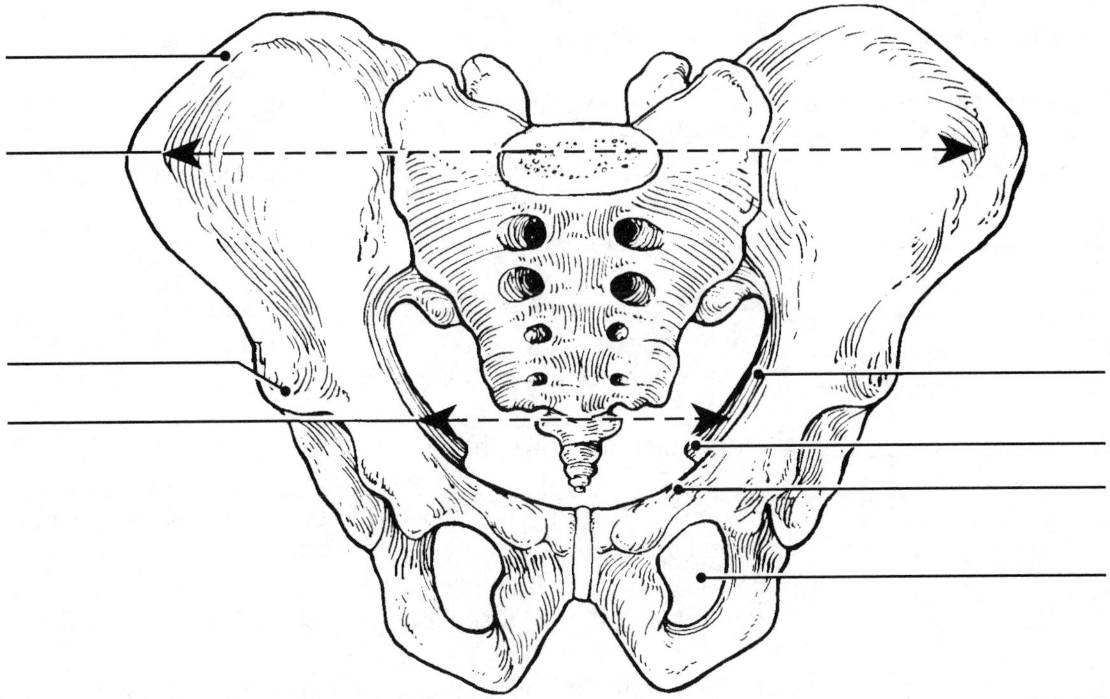

Figure 5–11

1. _____

2. _____

3. _____

26. Circle the term that does not belong in each of the following groupings.

1. Tibia   Ulna   Fibula   Femur

2. Skull   Rib cage   Vertebral column   Pelvis

3. Ischium   Scapula   Ilium   Pubis

4. Mandible   Frontal bone   Temporal bone   Occipital bone

5. Calcaneus   Tarsals   Carpals   Talus

27. Using the key choices, identify the bone names and markings, according to the descriptions that follow. Insert the appropriate key term(s) or letter(s) in the answer blanks.

Key Choices

A. Acetabulum	I. Ilium	Q. Patella
B. Calcaneus	J. Ischial tuberosity	R. Pubic symphysis
C. Femur	K. Ischium	S. Pubis
D. Fibula	L. Lateral malleolus	T. Sacroiliac joint
E. Gluteal tuberosity	M. Lesser sciatic notch	U. Talus
F. Greater sciatic notch	N. Medial malleolus	V. Tarsals
G. Greater and lesser trochanters	O. Metatarsals	W. Tibia
H. Iliac crest	P. Obturator foramen	X. Tibial tuberosity

_____ 1. Fuse to form the coxal bone (hip bone)

_____ 2. Receives the weight of the body when sitting

_____ 3. Point where the coxal bones join anteriorly

_____ 4. Upper margin of iliac bones

_____ 5. Deep socket in the coxal bone (hip bone) that receives the head of the thigh bone

_____ 6. Point where the axial skeleton attaches to the pelvic girdle

_____ 7. Longest bone in body; articulates with the coxal bone

_____ 8. Lateral bone of the leg

_____ 9. Medial bone of the leg

_____ 10. Bones forming the knee joint

_____ 11. Point where the patellar ligament attaches

_____ 12. Kneecap

_____ 13. Shinbone

_____ 14. Distal process on medial tibial surface

_____ 15. Process forming the outer ankle

_____ 16. Heel bone

_____ 17. Bones of the ankle

_____ 18. Bones forming the instep of the foot

_____ 19. Opening in a coxal bone (hip bone) formed by the pubic and ischial rami

_____ 20. Sites of muscle attachment on the proximal end of the femur

_____ 21. Tarsal bone that articulates with the tibia

28. For each of the following statements that is true, insert *T* in the answer blank. If any of the statements are false, correct the underlined term by inserting the correct term in the answer blank.

_____ 1. The <u>pectoral</u> girdle is formed by the articulation of the coxal bones (hip bones) and the sacrum.

_____ 2. Bones present in both the hand and the foot are <u>carpals</u>.

_____ 3. The tough, fibrous connective tissue covering of a bone is the <u>periosteum</u>.

_____ 4. The point of fusion of the three bones forming a coxal bone is the <u>glenoid cavity</u>.

_____ 5. The large nerve that must be avoided when giving injections into the buttock muscles is the <u>femoral</u> nerve.

_____ 6. The long bones of a fetus are constructed of <u>hyaline</u> cartilage.

_____ 7. Bones that provide the most protection to the abdominal viscera are the <u>ribs</u>.

_____ 8. The largest foramen in the skull is the <u>foramen magnum</u>.

_____ 9. The intercondylar fossa, greater trochanter, and tibial tuberosity are all bone markings of the <u>humerus</u>.

_____ 10. The first major event of fracture healing is <u>hematoma formation</u>.

_____ 11. An exaggerated thoracic curvature known as "dowager's hump" is an abnormal condition called <u>scoliosis</u>.

29. The bones of the thigh and the leg are shown in Figure 5–12. Identify each and put your answers in the blanks labeled A, B, and C. Select different colors for the lower limb bones listed below and use them to color in the coding circles and corresponding bones on the diagram. Complete the illustration by inserting the terms indicating bone markings at the ends of the appropriate leader lines in the figure.

◯ Femur

Head of femur

Intercondylar eminence

Tibial tuberosity

◯ Tibia

Anterior border of tibia

Lesser trochanter

Greater trochanter

◯ Fibula

Head of fibula

Medial malleolus

Lateral malleolus

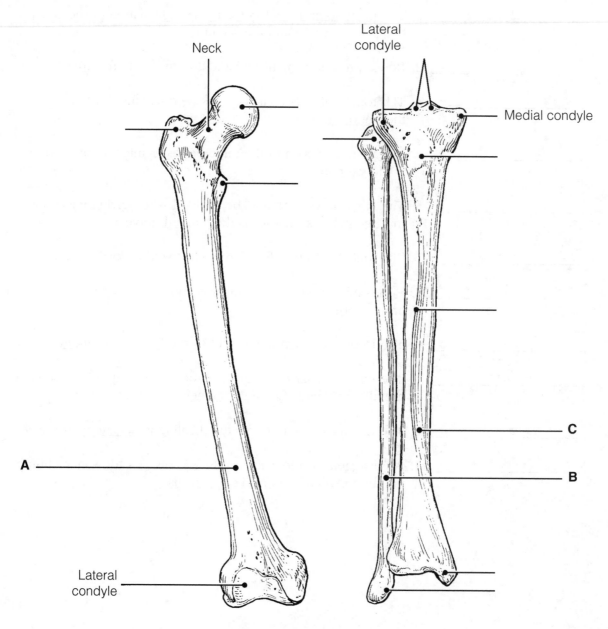

Figure 5–12

30. Figure 5–13 is a diagram of the articulated skeleton. Identify all bones or groups of bones by writing the correct labels at the end of the leader lines. Then, select two different colors for the bones of the axial and appendicular skeletons and use them to color in the coding circles and corresponding structures in the diagram.

◯ Axial skeleton   ◯ Appendicular skeleton

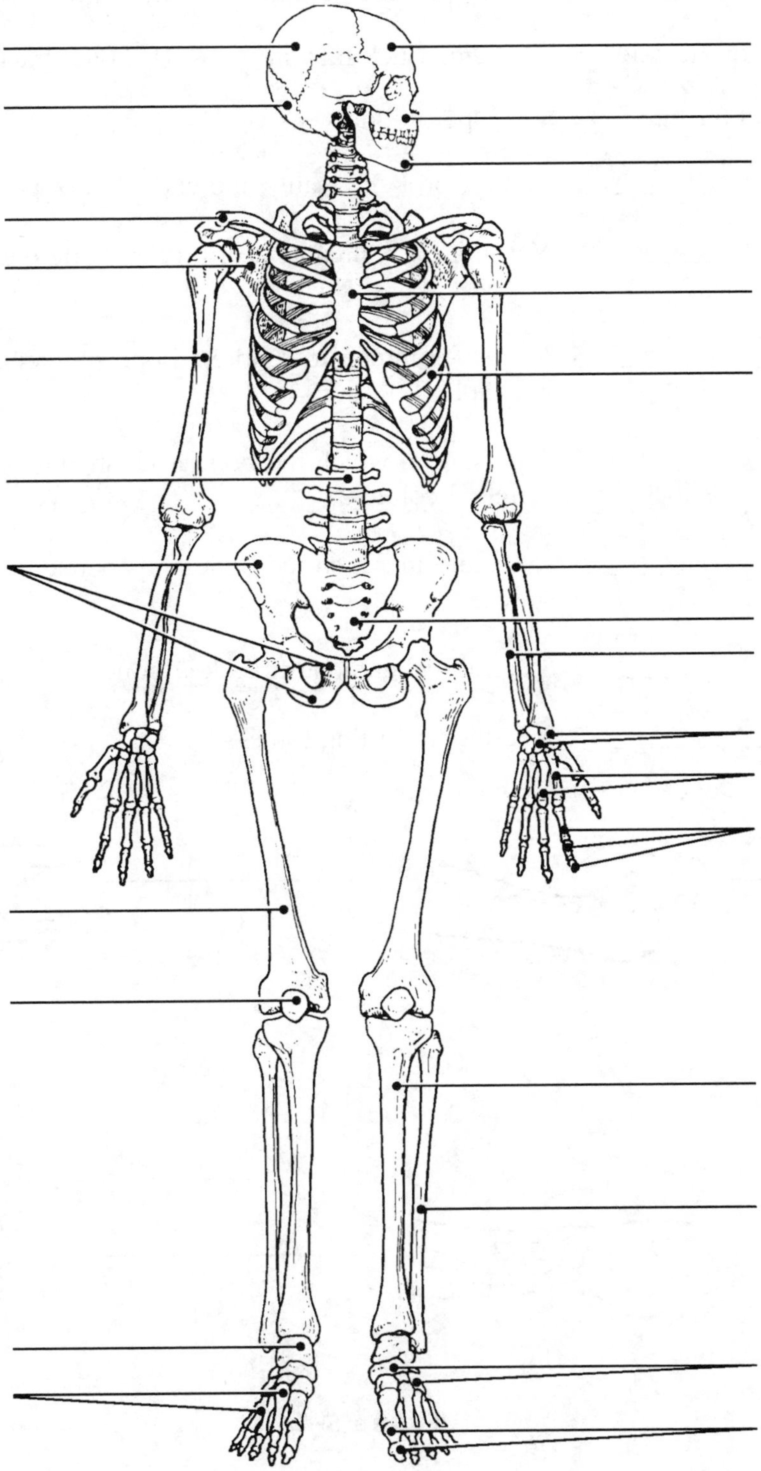

Figure 5–13

BONE FRACTURES

31. Using the key choices, identify the fracture (fx) types shown in Figure 5–14 and the fracture types and treatments described below. Enter the appropriate key letter or term in each answer blank.

Key Choices

A. Closed reduction   D. Depressed fracture   G. Simple fracture

B. Compression fracture   E. Greenstick fracture   H. Spiral fracture

C. Compound fracture   F. Open reduction

_____ 1. Bone is broken cleanly; the ends do not penetrate the skin

_____ 2. Nonsurgical realignment of broken bone ends and splinting of bone

_____ 3. A break common in children; bone splinters, but break is incomplete

_____ 4. A fracture in which the bone is crushed; common in the vertebral column

_____ 5. A fracture in which the bone ends penetrate through the skin surface

_____ 6. Surgical realignment of broken bone ends

_____ 7. A result of twisting forces

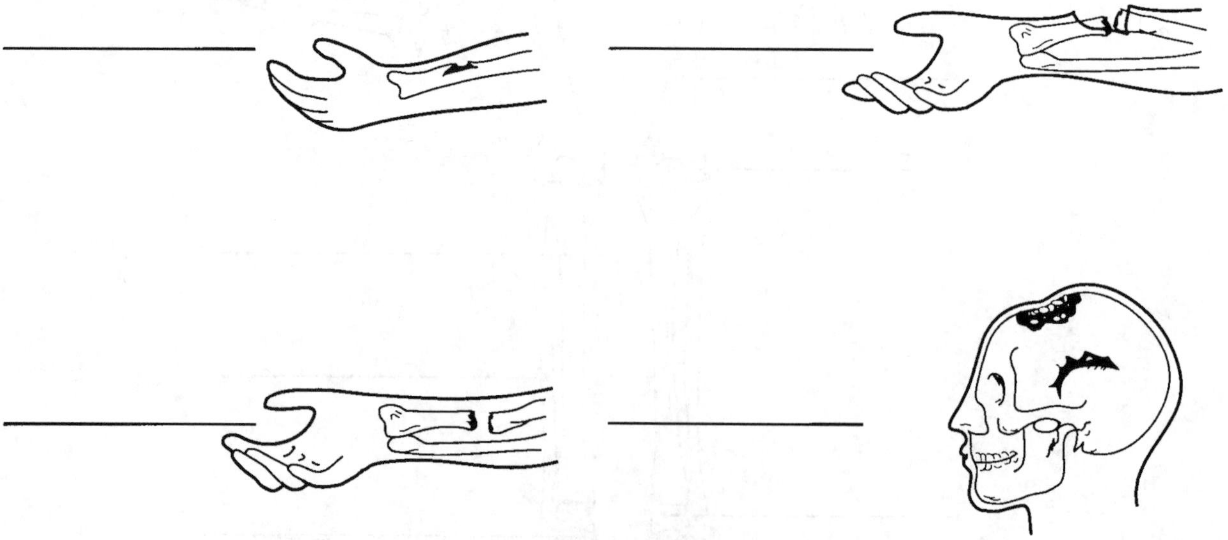

Figure 5–14

32. For each of the following statements that is true about bone breakage and the repair process, insert *T* in the answer blank. For false statements, correct the underlined terms by inserting the correct term in the answer blank.

_____ 1. A <u>hematoma</u> usually forms at a fracture site.

_____ 2. Deprived of nutrition, <u>osteocytes</u> at the fracture site die.

_____ 3. Nonbony debris at the fracture site is removed by <u>osteoclasts</u>.

_____ 4. Growth of a new capillary supply into the region produces <u>granulation tissue</u>.

_____ 5. Osteoblasts from the <u>medullary cavity</u> migrate to the fracture site.

_____ 6. The <u>fibrocartilage callus</u> is the first repair mass to splint the broken bone.

_____ 7. The bony callus is initially composed of <u>compact</u> bone.

JOINTS

33. Figure 5–15 shows the structure of a typical diarthrotic joint. Select different colors to identify each of the following areas and use them to color the coding circles and the corresponding structures on the figure. Then, complete the statements below the figure.

◯ Articular cartilage of bone ends

◯ Fibrous capsule

◯ Synovial membrane

◯ Joint cavity

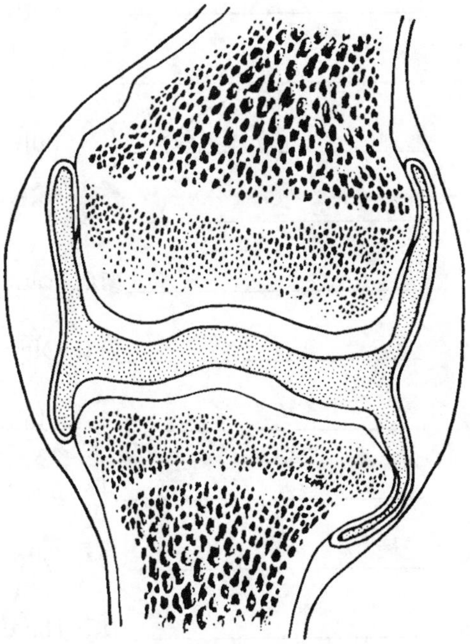

Figure 5–15

1. _____ The lubricant that minimizes friction and abrasion of joint surfaces is __(1)__.

2. _____ The resilient substance that keeps bone ends from crushing when compressed is __(2)__.

3. _____ __(3)__, which reinforce the fibrous capsule, help to prevent dislocation of the joint.

34. For each joint described below, select an answer from Key A. Then, if the Key A selection *is other than C* (a synovial joint), see if you can classify the joint further by making a choice from Key B.

Key Choices

Key A: A. Cartilaginous   Key B: 1. Epiphyseal disk

            B. Fibrous   2. Suture

            C. Synovial   3. Symphysis

_____ 1. Has amphiarthrotic and synarthrotic examples

_____ 2. All have a fibrous capsule lined with synovial membrane surrounding a joint cavity

_____ 3. Bone regions united by fibrous connective tissue

_____ 4. Joints between skull bones

_____ 5. Joint between the atlas and axis

_____ 6. Hip, elbow, and knee

_____ 7. All examples are diarthroses

_____ 8. Pubic symphysis

_____ 9. All are reinforced by ligaments

_____10. Joint providing the most protection to underlying structures

_____11. Often contains a fluid-filled cushion

_____12. Child's long-bone growth plate made of hyaline cartilage

_____13. Most joints of the limbs

_____14. Often associated with bursae

_____15. Have the greatest mobility

35. Which structural joint type is *not* commonly found in the axial skeleton and why not?

_____

_____

_____

Homeostatic Imbalances of Bones and Joints

36. For each of the following statements that is true, enter *T* in the answer blank. For each false statement, correct the underlined words by writing the correct words in the answer blank.

_____ 1. In a sprain, the ligaments reinforcing a joint are excessively stretched or torn.

_____ 2. Age-related erosion of articular cartilages and formation of painful bony spurs are characteristic of gouty arthritis.

_____ 3. Chronic arthritis usually results from bacterial invasion.

_____ 4. Healing of a partially torn ligament is slow because its hundreds of fibrous strands are poorly aligned.

_____ 5. Rheumatoid arthritis is an autoimmune disease.

_____ 6. High levels of uric acid in the blood may lead to rheumatoid arthritis.

_____ 7. A "soft" bone condition in children, usually caused by a lack of calcium or vitamin D in the diet, is called osteomyelitis.

_____ 8. Atrophy and thinning of bone owing to hormonal changes or inactivity (generally in the elderly) is called osteoporosis.

DEVELOPMENTAL ASPECTS OF THE SKELETON

37. Using the key choices, identify the body systems that relate to bone tissue viability. Enter the appropriate key terms or letters in the answer blanks.

Key Choices

A. Endocrine   C. Muscular   E. Reproductive

B. Integumentary   D. Nervous   F. Urinary

_____ 1. Conveys the sense of pain in bone and joints

_____ 2. Activates vitamin D for proper calcium usage

_____ 3. Regulates uptake and release of calcium by bones

_____ 4. Increases bone strength and viability by pulling action

_____ 5. Influences skeleton proportions and adolescent growth of long bones

_____ 6. Provides vitamin D for proper calcium absorption

38. Complete the following statements concerning fetal and infant skeletal development. Insert the missing words in the answer blanks.

_____ 1.

_____ 2.

_____ 3.

_____ 4.

_____ 5.

_____ 6.

_____ 7.

_____ 8.

_____ 9.

"Soft spots," or membranous joints called __(1)__ in the fetal skull, allow the skull to be __(2)__ slightly during birth passage. They also allow for continued brain __(3)__ during the later months of fetal development and early infancy. Eventually these soft spots are replaced by immovable joints called __(4)__.

The two spinal curvatures well developed at birth are the __(5)__ and __(6)__ curvatures. Because they are present at birth, they are called __(7)__ curvatures. The secondary curvatures develop as the baby matures. The __(8)__ curvature develops as the baby begins to lift his or her head. The __(9)__ curvature matures when the baby begins to walk or assume the upright posture.

INCREDIBLE JOURNEY

A Visualization Exercise for the Skeletal System

. . . stalagmite- and stalactite-like structures that surround you. . . .
Since the texture is so full of holes. . . .

39. Where necessary, complete statements by inserting the missing words in the answer blanks.

_____ 1.

_____ 2.

_____ 3.

_____ 4.

_____ 5.

_____ 6.

For this journey you are miniaturized and injected into the interior of the largest bone of your host's body, the __(1)__. Once inside this bone, you look around and find yourself examining the stalagmite- and stalactite-like structures that surround you. Although you feel as if you are in an underground cavern, you know that it has to be bone. Since the texture is so full of holes, it obviously is __(2)__ bone. Although the arrangement of these bony spars seems to be haphazard, as if someone randomly dropped straws, they are precisely arranged to resist points of __(3)__. All about you is frantic, hurried activity. Cells are dividing rapidly, nuclei are being ejected, and disklike cells are appearing. You decide that these disklike cells are __(4)__ and that this is the __(5)__ cavity. As you explore further, strolling along the edge of the cavity, you spot many tunnels leading into the solid bony area on which you are walking. Walking into one of these drainpipe-like openings, you notice that it contains a glistening white ropelike structure (a __(6)__, no doubt) and blood vessels running the length of the tube. You eventually come to a point in the channel where the

_____ 7.

_____ 8.

_____ 9.

_____ 10.

_____ 11.

_____ 12.

horizontal passageway joins with a vertical passage that runs with the longitudinal axis of the bone. This is obviously a __(7)__ canal. Because you would like to see how nutrients are brought into __(8)__ bone, you decide to follow this channel. Reasoning that there is no way you can possibly scale the slick walls of the channel, you leap and grab onto a white cord hanging down its length. Because it is easier to slide down than to try to climb up the cord, you begin to lower yourself, hand-over-hand. During your descent, you notice small openings in the wall, which are barely large enough for you to wriggle through. You conclude that these are the __(9)__ that connect all the __(10)__ to the nutrient supply in the central canal. You decide to investigate one of these tiny openings and begin to swing on your cord, trying to get a foothold on one of the openings. After managing to anchor yourself and squeezing into an opening, you use a flashlight to illuminate the passageway in front of you. You are startled by a giant cell with many dark nuclei. It appears to be plastered around the entire lumen directly ahead of you. As you watch this cell, the bony material beneath it, the __(11)__ , begins to liquefy. The cell apparently is a bone-digesting cell, or __(12)__ , and because you are unsure whether or not its enzymes can also liquefy you, you slither backwards hurriedly and begin your trek back to your retrieval site.

AT THE CLINIC

40. Antonio is hit in the face with a football during practice. An X-ray reveals multiple fractures of the bones around an orbit. Name the bones that form margins of the orbit.

41. Mrs. Bruso, a woman in her 80s, is brought to the clinic with a fractured hip. X-rays reveal compression fractures in her lower vertebral column and extremely low bone density in her vertebrae, coxal bones (hip bones), and femurs. What are the condition, cause, and treatment?

42. Jack, a young man, is treated at the clinic for an accident in which he hit his forehead. When he returns for a checkup, he complains that he can't smell anything. A hurried X-ray of his head reveals a fracture. What part of which bone was fractured to cause his loss of smell?

43. A middle-aged woman comes to the clinic complaining of stiff, painful joints and increasing immobility of her finger joints. A glance at her hands reveals knobby, deformed knuckles. For what condition will she be tested?

44. At his 94th birthday party, James was complimented on how good he looked and was asked about his health. He replied, "I feel good most of the time, but some of my joints ache and are stiff, especially my knees, hips, and lower back, and especially in the morning when I wake up." A series of X-rays and an MRI scan taken a few weeks earlier had revealed that the articular cartilages of these joints were rough and flaking off, and bone spurs (overgrowths) were present at the ends of some of James's bones. What is James's probable condition?

45. Janet, a 10-year-old girl, is brought to the clinic after falling out of a tree. An X-ray shows she has small fractures of the transverse processes of T_3 to T_5 on the right side. Janet will be watched for what abnormal spinal curvature over the next several years?

46. The serving arm of many tennis players is often significantly larger (thicker) than the other arm. Explain this phenomenon.

47. Jerry is giving cardiopulmonary resuscitation to Ms. Jackson, an elderly woman who has just been rescued from the waters of Cape Cod Bay. What bone is he compressing?

48. Rita's bone density scan revealed she has osteoporosis. Her physician prescribed a drug that inhibits osteoclast activity. Explain this treatment.

✓ THE FINALE: MULTIPLE CHOICE

49. Select the best answer or answers from the choices given.

1. Important bone functions include:

   A. support of the pelvic organs

   B. protection of the brain

   C. provision of levers for movement of the limbs

   D. protection of the skin and limb musculature

   E. storage of water

2. A passageway connecting neighboring osteocytes in an osteon is a:

   A. central canal   D. canaliculus

   B. lamella   E. perforating canal

   C. lacuna

3. What is the earliest event (of those listed) in endochondral ossification?

   A. Ossification of proximal epiphysis

   B. Appearance of the epiphyseal plate

   C. Invasion of the shaft by the periosteal bud

   D. Cavitation of the cartilage shaft

   E. Formation of secondary ossification centers

4. The growth spurt of puberty is triggered by:

   A. high levels of sex hormones

   B. the initial, low levels of sex hormones

   C. growth hormone

   D. parathyroid hormone

   E. calcitonin

5. Deficiency of which of the following hormones will cause dwarfism?

   A. Growth hormone

   B. Sex hormones

   C. Thyroid hormones

   D. Calcitonin

   E. Parathyroid hormone

6. Women suffering from osteoporosis are frequent victims of _____ fractures of the vertebrae.

   A. compound   D. compression

   B. spiral   E. depression

   C. comminuted

7. Which of the following bones are part of the axial skeleton?

   A. Vomer   D. Parietal

   B. Clavicle   E. Coxal bone (hip bone)

   C. Sternum

8. A blow to the cheek is most likely to break what superficial bone or bone part?

   A. Superciliary arches

   B. Zygomatic process

   C. Mandibular ramus

   D. Styloid process

9. Which of the following are part of the sphenoid?

   A. Crista galli   D. Pterygoid process

   B. Sella turcica   E. Lesser wings

   C. Petrous portion

10. Structural characteristics of *all* cervical vertebrae are:

    A. small body

    B. bifid spinous process

    C. transverse foramina

    D. small vertebral foramen

    E. costal facets

11. Which of the following bones exhibit a styloid process?

    A. Hyoid   D. Radius

    B. Temporal   E. Ulna

    C. Humerus

12. Coxal bone (hip bone) markings include:

    A. ala   D. pubic ramus

    B. sacral hiatus   E. fovea capitis

    C. gluteal surface

13. Cartilaginous joints include:

    A. syndesmoses   C. synostoses

    B. symphyses   D. synchondroses

14. Considered to be part of a synovial joint are:

    A. bursae   C. tendon sheath

    B. articular cartilage   D. capsular ligaments

15. Abduction is:

    A. moving the right arm out to the right

    B. spreading out the fingers

    C. wiggling the toes

    D. moving the sole of the foot laterally

16. In comparing two joints of the same type, what characteristic(s) would you use to determine strength and flexibility?

    A. Depth of the depression of the concave bone of the joint

    B. Snugness of fit of the bones

    C. Size of bone projections for muscle attachments

    D. Presence of menisci

17. Which of the following joints has the greatest freedom of movement?

    A. Interphalangeal

    B. Saddle joint of thumb

    C. Distal tibiofibular

    D. Coxal (hip)

18. Which specific joint does the following description identify? "Articular surfaces are deep and secure, multiaxial; capsule heavily reinforced by ligaments; labrum helps prevent dislocation; the first joint to be built artificially; very stable."

    A. Elbow   C. Knee

    B. Hip   D. Shoulder

19. An autoimmune disease resulting in inflammation and eventual fusion of diarthrotic joints is:

    A. gout

    B. rheumatoid arthritis

    C. degenerative joint disease

    D. pannus

20. Plane joints allow:

    A. pronation   C. rotation

    B. flexion   D. gliding

21. Movements made in chewing food are:

    A. Flexion   D. Depression

    B. Extension   E. Opposition

    C. Elevation

22. Which of the following bones are *not* paired?

    A. Parietal   D. Pubis

    B. Frontal   E. Calcaneus

    C. Sternum

6 THE MUSCULAR SYSTEM

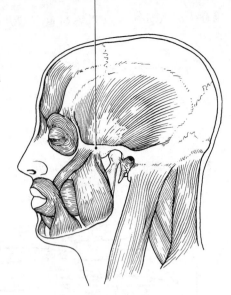

Muscles, the specialized tissues that facilitate body movement, make up about 40% of body weight. Most body muscle is the voluntary type, called skeletal muscle because it is attached to the bony skeleton. Skeletal muscle contributes to body contours and shape, and it composes the organ system called the muscular system. These muscles allow you to grin, frown, run, swim, shake hands, swing a hammer, and to otherwise manipulate your environment. The balance of body muscle consists of smooth and cardiac muscles, which form the bulk of the walls of hollow organs and the heart. Smooth and cardiac muscles are involved in the transport of materials within the body.

Study activities in this chapter deal with microscopic and gross structure of muscle, identification of voluntary muscles, body movements, and important understandings of muscle physiology.

OVERVIEW OF MUSCLE TISSUES

1. Nine characteristics of muscle tissue are listed below and on page 104. Identify the muscle tissue type described by choosing the correct response(s) from the key choices. Enter the appropriate term(s) or letter(s) of the key choice in the answer blank.

Key Choices

A. Cardiac   B. Smooth   C. Skeletal

_____ 1. Involuntary

_____ 2. Banded appearance

_____ 3. Longitudinally and circularly arranged layers

_____ 4. Dense connective tissue packaging

_____ 5. Figure eight packaging of the cells

_____ 6. Coordinated activity to act as a pump

→

103

_____ 7. Moves bones and the facial skin

_____ 8. Referred to as the muscular system

_____ 9. Voluntary

2. Identify the type of muscle in each of the illustrations in Figure 6–1. Color the diagrams as you wish.

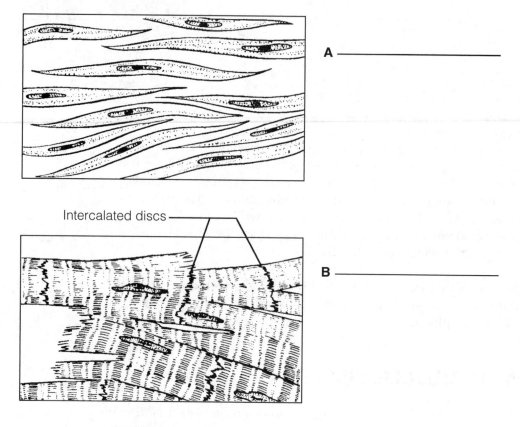

A _____

Intercalated discs

B _____

Figure 6–1

3. Regarding the functions of muscle tissues, circle the term in each of the groupings that does not belong with the other terms.

1. Urine   Foodstuffs   Bones   Smooth muscle

2. Heart   Cardiac muscle   Blood pump   Promotes labor during birth

3. Excitability   Response to a stimulus   Contractility   Action potential

4. Ability to shorten   Contractility   Pulls on bones   Stretchability

5. Maintains posture   Movement   Promotes growth   Generates heat

MICROSCOPIC ANATOMY OF SKELETAL MUSCLE

4. First, identify the structures in Column B by matching them with the descriptions in Column A. Enter the correct letters (or terms if desired) in the answer blanks. Then, select a different color for each of the terms in Column B that has a color-coding circle and color in the structures on Figure 6–2.

Column A

_____ 1. Connective tissue surrounding a fascicle

_____ 2. Connective tissue ensheathing the entire muscle

_____ 3. Contractile unit of muscle

_____ 4. A muscle cell

_____ 5. Thin connective tissue investing each muscle cell

_____ 6. Plasma membrane of the muscle cell

_____ 7. A long, filamentous organelle found within muscle cells that has a banded appearance

_____ 8. Actin- or myosin-containing structure

_____ 9. Cordlike extension of connective tissue beyond the muscle, serving to attach it to the bone

_____ 10. A discrete bundle of muscle cells

Column B

A. Endomysium ○

B. Epimysium ○

C. Fascicle

D. Fiber ○

E. Myofilament

F. Myofibril ○

G. Perimysium ○

H. Sarcolemma

I. Sarcomere

J. Sarcoplasm

K. Tendon ○

Figure 6–2

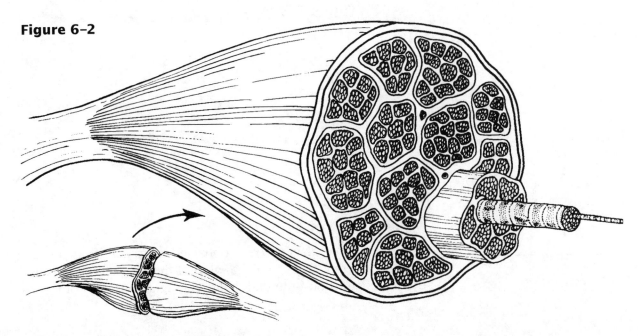

5. Figure 6–3 is a diagrammatic representation of a small portion of a relaxed muscle cell (bracket indicates the portion enlarged). First, select different colors for the structures listed below. Use them to color the coding circles and corresponding structures on Figure 6–3. Then bracket and label an A band, an I band, and a sarcomere. When you have finished, draw a contracted sarcomere in the space beneath the figure and label the same structures, as well as the light and dark bands.

◯ Myosin   ◯ Actin filaments   ◯ Z disc

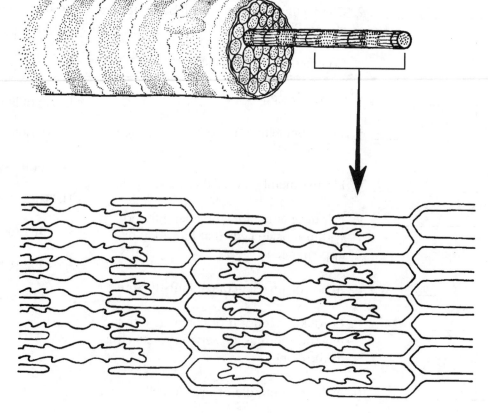

Figure 6–3

_____ 1. Looking at your diagram of a contracted sarcomere from a slightly different angle, which region of the sarcomere shortens during contraction—the dark band, the light band, or both?

SKELETAL MUSCLE ACTIVITY

6. Complete the following statements relating to the neuromuscular junction. Insert the correct answers in the numbered answer blanks.

_____ 1.

_____ 2.

_____ 3.

_____ 4.

_____ 5.

_____ 6.

A motor neuron and all of the skeletal muscle cells it stimulates is called a __(1)__ . The axon of each motor neuron has numerous endings called __(2)__ . The actual gap between an axonal ending and the muscle cell is called a __(3)__ . Within the axonal endings are many small vesicles containing a neurotransmitter substance called __(4)__ .

When the __(5)__ reaches the ends of the axon, the neurotransmitter is released, and it diffuses to the muscle cell membrane to combine with receptors there. Binding of the neurotransmitters with muscle membrane receptors causes the membrane to become permeable to sodium, resulting in the influx of sodium ions and __(6)__ of the membrane. Then contraction of the muscle cell occurs.

7. Figure 6–4 shows the components of a neuromuscular junction. Identify the parts by coloring the coding circles and the corresponding structures in the diagram. Add small arrows to indicate the location of the acetylcholine (ACh) receptors and label appropriately.

◯ Mitochondrion   ◯ T tubule   ◯ Sarcomere

◯ Synaptic vesicles   ◯ Synaptic cleft   ◯ Junctional folds

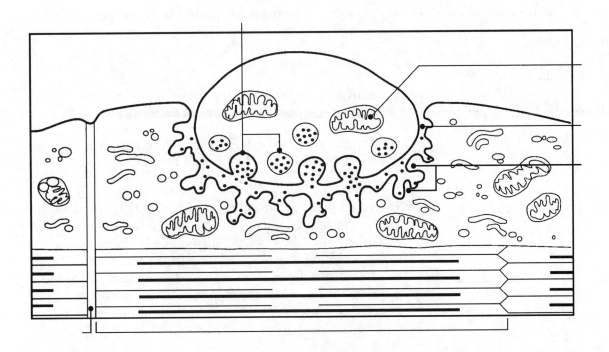

Figure 6–4

8. Number the following statements in their proper sequence to describe the contraction mechanism in a skeletal muscle cell. The first step has already been identified as number 1.

___1___   1. ACh is released into the neuromuscular junction by the axonal terminal.

_____   2. The action potential, carried deep into the cell, causes the sarcoplasmic reticulum to release calcium ions.

_____   3. The muscle cell relaxes and lengthens.

_____   4. ACh diffuses across the neuromuscular junction and binds to receptors on the sarcolemma.

_____   5. The calcium ion concentration at the myofilaments increases; the myofilaments slide past one another, and the cell shortens.

_____   6. Depolarization occurs, and the action potential is generated.

_____   7. As calcium is actively reabsorbed into the sarcoplasmic reticulum, its concentration at the myofilaments decreases.

9. The following incomplete statements refer to a muscle cell in the resting, or polarized, state just before stimulation. Complete each statement by choosing the correct response from the key choices and entering the appropriate letter in the answer blanks.

Key Choices

A. Na^+ diffuses out of the cell

B. K^+ diffuses out of the cell

C. Na^+ diffuses into the cell

D. K^+ diffuses into the cell

E. Inside the cell

F. Outside the cell

G. Relative ionic concentrations on the two sides of the membrane during rest

H. Electrical conditions

I. Activation of the sodium-potassium pump, which moves K^+ into the cell and Na^+ out of the cell

J. Activation of the sodium-potassium pump, which moves Na^+ into the cell and K^+ out of the cell

_____ 1.

_____ 2.

_____ 3.

_____ 4.

_____ 5.

_____ 6.   _____ 7.

There is a greater concentration of Na^+ __(1)__ , and there is a greater concentration of K^+ __(2)__ . When the stimulus is delivered, the permeability of the membrane is changed, and __(3)__ , initiating the depolarization of the membrane. Almost as soon as the depolarization wave begins, a repolarization wave follows it across the membrane. This occurs as __(4)__ . Repolarization restores the __(5)__ of the resting cell membrane. The __(6)__ is (are) reestablished by __(7)__ .

10. Complete the following statements by choosing the correct response from the key choices and entering the appropriate letter or term in the answer blanks.

Key Choices

A. Fatigue

B. Isotonic contraction

C. Muscle cell

D. Muscle tone

E. Isometric contraction

F. Whole muscle

G. Fused tetanus

H. Few motor units

I. Many motor units

J. Repolarization

K. Depolarization

L. Unfused tetanus

_____ 1. _____ is a continuous contraction that shows no evidence of relaxation.

_____ 2. A(n) _____ is a contraction in which the muscle shortens and work is done.

_____ 3. To accomplish a strong contraction, _____ are stimulated at a rapid rate.

_____ 4. When a weak but smooth muscle contraction is desired, _____ are stimulated at a rapid rate.

_____ 5. When a muscle is being stimulated but is not able to respond because of "oxygen deficit," the condition is called _____.

_____ 6. A(n) _____ is a contraction in which the muscle does not shorten, but tension in the muscle keeps increasing.

11. The terms in the key refer to the three ways that muscle cells replenish their ATP supplies. Select the term(s) that best apply to the conditions described and insert the correct key letter(s) in the answer blanks.

Key Choices

A. Coupled reaction of creatine phosphate (CP) and ADP

B. Anaerobic glycolysis   C. Aerobic respiration

_____   1. Accompanied by lactic acid formation

_____   2. Supplies the highest ATP yield per glucose molecule

_____   3. Involves the simple transfer of a phosphate group

_____   4. Requires no oxygen

_____   5. The slowest ATP regeneration process

_____   6. Produces carbon dioxide and water

_____   7. The energy mechanism used in the second hour of running in a marathon

_____   8. Used when the oxygen supply is inadequate over time

_____   9. Good for a sprint

12. Briefly describe how you can tell when you are repaying the oxygen deficit.

_____

_____

13. Which of the following occur within a muscle cell during oxygen deficit?
Place a check (✓) by the correct choices.

_____ 1. Decreased ATP _____ 5. Increased oxygen

_____ 2. Increased ATP _____ 6. Decreased carbon dioxide

_____ 3. Increased lactic acid _____ 7. Increased carbon dioxide

_____ 4. Decreased oxygen _____ 8. Increased glucose

MUSCLE MOVEMENTS, TYPES, AND NAMES

14. Relative to general terminology concerning muscle activity, first label the
following structures on Figure 6–5: insertion, origin, tendon, resting muscle,
and contracting muscle. Next, identify the two structures named below by
choosing different colors for the coding circles and the corresponding
structures in the figure.

◯ Movable bone

◯ Immovable bone

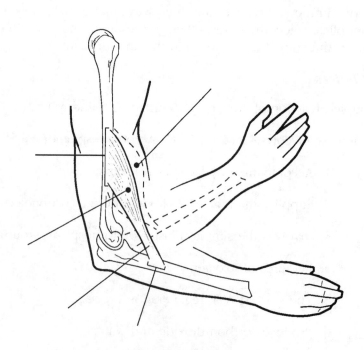

Figure 6–5

15. Complete the following statements. Insert your answers in the answer blanks.

_____ 1.

_____ 2.

_____ 3.

_____ 4.

_____ 5.

_____ 6.

_____ 7.

_____ 8.

_____ 9.

_____10.

_____11.

_____12.

_____13.

_____14.

1. Standing on your toes as in ballet is __(1)__ of the foot. Walking on your heels is __(2)__.

Winding up for a pitch (as in baseball) can properly be called __(3)__. To keep your seat when riding a horse, the tendency is to __(4)__ your thighs.

In running, the action at the hip joint is __(5)__ in reference to the leg moving forward and __(6)__ in reference to the leg in the posterior position. When kicking a football, the action at the knee is __(7)__. In climbing stairs, the hip and knee of the forward leg are both __(8)__. You have just touched your chin to your chest; this is __(9)__ of the neck.

Using a screwdriver with a straight arm requires __(10)__ of the arm. Consider all the movements of which the arm is capable. One often used for strengthening all the upper arm and shoulder muscles is __(11)__.

Moving the head to signify "no" is __(12)__. Action that moves the distal end of the radius across the ulna is __(13)__. Raising the arms laterally away from the body is called __(14)__ of the arms.

16. The terms provided in the key choices are often used to describe the manner in which muscles interact with other muscles. Select the key terms that apply to the following definitions and insert the correct letter or term in the answer blanks.

Key Choices

A. Antagonist   B. Fixator   C. Prime mover   D. Synergist

_____ 1. Agonist

_____ 2. Postural muscles for the most part

_____ 3. Stabilizes a joint so that the prime mover can act at more distal joints

_____ 4. Performs the same movement as the prime mover

_____ 5. Reverses and/or opposes the action of a prime mover

_____ 6. Immobilizes the origin of a prime mover

17. Several criteria are applied to the naming of muscles. These are provided in Column B. Identify which criteria pertain to the muscles listed in Column A and enter the correct letter(s) in the answer blank.

Column A	Column B
_____ 1. Gluteus maximus	A. Action of the muscle
_____ 2. Adductor magnus	B. Shape of the muscle
_____ 3. Biceps femoris	C. Location of the muscle's origin and/or insertion
_____ 4. Transversus abdominis	D. Number of origins
_____ 5. Extensor carpi ulnaris	E. Location of muscle relative to a bone or body region
_____ 6. Trapezius	F. Direction in which the muscle fibers run relative to some imaginary line
_____ 7. Rectus femoris	
_____ 8. External oblique	G. Relative size of the muscle

GROSS ANATOMY OF THE SKELETAL MUSCLES

Muscles of the Head

18. Identify the major muscles described in Column A by choosing a response from Column B. Enter the correct letter in the answer blank. Select a different color for each muscle described and color in the coding circle and corresponding muscle on Figure 6–6.

Column A	Column B
◯ _____ 1. Used to show you're happy	A. Buccinator
◯ _____ 2. Used to suck in your cheeks	B. Frontalis
◯ _____ 3. Used in winking	C. Masseter
◯ _____ 4. Wrinkles the forehead horizontally	D. Orbicularis oculi
◯ _____ 5. The "kissing" muscle	E. Orbicularis oris
◯ _____ 6. Prime mover of jaw closure	F. Sternocleidomastoid
◯ _____ 7. Synergist muscle for jaw closure	G. Temporalis
◯ _____ 8. Prime mover of head flexion; a two-headed muscle	H. Trapezius
	I. Zygomaticus

Zygomatic bone

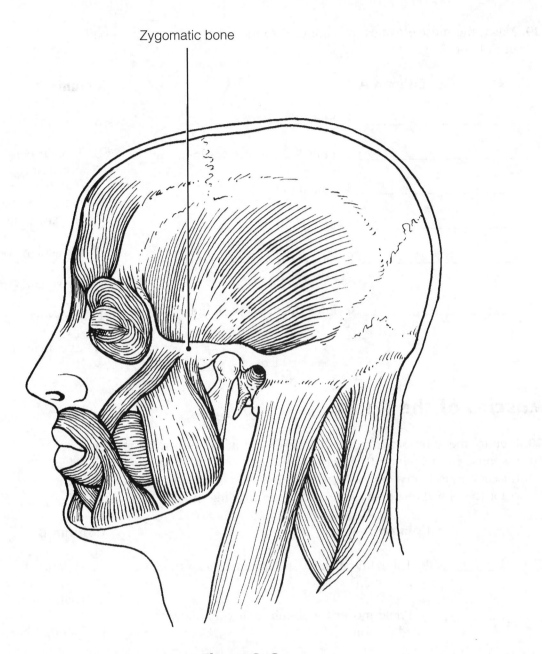

Figure 6-6

19. Match the muscle names in Column B to the facial muscles described in Column A.

Column A	Column B
_____ 1. Squints the eyes	A. Buccinator
_____ 2. Pulls the eyebrows superiorly	B. Frontal belly of the epicranius
_____ 3. Smiling muscle	C. Occipital belly of the epicranius
_____ 4. Puckers the lips	
_____ 5. Draws the corners of the lips downward	D. Orbicularis oculi
	E. Orbicularis oris
_____ 6. Pulls the scalp posteriorly	F. Platysma
	G. Zygomaticus

Muscles of the Trunk

20. Identify the anterior trunk muscles described in Column A by choosing a response from Column B. Enter the correct letter in the answer blank. Then, for each muscle description that has a color-coding circle, select a different color to color the coding circle and corresponding muscle on Figure 6–7.

Column A	Column B
○ _____ 1. The name means "straight muscle of the abdomen"	A. Deltoid
○ _____ 2. Prime mover for shoulder flexion and adduction	B. Diaphragm
○ _____ 3. Prime mover for shoulder abduction	C. External intercostal
○ _____ 4. Part of the abdominal girdle; forms the external lateral walls of the abdomen	D. External oblique
○ _____ 5. Acting alone, each muscle of this pair turns the head toward the opposite shoulder	E. Internal intercostal
	F. Internal oblique
_____ 6. and 7. Besides the two abdominal muscles (pairs) named above, two muscle pairs that help form the natural abdominal girdle	G. Latissimus dorsi
	H. Pectoralis major
_____ 8. Deep muscles of the thorax that promote the inspiratory phase of breathing	I. Rectus abdominis
	J. Sternocleidomastoid
_____ 9. An unpaired muscle that acts with the muscles named immediately above to accomplish inspiration	K. Transversus abdominis

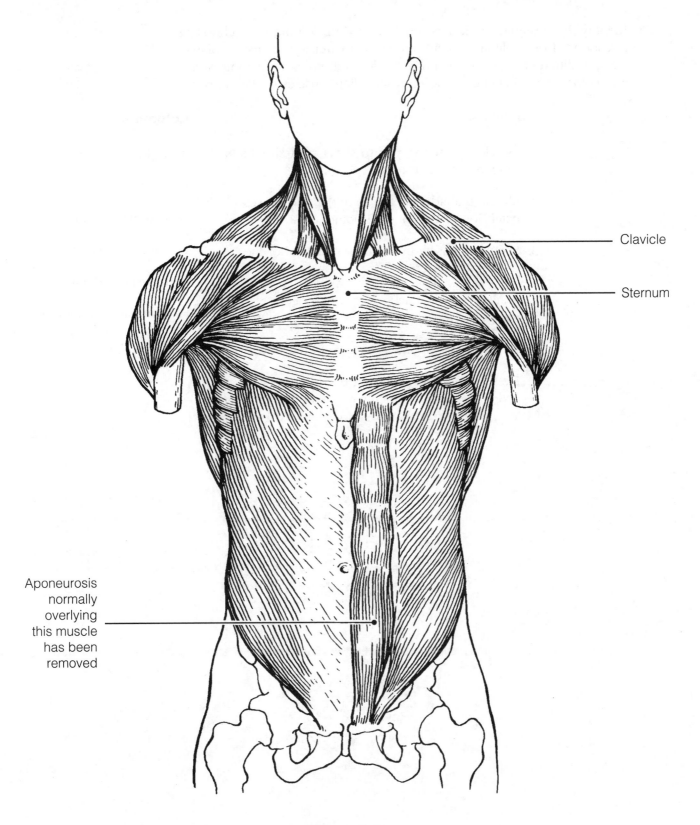

Clavicle

Sternum

Aponeurosis
normally
overlying
this muscle
has been
removed

Figure 6–7

21. Identify the posterior trunk muscles described in Column A by choosing a response from Column B. Enter the correct letter in the answer blank. Select a different color for each muscle description with a coding circle and color the coding circles and corresponding muscles on Figure 6–8.

Column A

○ _____ 1. Muscle that allows you to shrug your shoulders or extend your head

○ _____ 2. Muscle that adducts the shoulder and causes extension of the shoulder joint

○ _____ 3. Shoulder muscle that is the antagonist of the muscle just described

_____ 4. Prime mover of back extension; a deep composite muscle consisting of three columns

_____ 5. Large paired superficial muscle of the lower back

○ _____ 6. Fleshy muscle forming part of the posterior abdominal wall that helps maintain upright posture

Column B

A. Deltoid

B. Erector spinae

C. External oblique

D. Gluteus maximus

E. Latissimus dorsi

F. Quadratus lumborum

G. Trapezius

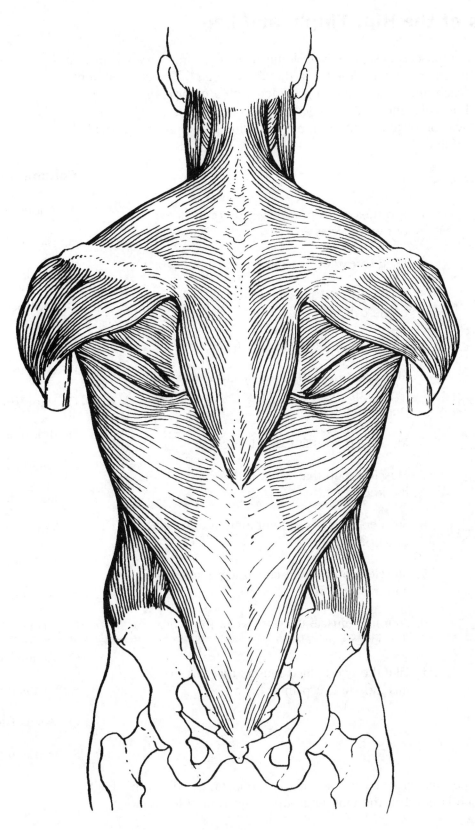

Figure 6-8

Muscles of the Hip, Thigh, and Leg

22. Identify the muscles described in Column A by choosing a response from Column B. Enter the correct letter in the answer blank. Select a different color for each muscle description provided with a color-coding circle, and use it to color the coding circles and corresponding muscles on Figure 6–9. Complete the illustration by labeling those muscles provided with leader lines.

Column A	Column B
_____ 1. Hip flexor, deep in pelvis; a composite of two muscles	A. Adductors
○ _____ 2. Used to extend the hip when climbing stairs	B. Biceps femoris
○ _____ 3. "Toe dancer's" muscle; a two-bellied muscle of the calf	C. Fibularis muscles
○ _____ 4. Inverts and dorsiflexes the foot	D. Gastrocnemius
○ _____ 5. Muscle group that allows you to draw your legs to the midline of your body, as when standing at attention	E. Gluteus maximus
	F. Gluteus medius
○ _____ 6. Muscle group that extends the knee	G. Hamstrings
○ _____ 7. Muscle group that extends the thigh and flexes the knee	H. Iliopsoas
○ _____ 8. Smaller hip muscle commonly used as an injection site	I. Quadriceps
○ _____ 9. Muscle group of the lateral leg; plantar flex and evert the foot	J. Rectus femoris
	K. Sartorius
○ _____ 10. Straplike muscle that is a weak thigh flexor; the "tailor's muscle"	L. Semimembranosus
	M. Semitendinosus
○ _____ 11. Like the two-bellied muscle that lies over it, this muscle is a plantar flexor	N. Soleus
	O. Tibialis anterior
	P. Vastus intermedius
	Q. Vastus lateralis
	R. Vastus medialis

23. What is the functional reason the muscle group on the dorsal leg (calf) is so much larger than the muscle group in the ventral leg region?

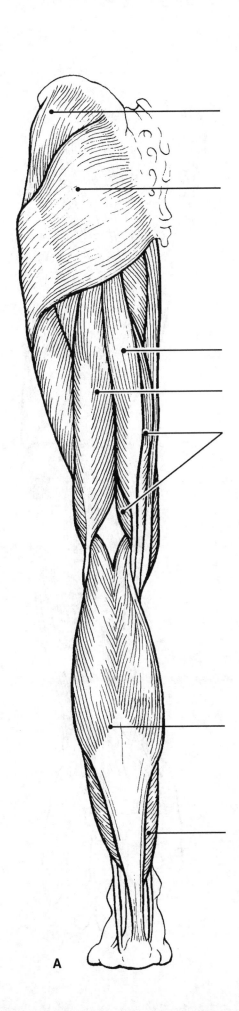

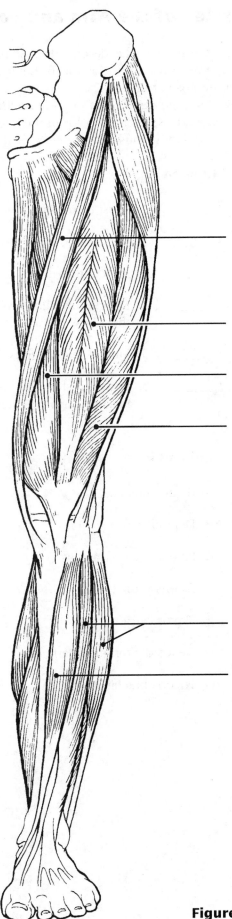

A

B

Figure 6-9

Muscles of the Arm and Forearm

24. Identify the muscles described in Column A by choosing a response from Column B. Enter the correct letter in the answer blank. Then select different colors for each muscle description provided with a color-coding circle and use them to color in the coding circles and corresponding muscles on Figure 6–10.

Column A

○ _____ 1. Wrist flexor that follows the ulna

○ _____ 2. Muscle that extends the fingers

_____ 3. Muscle that flexes the fingers

○ _____ 4. Muscle that allows you to bend (flex) the elbow

○ _____ 5. Muscle that extends the elbow

○ _____ 6. Powerful shoulder abductor, used to raise the arm overhead

Column B

A. Biceps brachii

B. Deltoid

C. Extensor carpi radialis

D. Extensor digitorum

E. Flexor carpi ulnaris

F. Flexor digitorum superficialis

G. Triceps brachii

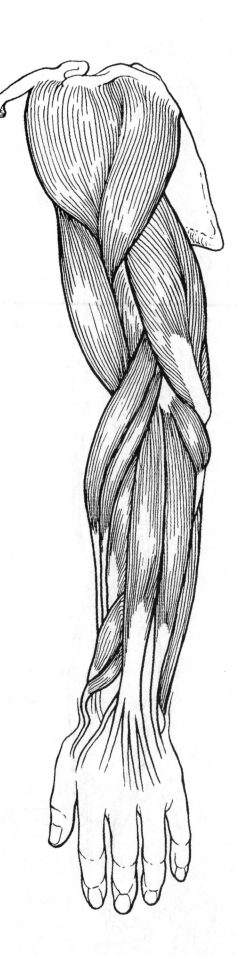

Figure 6–10

Chapter 6 The Muscular System **121**

General Body Muscle Review

25. Complete the following statements describing muscles. Insert the correct answers in the answer blanks.

_____ 1.

_____ 2.

_____ 3.

_____ 4.

_____ 5.

_____ 6.

_____ 7.

_____ 8.

_____ 9.

_____ 10.

_____ 11.

Three muscles— (1) , (2) , and (3) —are commonly used for intramuscular injections in adults.

The insertion tendon of the (4) group contains a large sesamoid bone, the patella.

The triceps surae insert in common into the (5) tendon.

The bulk of the tissue of a muscle tends to lie (6) to the part of the body it causes to move.

The extrinsic muscles of the hand originate on the (7) .

Most flexor muscles are located on the (8) aspect of the body; most extensors are located (9) . An exception to this generalization is the extensor–flexor musculature of the (10) .

The pectoralis major and deltoid muscles act synergistically to (11) the arm.

26. Circle the term that does not belong in each of the following groupings.

1. Vastus lateralis   Vastus medialis   Knee extension   Biceps femoris

2. Latissimus dorsi   Pectoralis major   Shoulder adduction   Antagonists

3. Buccinator   Frontalis   Masseter   Mastication   Temporalis

4. Vastus medialis   Rectus femoris   Iliacus   Origin on coxal bone

27. When kicking a football, at least three major actions of the lower limb are involved. Name the major muscles (or muscle groups) responsible for the following:

1. Flexing the hip joint: _____

2. Extending the knee: _____

3. Dorsiflexing the foot: _____

_____

28. Identify the numbered muscles in Figure 6–11 by placing the numbers in the blanks next to the following muscle names. Then select a different color for each muscle provided with a color-coding circle and color the coding circle and corresponding muscle in Figure 6–11.

◯ _____   1. Orbicularis oris

◯ _____   2. Pectoralis major

◯ _____   3. External oblique

◯ _____   4. Sternocleidomastoid

◯ _____   5. Biceps brachii

◯ _____   6. Deltoid

◯ _____   7. Vastus lateralis

◯ _____   8. Frontalis

◯ _____   9. Rectus femoris

◯ _____   10. Sartorius

◯ _____   11. Gracilis

◯ _____   12. Adductor group

◯ _____   13. Fibularis longus

◯ _____   14. Temporalis

◯ _____   15. Orbicularis oculi

◯ _____   16. Zygomaticus

◯ _____   17. Masseter

◯ _____   18. Vastus medialis

◯ _____   19. Tibialis anterior

◯ _____   20. Transversus abdominis

◯ _____   21. Rectus abdominis

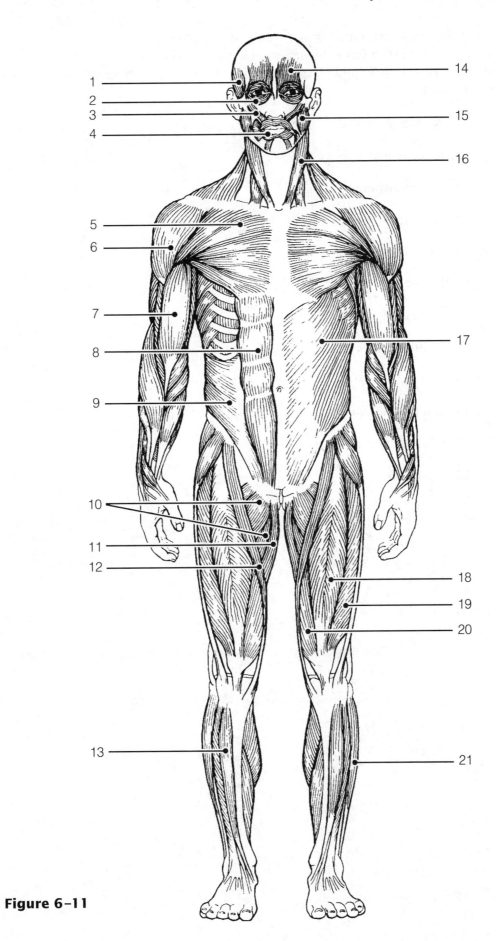

1
2
3
4

14

15

16

5
6

7

8

9

17

10

11
12

18
19
20

13

21

Figure 6–11

29. Identify each of the numbered muscles in Figure 6–12 by placing the numbers in the blanks next to the following muscle names. Then select different colors for each muscle and color the coding circles and corresponding muscles on Figure 6–12.

◯ _____   1. Adductor muscle

◯ _____   2. Gluteus maximus

◯ _____   3. Gastrocnemius

◯ _____   4. Latissimus dorsi

◯ _____   5. Deltoid

◯ _____   6. Semitendinosus

◯ _____   7. Soleus

◯ _____   8. Biceps femoris

◯ _____   9. Triceps brachii

◯ _____   10. External oblique

◯ _____   11. Gluteus medius

◯ _____   12. Trapezius

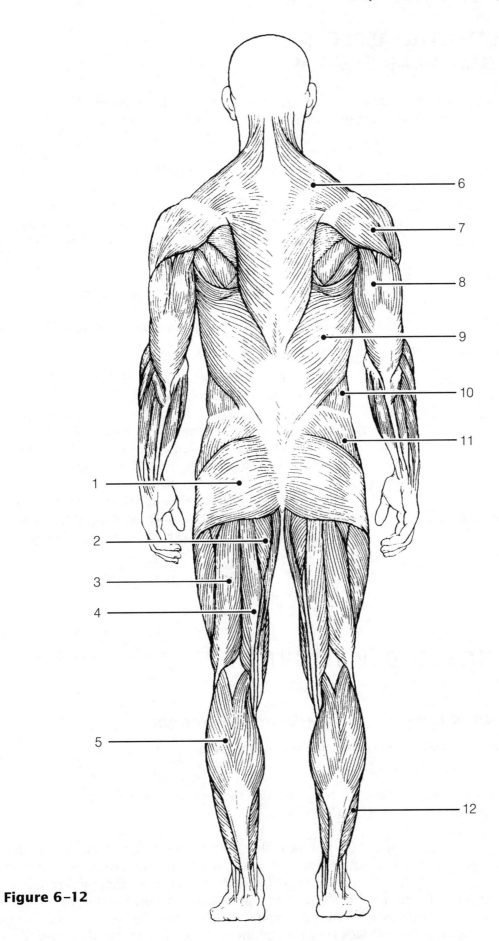

6

7

8

9

10

11

1

2

3

4

5

12

Figure 6-12

DEVELOPMENTAL ASPECTS OF THE MUSCULAR SYSTEM

30. Complete the following statements concerning the embryonic development of muscles and their functioning throughout life. Insert your answers in the answer blanks.

_____ 1.

_____ 2.

_____ 3.

_____ 4.

_____ 5.

_____ 6.

_____ 7.

_____ 8.

_____ 9.

_____ 10.

_____ 11.

_____ 12.

The first movement of the baby detected by the mother-to-be is called the __(1)__ .

An important congenital muscular disease that results in the degeneration of the skeletal muscles by young adulthood is called __(2)__ .

A baby's control over muscles progresses in a __(3)__ direction as well as a __(4)__ direction. In addition, __(5)__ muscular control (that is, waving of the arms) occurs before __(6)__ control (pincer grasp) does.

Muscles will ordinarily stay healthy if they are __(7)__ regularly; without normal stimulation they __(8)__ .

__(9)__ is a disease of the muscles, which results from some problem with the stimulation of muscles by ACh. The muscles become progressively weaker in this disease.

With age, our skeletal muscles decrease in mass; this leads to a decrease in body __(10)__ and in muscle __(11)__ . Muscle tissue that is lost is replaced by noncontractile __(12)__ tissue.

 INCREDIBLE JOURNEY

A Visualization Exercise for the Muscular System

As you straddle this structure, you wonder what is happening.

31. Where necessary, complete statements by inserting the missing words in the numbered spaces.

_____ 1.

On this incredible journey, you will be miniaturized and enter a skeletal muscle cell to observe the events that occur during muscle contraction. You prepare yourself by donning a wet suit and charging your ion detector. Then you climb into a syringe to prepare for injection. Your journey will begin when you see the gleaming connective tissue covering, the __(1)__ of a single muscle cell. Once injected, you monitor your descent through the epidermis and subcutaneous tissue. When you reach the muscle cell surface, you see that it is punctuated with pits at relatively

_____ 2.

_____ 3.

_____ 4.

_____ 5.

_____ 6.

_____ 7.

_____ 8.

_____ 9.

_____ 10.

regular intervals. Looking into the darkness and off in the distance, you can see a group of fibers ending close to a number of muscle cells. Considering that all of these fibers must be from the same motor neuron, this functional unit is obviously a __(2)__. You approach the fiber ending on your muscle cell and scrutinize the __(3)__ junction there. As you examine the junction, minute fluid droplets leave the nerve ending and attach to doughnut-shaped receptors on the muscle cell membrane. This substance released by the nerve ending must be __(4)__. Then, as a glow falls over the landscape, your ion detector indicates ions are disappearing from the muscle cell exterior and entering the muscle pits. The needle drops from high to low as the __(5)__ ions enter the pits from the watery fluid outside. You should have expected this, because these ions must enter to depolarize the muscle cells and start the __(6)__.

Next, you begin to explore one of the surface pits. As the muscle jerks into action, you topple deep into the pit. Sparkling electricity lights up the wall on all sides. You grasp for a handhold. Finally successful, you pull yourself laterally into the interior of the muscle cell and walk carefully along what seems to be a log. Then, once again, you notice an eerie glow as your ion detector reports that __(7)__ ions are entering the cytoplasm rapidly. The "log" you are walking on "comes to life" and begins to slide briskly in one direction. Unable to keep your balance, you fall. As you straddle this structure, you wonder what is happening. On all sides, cylindrical structures—such as the one you are astride—are moving past other similar but larger structures. Suddenly you remember, these are the __(8)__ myofilaments that slide past the __(9)__ myofilaments during muscle contraction.

Seconds later, the forward movement ends, and you begin to journey smoothly in the opposite direction. The ion detector now indicates low __(10)__ ion levels. Because you cannot ascend the smooth walls of one of the entry pits, you climb from one myofilament to another to reach the underside of the sarcolemma. Then you travel laterally to enter a pit close to the surface and climb out onto the cell surface. Your journey is completed, and you prepare to leave your host once again.

AT THE CLINIC

32. Pete, who has been moving furniture all day, arrives at the clinic complaining of painful spasms in his back. He reports having picked up a heavy table by stooping over. What muscle group has Pete probably strained, and why are these muscles at risk when one lifts objects improperly?

33. During an overambitious workout, a high school athlete pulled some muscles by forcing his knee into extension when his hip was already fully flexed. What muscles did he pull?

34. An emergency appendectomy is performed on Mr. Geiger. The incision was made at the lateral edge of the right iliac abdominopelvic region. Was his rectus abdominis cut?

35. Susan, a massage therapist, was giving Mr. Graves a back rub. What two broad superficial muscles of the back were receiving most of her attention?

36. Mrs. Sanchez says that her 6-year-old son seems to be unusually clumsy and tires easily. The doctor notices that his calf muscles appear to be normal in size. If anything, they seem a bit enlarged rather than wasted. For what condition must the boy be checked? What is the prognosis?

37. People with chronic back pain occasionally get relief from a tummy tuck. How does this help?

38. Gregor, who works at a pesticide factory, comes to the clinic complaining of muscle spasms that interfere with his movement and breathing. A blood test shows that he has been contaminated with organophosphate pesticide, which is an acetylcholinesterase inhibitor. How would you explain to Gregor what this means?

39. While riding an unusually large horse, Chao Jung had to spread her thighs wide to span its back, and she pulled the muscles in her medial thighs. Which muscles were these?

40. Do all muscles attach to bone? If not, what else do they attach to?

✓ THE FINALE: MULTIPLE CHOICE

41. Select the best answer or answers from the choices given.

1. Select the type of muscle tissue that fits the following description: self-excitable, pacemaker cells, gap junctions, limited sarcoplasmic reticulum.

   A. Skeletal muscle   C. Smooth muscle

   B. Cardiac muscle   D. Involuntary muscle

2. Skeletal muscle is *not* involved in:

   A. movement of skin

   B. propulsion of a substance through a body tube

   C. heat production

   D. inhibition of body movement

3. Which of the following are part of a thin myofilament?

   A. ATP-binding site   C. Globular actin

   B. Regulatory proteins   D. Calcium

4. The movement of thin filaments toward the sarcomere is called:

   A. cocking of the myosin heads

   B. repolarization of the T tubules

   C. the power stroke

   D. the action potential

5. Transmission of the stimulus at the neuromuscular junction involves:

   A. synaptic vesicles   C. ACh

   B. sarcolemma   D. axon terminal

6. Your ability to lift that heavy couch would be increased by which type of exercise?

   A. Aerobic   C. Resistance

   B. Endurance   D. Swimming

7. Which of the following activities depends most on anaerobic metabolism?

   A. Jogging

   B. Swimming a race

   C. Sprinting

   D. Running a marathon

8. The first energy source used to regenerate ATP when muscles are extremely active is:

   A. fatty acids   C. creatine phosphate

   B. glucose   D. pyruvic acid

9. Head muscles that insert on a bone include the:

   A. zygomaticus   C. buccinator

   B. masseter   D. temporalis

10. Lateral flexion of the torso involves:

    A. erector spinae

    B. rectus abdominis

    C. quadratus lumborum

    D. external oblique

11. Muscles attached to the vertebral column include:

    A. quadratus lumborum

    B. external oblique

    C. diaphragm

    D. latissimus dorsi

12. Muscles that help stabilize the scapula and shoulder joint include:

    A. triceps brachii   C. trapezius

    B. biceps brachii   D. pectoralis major

13. Which of these thigh muscles causes movement at the hip joint?

    A. Rectus femoris

    B. Biceps femoris

    C. Vastus lateralis

    D. Semitendinosus

14. Leg muscles that can cause movement at the knee joint include:

    A. tibialis anterior

    B. fibularis longus

    C. gastrocnemius

    D. soleus

15. The main muscles used when doing chin-ups are:

    A. triceps brachii and pectoralis major

    B. infraspinatus and biceps brachii

    C. serratus anterior and external oblique

    D. latissimus dorsi and brachialis

16. The major muscles used in doing push-ups are:

    A. biceps brachii and brachialis

    B. supraspinatus and subscapularis

    C. coracobrachialis and latissimus dorsi

    D. triceps brachii and pectoralis major

17. Arm and leg muscles are arranged in antagonistic pairs. How does this affect their functioning?

    A. It provides a backup if one of the muscles is injured.

    B. One muscle of the pair pushes while the other pulls.

    C. A single neuron controls both of them.

    D. It allows the muscles to produce opposing movements.

18. Muscle A and muscle B are the same size, but muscle A is capable of much finer control than muscle B. Which of the following is likely to be true of muscle A?

    A. It is controlled by more neurons than muscle B.

    B. It contains fewer motor units than muscle B.

    C. It is controlled by fewer neurons than muscle B.

    D. Each of its motor units consists of more cells than the motor units of muscle B.

19. Binding sites for calcium are found on:

    A. thin filaments

    B. thick filaments

    C. myosin filaments

    D. actin filaments

7 THE NERVOUS SYSTEM

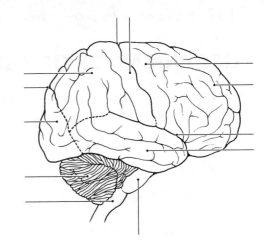

The nervous system is the master coordinating system of the body. Every thought, action, and sensation reflects its activity. The structures of the nervous system are described in terms of two principal divisions—the central nervous system (CNS) and the peripheral nervous system (PNS). The CNS (brain and spinal cord) interprets incoming sensory information and issues instructions based on past experience. The PNS (cranial and spinal nerves and ganglia) provides the communication lines between the CNS and the body's muscles, glands, and sensory receptors. The nervous system is also divided functionally in terms of motor activities into the somatic and autonomic divisions. It is important, however, to recognize that these classifications are made for the sake of convenience and that the nervous system acts in an integrated manner both structurally and functionally.

Student activities provided in this chapter review neuron anatomy and physiology, identify the various structures of the central and peripheral nervous system, consider reflex and sensory physiology, and summarize autonomic nervous system anatomy and physiology. Because every body system is controlled, at least in part, by the nervous system, these concepts are extremely important to understanding how the body functions as a whole.

1. List the three major functions of the nervous system.

1. _____

_____

2. _____

_____

3. _____

_____

ORGANIZATION OF THE NERVOUS SYSTEM

2. Choose the key responses that best correspond to the descriptions provided in the following statements. Insert the appropriate letter or term in the answer blanks.

Key Choices

A. Autonomic nervous system

C. Peripheral nervous system (PNS)

B. Central nervous system (CNS)

D. Somatic nervous system

_____ 1. Nervous system subdivision that is composed of the brain and spinal cord

_____ 2. Subdivision of the PNS that controls voluntary activities such as the activation of skeletal muscles

_____ 3. Nervous system subdivision that is composed of the cranial and spinal nerves and ganglia

_____ 4. Subdivision of the PNS that regulates the activities of the heart and smooth muscle, and of glands; it is also called the involuntary nervous system

_____ 5. A major subdivision of the nervous system that interprets incoming information and issues orders

_____ 6. A major subdivision of the nervous system that serves as communication lines, linking all parts of the body to the CNS

NERVOUS TISSUE—STRUCTURE AND FUNCTION

3. This exercise emphasizes the difference between neurons and neuroglia. Indicate which cell type is identified by the following descriptions. Insert the appropriate letter or term in the answer blanks.

Key Choices

A. Neurons

B. Neuroglia

_____ 1. Support, insulate, and protect cells

_____ 2. Demonstrate irritability and conductivity, and thus transmit electrical messages from one area of the body to another area

_____ 3. Release neurotransmitters

_____ 4. Are amitotic

_____ 5. Able to divide; therefore are responsible for most brain neoplasms

4. Relative to neuron anatomy, match the anatomical terms given in Column B
with the appropriate descriptions of functions provided in Column A. Place
the correct term or letter response in the answer blanks.

	Column A	**Column B**
_____	1. Releases neurotransmitters	A. Axon
_____	2. Conducts local electrical currents toward the cell body	B. Axon terminal
		C. Dendrite
_____	3. Increases the speed of impulse transmission	D. Myelin sheath
_____	4. Location of the nucleus	E. Neuron cell body
_____	5. Generally conducts impulses away from the cell body	F. Nissl bodies
_____	6. Clustered ribosomes and rough ER (endoplasmic reticulum)	

5. Certain activities or sensations are listed below. Using the key choices, select
the specific receptor type that would be activated by the activity or sensation
described. Insert the correct term(s) or letter response(s) in the answer blanks.
Note that more than one receptor type may be activated in some cases.

Key Choices

A. Bare nerve endings (pain)   C. Meissner's (tactile) corpuscle   E. Lamellated corpuscle

B. Golgi tendon organ   D. Muscle spindle

Activity or sensation	**Receptor type**
Walking on hot pavement	1. (Identify two) _____ and _____
Feeling a pinch	2. (Identify two) _____ and _____
Leaning on a shovel	3. _____
Muscle sensations when rowing a boat	4. (Identify two) _____ and _____
Feeling a caress	5. _____

6. Using the key choices, select the terms identified in the following descriptions by inserting the appropriate letter or term in the spaces provided.

Key Choices

A. Afferent neuron	F. Neuroglia	K. Proprioceptors
B. Association neuron (or interneuron)	G. Neurotransmitters	L. Schwann cells
C. Cutaneous sense organs	H. Nerve	M. Synapse
D. Efferent neuron	I. Nodes of Ranvier	N. Stimuli
E. Ganglion	J. Nuclei	O. Tract

_____ 1. Sensory receptors found in the skin, which are specialized to detect temperature, pressure changes, and pain

_____ 2. Specialized cells that myelinate the fibers of neurons found in the PNS

_____ 3. Junction or point of close contact between neurons

_____ 4. Bundle of nerve processes inside the CNS

_____ 5. Neuron, serving as part of the conduction pathway between sensory and motor neurons

_____ 6. Gaps in a myelin sheath

_____ 7. Collection of nerve cell bodies found outside the CNS

_____ 8. Neuron that conducts impulses away from the CNS to muscles and glands

_____ 9. Sensory receptors found in muscle and tendons that detect their degree of stretch

_____ 10. Changes, occurring within or outside the body, that affect nervous system functioning

_____ 11. Neuron that conducts impulses toward the CNS from the body periphery

_____ 12. Chemicals released by neurons that stimulate other neurons, muscles, or glands

7. Figure 7–1 is a diagram of a neuron. First, label the parts indicated on the
illustration by leader lines. Then choose different colors for each of the
structures listed below and use them to color in the coding circles and
corresponding structures in the illustration. Next, circle the term
in the list of three terms to the left of the diagram
that best describes this neuron's structural class.
Finally, draw arrows on the figure to indicate
the direction of impulse transmission
along the neuron's membrane.

◯ Axon

◯ Dendrites

◯ Cell body

◯ Myelin sheath

Unipolar

Bipolar

Multipolar

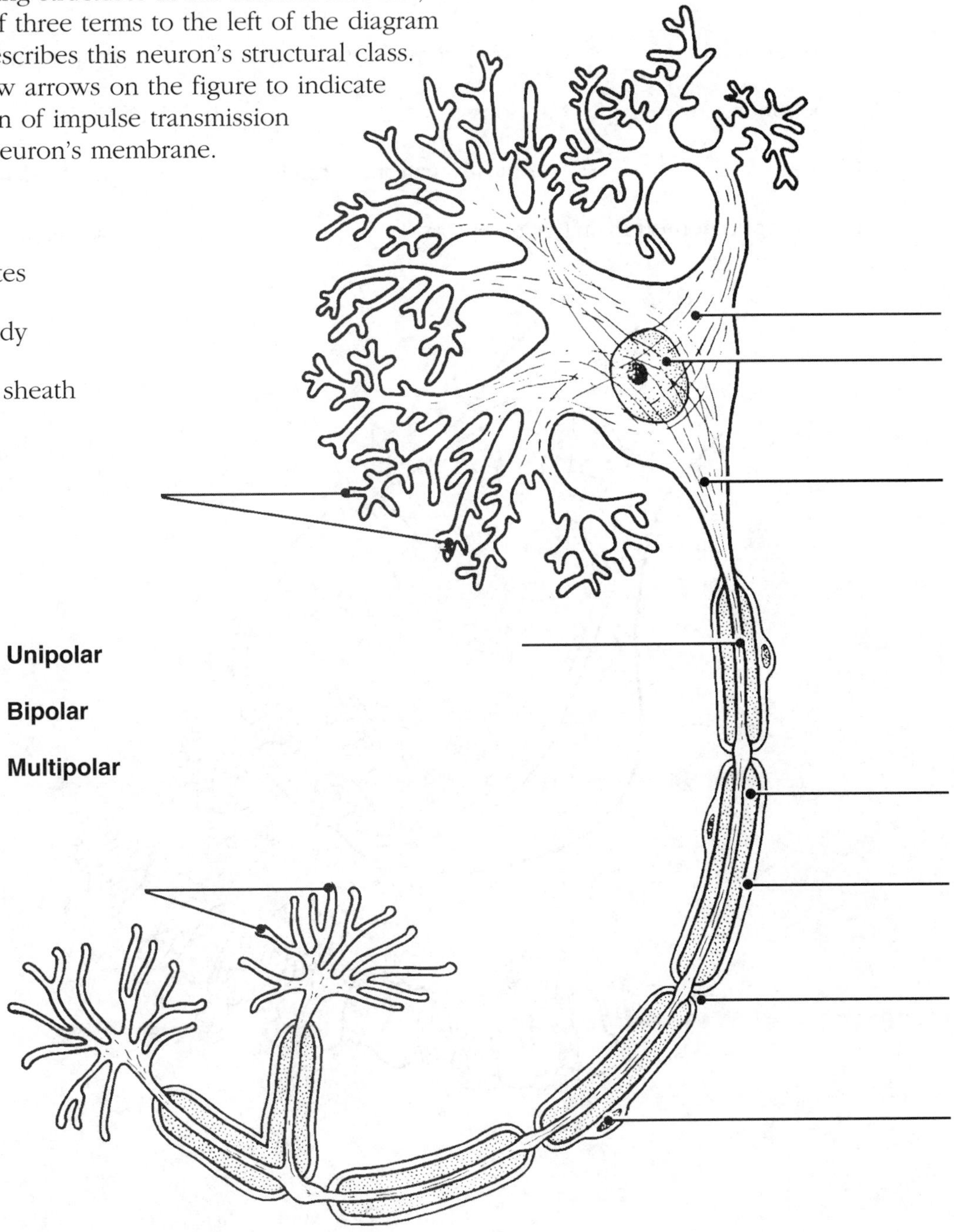

Figure 7–1

8. List in order the *minimum* elements in a reflex arc from the stimulus to
the activity of the effector. Place your responses in the answer blanks.

1. Stimulus

2. _____

3. _____

4. _____

5. Effector organ

9. In Figure 7–2, identify by coloring the following structures, which are typically part of a chemical synapse. Also, bracket the synaptic cleft, and identify the arrows showing (1) the direction of the presynaptic impulse and (2) the direction of net neurotransmitter movements.

- ○ Axon terminal
- ○ Postsynaptic membrane
- ○ Presynaptic membrane
- ○ Mitochondria
- ○ Na$^+$ ions
- ○ Ca^{2+} ions
- ○ K$^+$ ions
- ○ Chemically gated channels
- ○ Synaptic vesicles
- ○ Postsynaptic neurotransmitter receptors
- ○ Neurotransmitter molecules

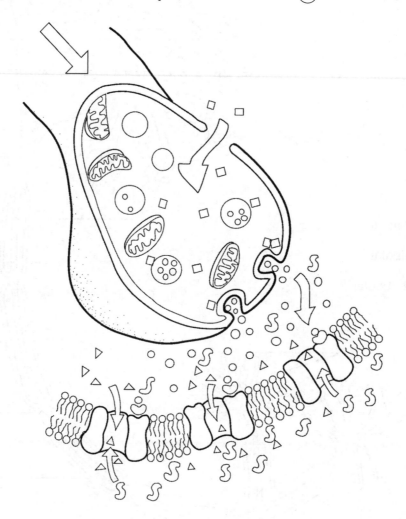

Figure 7–2

10. Using the key choices, identify the terms defined in the following statements. Place the correct term or letter response in the answer blanks.

Key Choices

A. Action potential   D. Potassium ions   G. Resting period

B. Depolarization   E. Refractory period   H. Sodium ions

C. Polarized   F. Repolarization   I. Sodium-potassium pump

_____ 1. Period of repolarization of the neuron during which it cannot respond to a second stimulus

_____ 2. State in which the resting potential is reversed as sodium ions rush into the neuron

_____ 3. Electrical condition of the plasma membrane of a resting neuron

_____ 4. Period during which potassium ions diffuse out of the neuron

_____ 5. Transmission of the depolarization wave along the neuron's membrane

_____ 6. The chief positive intracellular ion in a resting neuron

_____ 7. Process by which ATP is used to move sodium ions out of the cell and potassium ions back into the cell; completely restores the resting conditions of the neuron

_____ 8. State in which all voltage gated Na^+ and K^+ channels are closed

11. Using the key choices, identify the types of reflexes involved in each of the following situations.

Key Choices

A. Somatic reflex(es)   B. Autonomic reflex(es)

_____ 1. Patellar (knee-jerk) reflex

_____ 2. Pupillary light reflex

_____ 3. Effectors are skeletal muscles

_____ 4. Effectors are smooth muscle and glands

_____ 5. Flexor reflex

_____ 6. Regulation of blood pressure

_____ 7. Salivary reflex

12. Refer to Figure 7–3, showing a reflex arc, as you complete this exercise. First, briefly answer the following questions by inserting your responses in the spaces provided.

1. What is the stimulus? _____

2. What tissue is the effector? _____

3. How many synapses occur in this reflex arc? _____

Next, select different colors for each of the following structures and use them to color in the coding circles and corresponding structures in the diagram. Finally, draw arrows on the figure indicating the direction of impulse transmission through this reflex pathway.

○ Receptor region   ○ Interneuron

○ Afferent neuron   ○ Efferent neuron

○ Effector

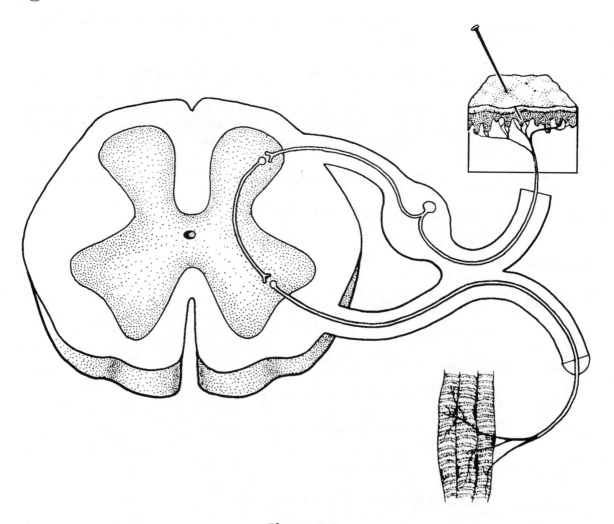

Figure 7–3

13. Circle the term that does not belong in each of the following groupings.

1. Astrocytes   Neurons   Oligodendrocytes   Microglia

2. K^+ enters the cell K^+ leaves the cell Repolarization Refractory period

3. Nodes of Ranvier Myelin sheath Nonmyelinated Saltatory conduction

4. Predictable response Voluntary act Involuntary act Reflex

5. Oligodendrocytes Schwann cells Myelin Microglia

6. Cutaneous receptors Free dendritic endings Stretch Pain and touch

7. Cell interior High Na^+ Low Na^+ High K^+

CENTRAL NERVOUS SYSTEM
Brain

14. Complete the following statements by inserting your answers in the answer blanks.

_____ 1.

_____ 2.

_____ 3.

_____ 4.

_____ 5.

The largest part of the human brain is the (paired) _(1)_ . The other major subdivisions of the brain are the _(2)_ and the _(3)_ . The cavities found in the brain are called _(4)_ . They contain _(5)_ .

15. Circle the terms indicating structures that are *not* part of the brain stem.

Cerebral hemispheres Midbrain Medulla

Pons Cerebellum Diencephalon

16. Complete the following statements by inserting your answers in the answer blanks.

_____ 1.

_____ 2.

_____ 3.

_____ 4.

_____ 5.

A _(1)_ is an elevated ridge of cerebral cortex tissue. The convolutions seen in the cerebrum are important because they increase the _(2)_ . Gray matter is composed of _(3)_ . White matter is composed of _(4)_ , which provide for communication between different parts of the brain as well as with lower CNS centers. The lentiform nucleus, the caudate, and other nuclei are collectively called the _(5)_ .

17. Figure 7–4 is a diagram of the right lateral view of the human brain. First, match the letters on the diagram with the following list of terms and insert the appropriate letters in the answer blanks. Then, select different colors for each of the areas of the brain provided with a color-coding circle and use them to color in the coding circles and corresponding structures in the diagram. If an identified area is part of a lobe, use the color you selected for the lobe but use *stripes* for that area.

_____ 1. ◯ Frontal lobe

_____ 2. ◯ Parietal lobe

_____ 3. ◯ Temporal lobe

_____ 4. ◯ Precentral gyrus

_____ 5. Parieto-occipital fissure

_____ 6. ◯ Postcentral gyrus

_____ 7. Lateral sulcus

_____ 8. Central sulcus

_____ 9. ◯ Cerebellum

_____ 10. ◯ Medulla

_____ 11. ◯ Occipital lobe

_____ 12. ◯ Pons

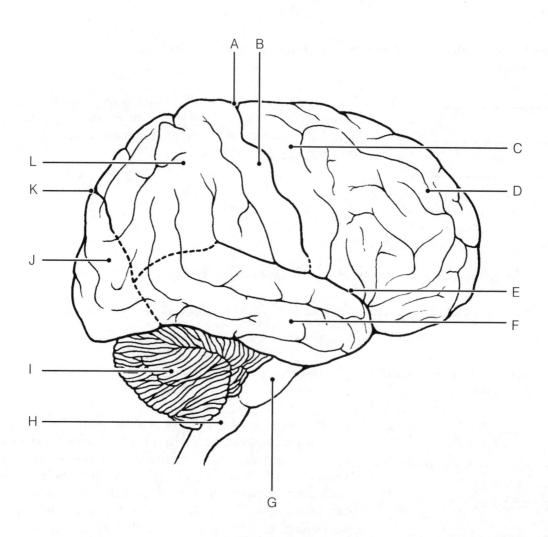

Figure 7–4

18. Figure 7–5 is a diagram of the sagittal view of the human brain. First, match the letters on the diagram with the following list of terms and insert the appropriate letter in each answer blank. Then, color the brainstem areas blue and the areas where cerebrospinal fluid is found yellow.

_____ 1. Cerebellum

_____ 2. Cerebral aqueduct

_____ 3. Cerebral hemisphere

_____ 4. Cerebral peduncle

_____ 5. Choroid plexus

_____ 6. Corpora quadrigemina

_____ 7. Corpus callosum

_____ 8. Fourth ventricle

_____ 9. Mammillary body

_____ 10. Medulla oblongata

_____ 11. Optic chiasma

_____ 12. Pineal body

_____ 13. Pituitary gland

_____ 14. Pons

_____ 15. Thalamus

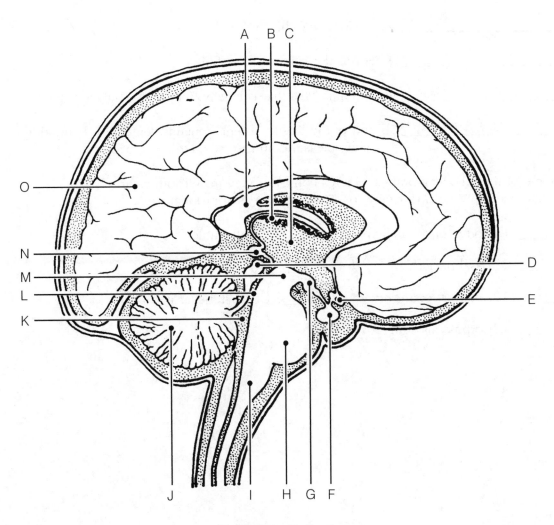

Figure 7–5

19. Referring to the brain areas listed in Exercise 18, match the appropriate brain structures with the following descriptions. Insert the correct terms in the answer blanks.

_____ 1. Site of regulation of water balance and body temperature

_____ 2. Contains reflex centers involved in regulating respiratory rhythm in conjunction with lower brainstem centers

_____ 3. Responsible for the regulation of posture and coordination of skeletal muscle movements

_____ 4. Important relay station for afferent fibers traveling to the sensory cortex for interpretation

_____ 5. Contains autonomic centers, which regulate blood pressure and respiratory rhythm, as well as coughing and sneezing centers

_____ 6. Large fiber tract connecting the cerebral hemispheres

_____ 7. Connects the third and fourth ventricles

_____ 8. Encloses the third ventricle

_____ 9. Forms the cerebrospinal fluid

_____ 10. Midbrain area that is largely fiber tracts; bulges anteriorly

_____ 11. Part of the limbic system; contains centers for many drives (rage, pleasure, hunger, sex, etc.)

20. Some of the following brain structures consist of gray matter; others are white matter. Write _G_ (for gray) or _W_ (for white) as appropriate.

_____ 1. Cortex of cerebellum   _____ 5. Pyramids

_____ 2. Basal nuclei   _____ 6. Thalamic nuclei

_____ 3. Anterior commisure   _____ 7. Cerebellar peduncle

_____ 4. Corpus callosum

21. Figure 7–6 illustrates a "see-through" brain showing the positioning of the ventricles and connecting canals or apertures. Correctly identify all structures having leader lines by using the key choices provided below. One of the lateral ventricles has already been identified. Color the spaces filled with cerebrospinal fluid blue.

Key Choices

A. Anterior horn   D. Fourth ventricle   G. Lateral aperture

B. Central canal   E. Inferior horn   H. Third ventricle

C. Cerebral aqueduct   F. Interventricular foramen

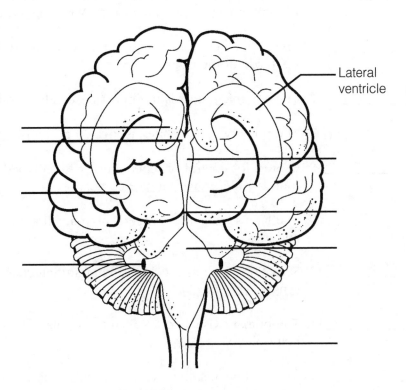

Lateral
ventricle

Figure 7–6

22. If a statement is true, write the letter *T* in the answer blank. If a statement is false, correct the <u>underlined</u> word(s) and write the correct word(s) in the answer blank.

_____ 1. The primary somatosensory area of the cerebral hemisphere(s) is found in the <u>precentral</u> gyrus.

_____ 2. Cortical areas involved in audition are found in the <u>occipital</u> lobe.

_____ 3. The primary motor area in the <u>temporal</u> lobe is involved in the initiation of voluntary movements.

_____ 4. The specialized motor speech area is located at the base of the precentral gyrus in an area called <u>Wernicke's</u> area.

_____ 5. The right cerebral hemisphere receives sensory input from the <u>right</u> side of the body.

_____ 6. The <u>pyramidal</u> tract is the major descending voluntary motor tract.

_____ 7. The primary motor cortex is located in the <u>postcentral</u> gyrus.

_____ 8. Centers for control of repetitious or stereotyped motor skills are found in the <u>primary motor</u> cortex.

_____ 9. The largest parts of the motor homunculi are the lips, tongue, and <u>toes</u>.

_____10. Sensations such as touch and pain are integrated in the <u>primary sensory cortex</u>.

_____11. The primary visual cortex is in the <u>frontal</u> lobe of each cerebral hemisphere.

_____12. In most humans, the area that controls the comprehension of language is located in the <u>left</u> cerebral hemisphere.

_____13. A <u>flat</u> electroencephalogram (EEG) is evidence of clinical death.

_____14. Beta waves are recorded when an individual is awake and <u>relaxed</u>.

Protection of the CNS

23. Identify the meningeal (or associated) structures described here.

_____ 1. Outermost covering of the brain, composed of tough fibrous connective tissue

_____ 2. Innermost covering of the brain; delicate and vascular

_____ 3. Structures that return cerebrospinal fluid to the venous blood in the dural sinuses

_____ 4. Middle meningeal layer; like a cobweb in structure

_____ 5. Its outer layer forms the periosteum of the skull

24. Figure 7–7 shows a frontal view of the meninges of the brain at the level of the superior sagittal (dural) sinus. First, label the _arachnoid villi_ on the figure. Then, select different colors for each of the following structures and use them to color the coding circles and corresponding structures in the diagram.

◯ Dura mater   ◯ Pia mater

◯ Arachnoid mater   ◯ Subarachnoid space

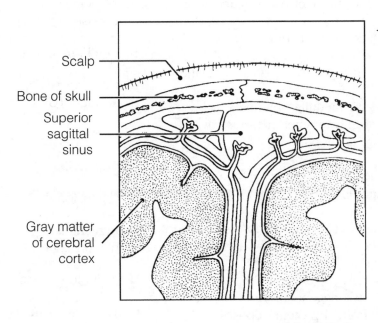
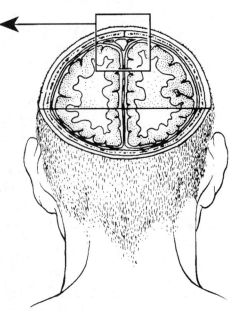

Figure 7–7

25. Complete the following statements by inserting your answers in the answer blanks.

_____ 1.

_____ 2.

_____ 3.

_____ 4.

_____ 5.

_____ 6.

_____ 7.

1. Cerebrospinal fluid is formed by capillary knots called __(1)__, which hang into the __(2)__ of the brain. Ordinarily, cerebrospinal fluid flows from the lateral ventricles to the third ventricle and then through the __(3)__ to the fourth ventricle. Some of the fluid continues down the __(4)__ of the spinal cord, but most of it circulates into the __(5)__ by passing through three tiny openings in the walls of the __(6)__. As a rule, cerebrospinal fluid is formed and drained back into the venous blood at the same rate. If its drainage is blocked, a condition called __(7)__ occurs, which results in increased pressure on the brain.

Brain Dysfunctions

26. Match the brain disorders listed in Column B with the conditions described in Column A. Place the correct answers in the answer blanks.

Column A

_____ 1. Slight and transient brain injury

_____ 2. Traumatic injury that destroys brain tissue

_____ 3. Total nonresponsiveness to stimulation

_____ 4. May cause medulla oblongata to be wedged into foramen magnum by pressure of blood

_____ 5. After head injury, retention of water by brain

_____ 6. Results when a brain region is deprived of blood or exposed to prolonged ischemia

_____ 7. Progressive degeneration of the brain with abnormal protein deposits

_____ 8. Autoimmune disorder with extensive demyelination

_____ 9. A ministroke; fleeting symptoms of a CVA

Column B

A. Alzheimer's disease

B. Cerebral edema

C. Cerebrovascular accident (CVA)

D. Coma

E. Concussion

F. Contusion

G. Intracranial hemorrhage

H. Multiple sclerosis

I. Transient ischemic attack (TIA)

Spinal Cord

27. Complete the following statements by inserting your responses in the answer blanks.

_____ 1.

_____ 2.

_____ 3.

_____ 4.

_____ 5.

_____ 6.

_____ 7.

_____ 8.

_____ 9.

The spinal cord extends from the __(1)__ of the skull to the __(2)__ region of the vertebral column. The meninges, which cover the spinal cord, extend more inferiorly to form a sac from which cerebrospinal fluid can be withdrawn without damage to the spinal cord. This procedure is called a __(3)__. __(4)__ pairs of spinal nerves arise from the cord. Of these, __(5)__ pairs are cervical nerves, __(6)__ pairs are thoracic nerves, __(7)__ pairs are lumbar nerves, and __(8)__ pairs are sacral nerves. The tail-like collection of spinal nerves at the inferior end of the spinal cord is called the __(9)__.

28. Using the key choices, select the appropriate terms to respond to the following descriptions referring to spinal cord anatomy. Place the correct term or letter in the answer blanks.

Key Choices

A. Afferent

B. Efferent

C. Both afferent and efferent

D. Association neurons (interneurons)

_____ 1. Neuron type found in the dorsal horn

_____ 2. Neuron type found in the ventral horn

_____ 3. Neuron type in a dorsal root ganglion

_____ 4. Fiber type in the ventral root

_____ 5. Fiber type in the dorsal root

_____ 6. Fiber type in a spinal nerve

_____ 7. Fiber type in the anterior ramus

_____ 8. Damage to this fiber type would lead to a loss of sensory function

_____ 9. Damage to this fiber type results in a loss of motor function

29. Figure 7–8 is a cross-sectional view of the spinal cord. First, identify the areas listed in the key choices by inserting the correct letters next to the appropriate leader lines on parts A and B of the figure. Then, color the bones of the vertebral column in part B gold.

Key Choices

A. Central canal   E. Dorsal root   I. Ventral horn

B. Columns of white matter   F. Dorsal root ganglion   J. Ventral root

C. Conus medullaris   G. Filum terminale

D. Dorsal horn   H. Spinal nerve

On part A, color the butterfly-shaped gray matter gray, and color the spinal nerves and roots yellow. Finally, select different colors to identify the following structures and use them to color the figure.

◯ Pia mater   ◯ Dura mater   ◯ Arachnoid mater

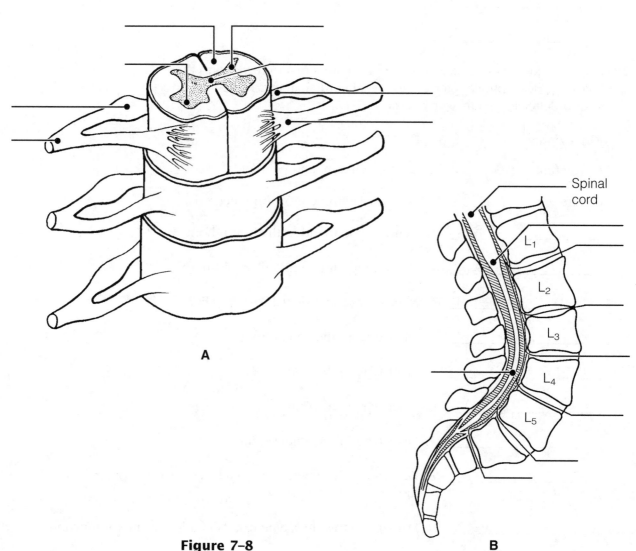

A

Spinal cord

L₁

L₂

L₃

L₄

L₅

Figure 7–8   B

30. On the ascending pathway shown in Figure 7–9, circle all synapse sites and use the terms listed below to identify all structures provided with leader lines. Then, select different colors for each of the terms and use them to color in the coding circles and corresponding structures in the illustration.

○ Sensory cortex   ○ Sensory receptor   ○ Thalamus

○ Sensory homunculus   ○ Spinal cord

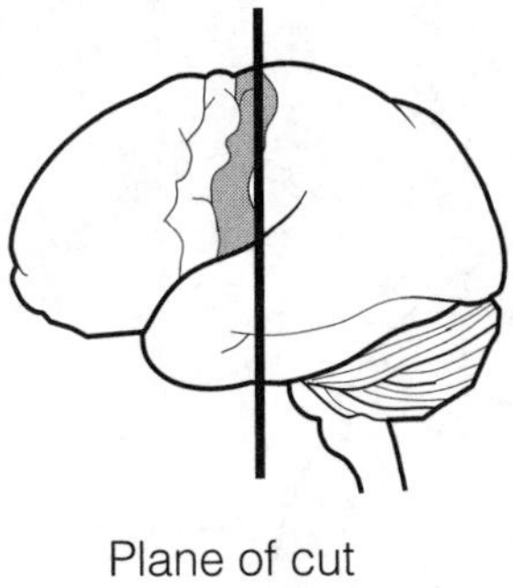

Plane of cut

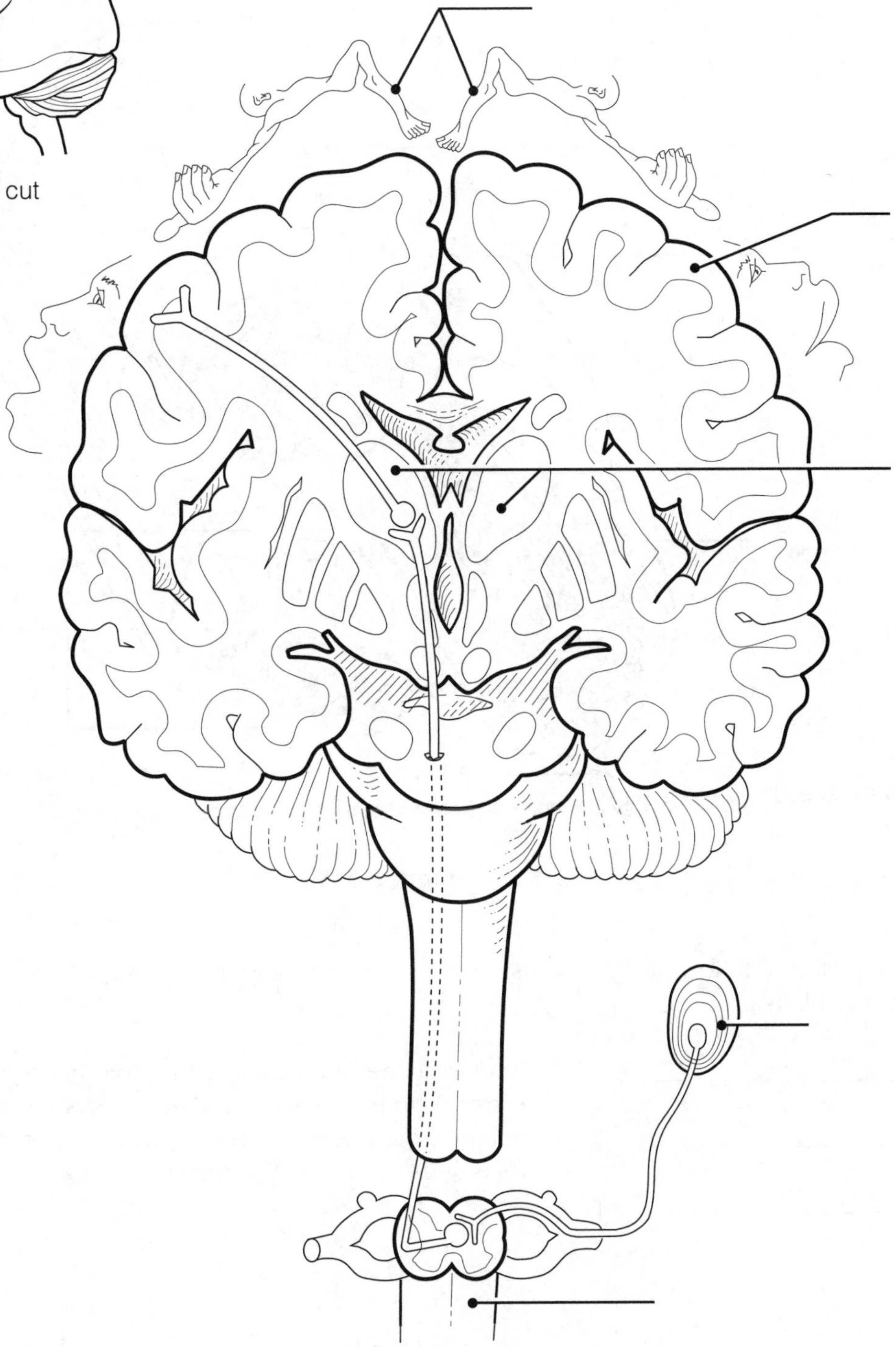

Figure 7–9

PERIPHERAL NERVOUS SYSTEM
Structure of a Nerve

31. Figure 7–10 is a diagrammatic view of a nerve wrapped in its connective tissue coverings. Select different colors to identify the following structures and use them to color the coding circles and corresponding structures in the figure. Then, label each of the sheaths indicated by leader lines on the figure.

◯ Endoneurium   ◯ Perineurium   ◯ Epineurium

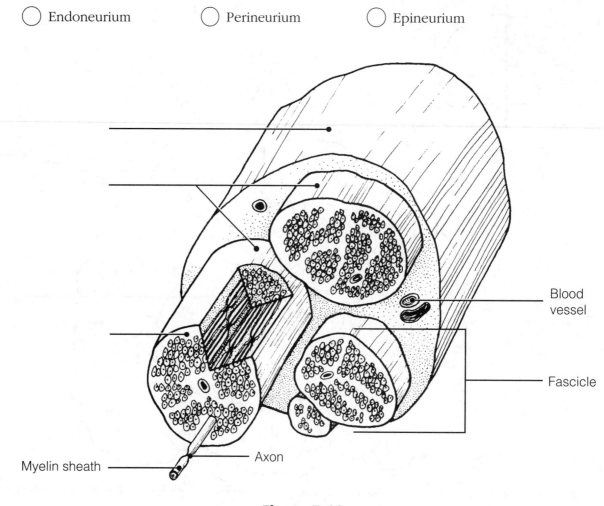

Figure 7–10

32. Complete the following statements by inserting your responses in the answer blanks.

_____ 1.

_____ 2.

_____ 3.

Another name for a bundle of nerve fibers is __(1)__. Nerves carrying both sensory and motor fibers are called __(2)__ nerves, whereas those carrying just sensory fibers are referred to as sensory, or __(3)__, nerves.

Cranial Nerves

33. The 12 pairs of cranial nerves are indicated by leader lines in Figure 7–11. First, label each by name and Roman numeral on the figure and then color each nerve with a different color.

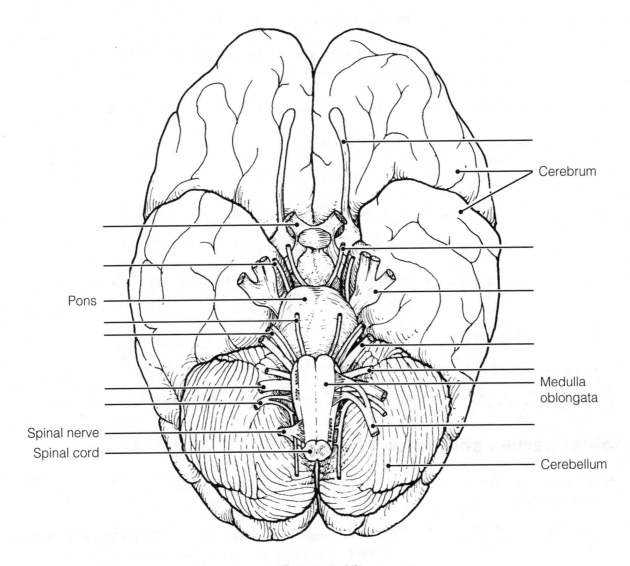

Cerebrum

Pons

Medulla oblongata

Spinal nerve

Spinal cord

Cerebellum

Figure 7–11

34. Provide the name and number of the cranial nerves involved in each of the following activities, sensations, or disorders. Insert your response in the answer blanks.

_____ 1. Shrugging the shoulders

_____ 2. Smelling a flower

_____ 3. Raising the eyelids and focusing the lens of the eye for accommodation; constriction of the eye pupils

_____ 4. Slows the heart; increases the mobility of the digestive tract

_____ 5. Involved in smiling

_____ 6. Involved in chewing gum

_____ 7. Listening to music; seasickness

_____ 8. Secretion of saliva; tasting well-seasoned food

_____ 9. Involved in "rolling" the eyes (three nerves—provide numbers only)

_____ 10. Feeling a toothache

_____ 11. Reading *Tennis* magazine or this study guide

_____ 12. Purely sensory (three nerves—provide numbers only)

Spinal Nerves and Nerve Plexuses

35. Complete the following statements by inserting your responses in the answer blanks.

_____ 1.

_____ 2.

_____ 3.

_____ 4.

The ventral rami of spinal nerves C_1 through T_1 and L_1 through S_4 take part in forming __(1)__, which serve the __(2)__ of the body. The ventral rami of T_1 through T_{12} run between the ribs to serve the __(3)__. The posterior rami of the spinal nerves serve the __(4)__.

36. Figure 7–12 is an anterior view of the principal nerves arising from the brachial plexus. Select five different colors and color the coding circles and the nerves listed below. Also, label each nerve by inserting its name at the appropriate leader line.

○ Axillary nerve

○ Musculocutaneous nerve

○ Median nerve

○ Radial nerve

○ Ulnar nerve

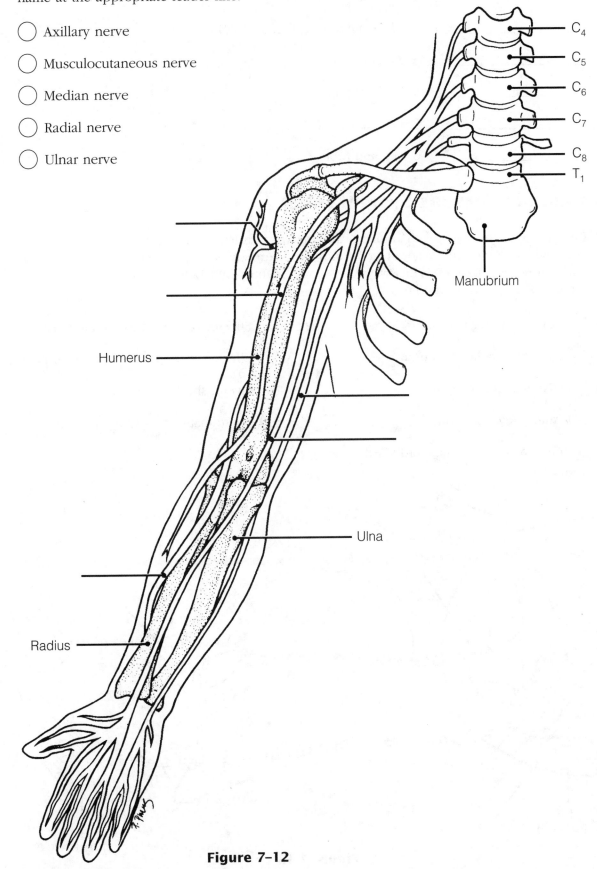

C_4

C_5

C_6

C_7

C_8

T_1

Manubrium

Humerus

Ulna

Radius

Figure 7–12

37. Name the major nerves that serve the following body areas. Insert your responses in the answer blanks.

_____ 1. Neck and shoulders (plexus only)

_____ 2. Abdominal wall (plexus only)

_____ 3. Anterior thigh

_____ 4. Diaphragm

_____ 5. Posterior thigh

_____ 6. Leg and foot (2)

Autonomic Nervous System (ANS)

38. Identify, by color coding and coloring, the following structures in Figure 7–13, which depicts the major anatomical differences between the somatic and autonomic motor divisions of the PNS. Also identify by labeling all structures provided with leader lines.

◯ Somatic motor neuron

◯ ANS preganglionic neuron

◯ ANS ganglionic neuron

◯ Autonomic ganglion

◯ Gray matter of spinal cord (CNS)

◯ Effector of the somatic motor neuron

◯ Effector of the autonomic motor neuron

◯ Myelin sheath

◯ White matter of spinal cord (CNS)

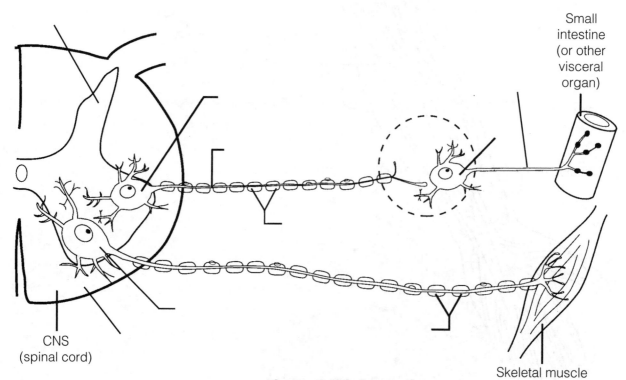

Small intestine (or other visceral organ)

CNS (spinal cord)

Skeletal muscle

Figure 7–13

39. The following table indicates a number of conditions. Use a check (✓) to show which division of the autonomic nervous system is involved in each condition. Then, respond to the true-to-life situation below the chart.

Condition	Sympathetic	Parasympathetic
1. Postganglionic axons secrete norepinephrine; adrenergic fibers		
2. Postganglionic axons secrete acetylcholine; cholinergic fibers		
3. Long preganglionic axon, short postganglionic axon		
4. Short preganglionic axon, long postganglionic axon		
5. Arises from cranial and sacral nerves		
6. Arises from spinal nerves T_1 to L_3		
7. Normally in control		
8. Fight-or-flight system		
9. Has more specific control		
10. Causes a dry mouth, dilates bronchioles		
11. Constricts eye pupils, decreases heart rate		

You are alone in your home late in the evening, and you hear an unfamiliar sound in your backyard. In the spaces provided, list four physiological events promoted by the sympathetic nervous system that would help you to cope with this rather frightening situation.

1. _____

2. _____

3. _____

4. _____

DEVELOPMENTAL ASPECTS OF THE NERVOUS SYSTEM

40. Complete the following statements by inserting your responses in the answer blanks.

_____ 1.

_____ 2.

_____ 3.

_____ 4.

_____ 5.

_____ 6.

_____ 7.

_____ 8.

Body temperature regulation is a problem in premature infants because the __(1)__ is not yet fully functional. Cerebral palsy involves crippling neuromuscular problems. It usually is a result of a lack of __(2)__ to the infant's brain during delivery. Normal maturation of the nervous system occurs in a __(3)__ direction, and fine control occurs much later than __(4)__ muscle control.

The sympathetic nervous system becomes less efficient as aging occurs, resulting in an inability to prevent sudden changes in __(5)__ when abrupt changes in position are made. The usual cause of decreasing efficiency of the nervous system as a whole is __(6)__. A change in intellect caused by a gradual decrease in oxygen delivery to brain cells is called __(7)__. Death of brain neurons, which results from a sudden cessation of oxygen delivery, is called a __(8)__.

INCREDIBLE JOURNEY

A Visualization Exercise for the Nervous System

You climb on the first cranial nerve you see. . . .

41. Where necessary, complete statements by inserting the missing words in the answer blanks.

_____ 1.

Nervous tissue is quite densely packed, and it is difficult to envision strolling through its various regions. Imagine instead that each of the various functional regions of the brain has a computerized room where you can observe what occurs in that particular area. Your assignment is to determine where you are at any given time during your journey through the nervous system.

You begin your journey after being miniaturized and injected into the warm pool of cerebrospinal fluid in your host's fourth ventricle. As you begin your stroll through the nervous tissue, you notice a huge area of branching white matter overhead. As you enter the first computer room you hear an announcement through the loudspeaker: "The pelvis is tipping too far posteriorly. Please correct. We are beginning to fall backward and will soon lose our balance." The computer responds immediately, decreasing impulses to the posterior hip muscles and increasing impulses to the anterior thigh muscles. "How is that, proprioceptor 1?" From this information, you determine that your first stop is the __(1)__.

_____ 2.

_____ 3.

_____ 4.

_____ 5.

_____ 6.

_____ 7.

_____ 8.

_____ 9.

_____ 10.

_____ 11.

_____ 12.

At the next computer room, you hear, "Blood pressure to head is falling; increase sympathetic nervous system stimulation of the blood vessels." Then, as it becomes apparent that your host has not only stood up but is going to run, you hear, "Increase rate of impulses to the heart and respiratory muscles. We are going to need more oxygen and a faster blood flow to the skeletal muscles of the legs." You recognize that this second stop must be the __(2)__.

Computer room 3 presents a problem. There is no loudspeaker here. Instead, incoming messages keep flashing across the wall, giving only bits and pieces of information. "Four hours since last meal: stimulate appetite center. Slight decrease in body temperature: initiate skin vasoconstriction. Mouth dry: stimulate thirst center. Oh, a stroke on the arm: stimulate pleasure center." Looking at what has been recorded here—appetite, temperature, thirst, and pleasure—you conclude that this has to be the __(3)__.

Continuing your journey upward toward the higher brain centers, finally you are certain that you have reached the cerebral cortex. The first center you visit is quiet, like a library with millions of "encyclopedias" of facts and recordings of past input. You conclude that this must be the area where __(4)__ are stored and that you are in the __(5)__ lobe. The next stop is close by. As you enter the computer center, you once again hear a loudspeaker: "Let's have the motor instructions to say 'tintinnabulation.' Hurry, we don't want them to think we're tongue-tied." This area is obviously __(6)__. Your final stop in the cerebral cortex is a very hectic center. Electrical impulses are traveling back and forth between giant neurons, sometimes in different directions and sometimes back and forth between a small number of neurons. Watching intently, you try to make some sense out of these interactions and suddenly realize that this _is_ what is happening here. The neurons _are_ trying to make some sense out of something, and this helps you decide that this must be the brain area where __(7)__ occurs in the __(8)__ lobe.

You hurry out of this center and retrace your steps back to the cerebrospinal fluid, deciding en route to observe a cranial nerve. You decide to pick one randomly and follow it to the organ it serves. You climb on to the first cranial nerve you see and slide down past the throat. Picking up speed, you quickly pass the heart and lungs and see the stomach and small intestine coming up fast. A moment later you land on the stomach and now you know that this wandering nerve has to be the __(9)__. As you look upward, you see that the nerve is traveling almost straight up and that you'll have to find an alternative route back to the cerebrospinal fluid. You begin to walk posteriorly until you find a spinal nerve, which you follow until you reach the vertebral column. You squeeze between two adjacent vertebrae to follow the nerve to the spinal cord. With your pocket knife you cut away the tough connective tissue covering the cord. Thinking that the __(10)__ covering deserves its name, you finally manage to cut an opening large enough to get through, and you return to the warm bath of cerebrospinal fluid that it encloses. At this point you are in the __(11)__, and from here you swim upward until you get to the lower brainstem. Once there, it should be an easy task to find the holes leading into the __(12)__ ventricle, where your journey began.

AT THE CLINIC

42. After surgery, patients are often temporarily unable to urinate, and bowel sounds are absent. Identify the division of the autonomic nervous system that is affected by anesthesia.

43. A brain tumor is found in a CT scan of Mr. Childs's head. The physician is assuming that it is not a secondary tumor (i.e., it did not spread from another part of the body) because an exhaustive workup has revealed no signs of cancer elsewhere in Mr. Childs's body. Is the brain tumor more likely to have developed from nerve tissue or from neuroglia? Why?

44. Amy, a high-strung teenager, was suddenly startled by a loud bang that sounded like a gunshot. Her heartbeat accelerated rapidly. When she realized that the noise was only a car backfiring, she felt greatly relieved but her heart kept beating heavily for several minutes more. Why does it take a long time to calm down after we are scared?

45. You have been told that the superior and medial part of the right precentral gyrus of your patient's brain has been destroyed by a stroke. What part of the body is the patient unable to move? On which side, right or left?

46. *Application of knowledge:* You have been given all of the information needed to identify the brain regions involved in the following situations. See how well your nervous system has integrated this information, and name the brain region (or condition) most likely to be involved in each situation. Place your responses in the answer blanks.

1. Following a train accident, a man with an obvious head injury was observed stumbling about the scene. An inability to walk properly and a loss of balance were quite obvious. What brain region was injured?

   _____

2. An elderly woman is admitted to the hospital to have a gallbladder operation. While she is being cared for, the nurse notices that she has trouble initiating movement and has a strange "pill-rolling" tremor of her hands. What cerebral area is most likely involved?

   _____

3. A child is brought to the hospital with a high temperature. The doctor states that the child's meninges are inflamed. What name is given to this condition?

   _____

4. A young woman is brought into the emergency room with extremely dilated pupils. Her friends state that she has overdosed on cocaine. What cranial nerve is stimulated by the drug?

   _____

5. A young man has just received serious burns, resulting from standing with his back too close to a bonfire. He is muttering that he never felt the pain. Otherwise, he would have smothered the flames by rolling on the ground. What part of his CNS might be malfunctioning?

   _____

6. An elderly gentleman has just suffered a stroke. He is able to understand verbal and written language, but when he tries to respond, his words are garbled. What cortical region has been damaged by the stroke?

   _____

7. A 12-year-old boy suddenly falls to the ground, having an epileptic seizure. He is rushed to the emergency room of the local hospital for medication. His follow-up care includes a recording of his brain waves to try to determine the area of the lesion. What is this procedure called?

   _____

47. Marie Nolin exhibits slow, tentative movements and a very unstable gait. Examination reveals she cannot touch her finger to her nose with eyes closed. What is the name of this condition and what part of her brain is damaged?

48. Which would be the more likely result of injury to the posterior side of the spinal cord only—paralysis or paresthesia (loss of sensory input)? Explain your answer.

49. While jogging in Riverside Park, Susan was confronted by an angry dog. What division of her ANS was activated as she turned tail and ran from the dog?

50. During action potential transmission, many ions cross the neuronal membrane at right angles to the membrane. What is it that travels *along* the membrane and acts as the signal?

51. Suppose you cut the little finger of your left hand. Would you expect that the cut might interfere with motor function, sensory function, or both? Explain your choice.

52. Bill's femoral nerve was crushed while clinicians tried to control bleeding from his femoral artery. This resulted in loss of function and sensation in his leg, which gradually returned over the course of a year. Which cells were important in his recovery?

53. As Melanie woke up, she stretched and quickly did 20 sit-ups before getting out of bed. As she brushed her teeth, the aroma of coffee stimulated her smell receptors and her stomach began to gurgle. Indicate the division of the nervous system involved in each of these activities or events.

✓ THE FINALE: MULTIPLE CHOICE

54. Select the best answer or answers from the choices given.

1. Bipolar neurons:
   A. are found in the head
   B. are always part of an afferent pathway
   C. have two dendrites
   D. have two axons

2. Which of the following skin cells would form a junction with a motor neuron?
   A. Keratinocyte
   B. Sudoriferous glandular epithelial cell
   C. Arrector pili muscle cell
   D. Fibroblast

3. A synapse between an axon terminal and a neuron cell body is called:
   A. axodendritic   C. axosomatic
   B. axoaxonic   D. axoneuronic

4. Which is an incorrect association of brain region and ventricle?
   A. Mesencephalon—third ventricle
   B. Cerebral hemispheres—lateral ventricles
   C. Pons—fourth ventricle
   D. Medulla—fourth ventricle

5. The pineal gland is located in the:
   A. hypophysis cerebri
   B. mesencephalon
   C. epithalamus
   D. corpus callosum

6. Which of the following is *not* part of the brainstem?
   A. Medulla   C. Pons
   B. Cerebellum   D. Midbrain

7. When neurons in Wernicke's area send impulses to neurons in Broca's area, the white matter tracts utilized are:
   A. commissural fibers
   B. projection fibers
   C. association fibers
   D. anterior funiculus

8. Functions that are at least partially overseen by the medulla are:

   A. regulation of the heart

   B. maintaining of equilibrium

   C. regulation of respiration

   D. visceral motor function

9. Which structures are directly involved with formation, circulation, and drainage of CSF?

   A. Ependymal cilia

   B. Ventricular choroid plexuses

   C. Arachnoid villi

   D. Serous layers of the dura mater

10. In an earthquake, which type of sensory receptor is most likely to sound the *first* alarm?

    A. Exteroceptor   C. Mechanoreceptor

    B. Visceroceptor   D. Proprioceptor

11. Cranial nerves that have some function in vision include the:

    A. trochlear   C. abducens

    B. trigeminal   D. facial

12. Eating difficulties would result from damage to the:

    A. mandibular division of trigeminal nerve

    B. facial nerve

    C. glossopharyngeal nerve

    D. vagus nerve

13. If the right trapezius and sternocleidomastoid muscles were atrophied, you would suspect damage to the:

    A. vagus nerve

    B. motor branches of the cervical plexus

    C. facial nerve

    D. accessory nerve

14. Which nerve stimulates muscles that flex the forearm?

    A. Ulnar   C. Radial

    B. Musculocutaneous   D. Median

15. Motor functions of arm, forearm, and fingers would be affected by damage to which one of these nerves?

    A. Radial   C. Ulnar

    B. Axillary   D. Median

16. An inability to extend the leg would result from a loss of function of the:

    A. lateral femoral cutaneous nerve

    B. ilioinguinal nerve

    C. saphenous branch of femoral nerve

    D. femoral nerve

Use the following choices to respond to questions 17–28:

    A. sympathetic division

    B. parasympathetic division

    C. both sympathetic and parasympathetic

    D. neither sympathetic nor parasympathetic

_____ 17. Typically has long preganglionic and short postganglionic fibers

_____ 18. Some fibers utilize gray rami communicantes

_____ 19. Courses through spinal nerves

_____ 20. Has splanchnic nerves

_____ 21. Courses through cranial nerves

_____ 22. Originates in cranial nerves

_____ 23. Effects enhanced by direct stimulation of a hormonal mechanism

_____ 24. Includes otic ganglion

_____ 25. Includes celiac ganglion

_____ 26. Hypoactivity of this division would lead to decrease in metabolic rate

_____ 27. Has widespread, long-lasting effects

_____ 28. Sets the tone for the heart

29. Which contains only motor fibers?

    A. Dorsal root   C. Ventral root

    B. Dorsal ramus   D. Ventral ramus

⑧ SPECIAL SENSES

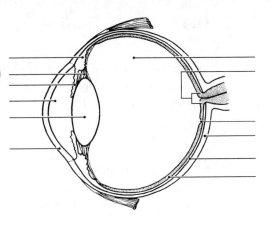

The body's sensory receptors react to stimuli or changes occurring both within the body and in the external environment. When triggered, these receptors send nerve impulses along afferent pathways to the brain for interpretation, thus allowing the body to assess and adjust to changing conditions so that homeostasis may be maintained.

The minute receptors of general sensation that react to touch—pressure, pain, temperature changes, and muscle tension—are widely distributed in the body. These are considered in Chapter 7. In contrast, receptors of the special senses—sight, hearing, equilibrium, smell, and taste—tend to be localized and in many cases are quite complex. The structure and function of the special sense organs are the subjects of the student activities in this chapter.

THE EYE AND VISION

1. Complete the following statements by inserting your responses in the answer blanks.

_____ 1.

_____ 2.

_____ 3.

_____ 4.

Attached to the eyes are the __(1)__ muscles that allow us to direct our eyes toward a moving object. The anterior aspect of each eye is protected by the __(2)__, which have eyelashes projecting from their edges. Closely associated with the lashes are oil-secreting glands called __(3)__ that help to lubricate the eyes. Inflammation of the mucosa lining the eyelids and covering the anterior part of the eyeball is called __(4)__.

2. Trace the pathway that the secretion of the lacrimal glands takes from the surface of the eye by assigning a number to each structure. (Note that #1 will be *closest* to the lacrimal gland.)

_____ 1. Lacrimal sac   _____ 3. Nasolacrimal duct

_____ 2. Nasal cavity   _____ 4. Lacrimal canals

3. Identify each of the eye muscles indicated by leader lines in Figure 8–1. Color code and color each muscle a different color. Then, in the blanks below, indicate the eye movement caused by each muscle.

◯ 1. Superior rectus _____

◯ 2. Inferior rectus _____

◯ 3. Superior oblique _____

◯ 4. Lateral rectus _____

◯ 5. Medial rectus _____

◯ 6. Inferior oblique _____

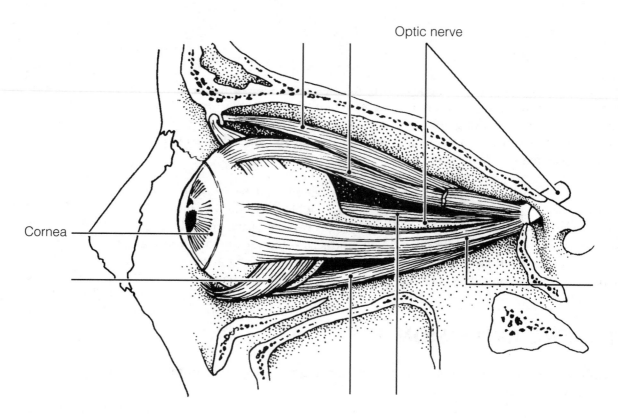

Figure 8–1

4. Three main accessory eye structures contribute to the formation of tears and/or aid in lubricating the eyeball. In the table, name each structure and then name its major secretory product. Indicate which of the secretions has antibacterial properties by circling that response.

Accessory eye structures	Secretory product
1.	
2.	
3.	

5. Match the terms provided in Column B with the appropriate descriptions in Column A. Insert the correct letter response or corresponding term in the answer blanks.

Column A

_____ 1. Light bending

_____ 2. Ability to focus for close vision (under 20 feet)

_____ 3. Normal vision

_____ 4. Inability to focus well on close objects; farsightedness

_____ 5. Reflex constriction of pupils when they are exposed to bright light

_____ 6. Clouding of the lens, resulting in loss of sight

_____ 7. Nearsightedness

_____ 8. Blurred vision, resulting from unequal curvatures of the lens or cornea

_____ 9. Condition of increasing pressure inside the eye, resulting from blocked drainage of aqueous humor

_____ 10. Medial movement of the eyes during focusing on close objects

_____ 11. Reflex constriction of the pupils when viewing close objects

_____ 12. Inability to see well in the dark; often a result of vitamin A deficiency

Column B

A. Accommodation

B. Accommodation pupillary reflex

C. Astigmatism

D. Cataract

E. Convergence

F. Emmetropia

G. Glaucoma

H. Hyperopia

I. Myopia

J. Night blindness

K. Photopupillary reflex

L. Refraction

6. The intrinsic eye muscles are under the control of which division of the nervous system? Circle the correct response.

1. Autonomic nervous system   2. Somatic nervous system

7. Complete the following statements by inserting your responses in the answer blanks.

_____ 1.

_____ 2.

_____ 3.

_____ 4.

_____ 5.   _____ 6.

A __(1)__ lens, like that of the eye, produces an image that is upside down and reversed from left to right. Such an image is called a __(2)__ image. In farsightedness, the light is focused __(3)__ the retina. The lens used to treat farsightedness is a __(4)__ lens. In nearsightedness, the light is focused __(5)__ the retina; it is corrected with a __(6)__ lens.

8. Using the key choices, identify the parts of the eye described in the following statements. Insert the correct term or letter response in the answer blanks.

Key Choices

A. ◯ Aqueous humor   F. ◯ Cornea   K. ◯ Retina

B.   Canal of Schlemm   G. ◯ Fovea centralis   L. ◯ Sclera

C. ◯ Choroid   H. ◯ Iris   M. ◯ Vitreous humor

D. ◯ Ciliary body   I. ◯ Lens

E. ◯ Ciliary zonule   J. ◯ Optic disk

_____ 1. Attaches the lens to the ciliary body

_____ 2. Fluid in the anterior segment that provides nutrients to the lens and cornea

_____ 3. The "white" of the eye

_____ 4. Area of retina that lacks photoreceptors

_____ 5. Contains muscle that controls the shape of the lens

_____ 6. Nutritive (vascular) layer of the eye

_____ 7. Drains the aqueous humor of the eye

_____ 8. Layer containing the rods and cones

_____ 9. Gel-like substance that helps to reinforce the eyeball

_____ 10. Heavily pigmented layer that prevents light scattering within the eye

_____ 11. _____ 12. Smooth muscle structures (intrinsic eye muscles)

_____ 13. Area of acute or discriminatory vision

_____ 14. _____ 15. Refractory media of the eye (#14–17)

_____ 16. _____ 17.

_____ 18. Most anterior part of the sclera—your "window on the world"

_____ 19. Pigmented "diaphragm" of the eye

9. Using the key choice terms given in Exercise 8, identify the structures indicated by leader lines on the diagram of the eye in Figure 8–2. Select different colors for all structures provided with a color-coding circle in Exercise 8, and then use them to color the coding circles and corresponding structures in the figure.

Figure 8-2

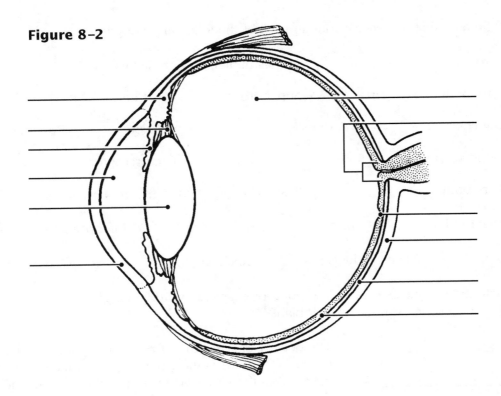

10. In the following table, circle the correct word under the vertical headings that describes events occurring within the eye during close and distant vision.

Vision	Ciliary muscle		Lens convexity		Degree of light refraction	
1. Distant	Relaxed	Contracted	Increased	Decreased	Increased	Decreased
2. Close	Relaxed	Contracted	Increased	Decreased	Increased	Decreased

11. Name in sequence the neural elements of the visual pathway, beginning with the retina and ending with the optic cortex.

Retina ➡ _____ ➡ _____ ➡

Synapse in thalamus ➡ _____ ➡ Optic cortex

12. Complete the following statements by inserting your responses in the answer blanks.

_____ 1.

_____ 2.

_____ 3.

_____ 4.

_____ 5.

_____ 6.

There are __(1)__ varieties of cones. One type responds most vigorously to __(2)__ light, another to __(3)__ light, and still another to __(4)__ light. The ability to see intermediate colors such as purple results from the fact that more than one cone type is being stimulated __(5)__. Lack of all color receptors results in __(6)__. Because this condition is sex linked, it occurs more commonly in __(7)__. Black and white, or dim light, vision is a function of the __(8)__.

_____ 7. _____ 8.

13. Circle the term that does not belong in each of the following groupings.

1. Choroid   Sclera   Vitreous humor   Retina

2. Ciliary body   Iris   Superior rectus   Choroid

3. Pupil constriction   Far vision   Accommodation   Bright light

4. Proprioceptors   Rods   Cones   Photoreceptors

5. Ciliary body   Iris   Suspensory ligaments   Lens

6. Inferior oblique   Iris   Superior rectus   Inferior rectus

7. Retina   Pigmented layer   Photoreceptors   Neural layer

14. Complete the statements concerning rod photopigment and physiology by writing your responses in the answer blanks.

_____ 1.

_____ 2.

_____ 3.

_____ 4.

_____ 5.   _____ 6.

The bent or kinked form of retinal is combined with a protein called __(1)__ to form the visual pigment called __(2)__. When light strikes the visual pigment, it straightens out and breaks down into its two components. This event is called __(3)__ because the purple color of the visual pigment changes to __(4)__ and finally becomes __(5)__ as retinal is converted all the way back to vitamin __(6)__.

THE EAR: HEARING AND BALANCE

15. Using the key choices, select the terms that apply to the following descriptions. Place the correct letter in the answer blanks.

Key Choices

A. Anvil (incus)   E. External acoustic meatus   I. Pinna   M. Tympanic membrane

B. Pharyngotympanic tube   F. Hammer (malleus)   J. Round window   N. Vestibule

C. Cochlea   G. Oval window   K. Semicircular canals

D. Endolymph   H. Perilymph   L. Stirrup (stapes)

____ 1. ____ 2. ____ 3. Structures composing the outer ear

____ 4. ____ 5. ____ 6. Structures composing the bony or osseous labyrinth

____ 7. ____ 8. ____ 9. Collectively called the ossicles

____ 10. ____ 11. Ear structures not involved with hearing

_____ 12. Allows pressure in the middle ear to be equalized with the atmospheric pressure

_____ 13. Vibrates as sound waves hit it; transmits the vibrations to the ossicles

_____ 14. Contains the organ of Corti

_____ 15. Connects the nasopharynx and the middle ear

_____ 16. _____ 17. Contain receptors for the sense of equilibrium

_____ 18. Transmits the vibrations from the stirrup to the fluid in the inner ear

_____ 19. Fluid that bathes the sensory receptors of the inner ear

_____ 20. Fluid contained within the osseous labyrinth, which bathes the membranous labyrinth

16. Figure 8–3 is a diagram of the ear. Use anatomical terms (as needed) from the key choices in Exercise 15 to correctly identify all structures in the figure provided with leader lines. Color all external ear structures yellow; color the ossicles red; color the equilibrium areas of the inner ear green; and color the internal ear structures involved with hearing blue.

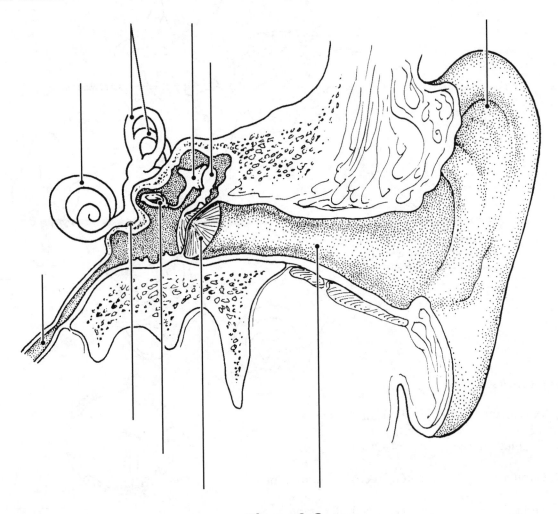

Figure 8–3

17. Sound waves hitting the eardrum set it into vibration. Trace the pathway through which vibrations and fluid currents travel to finally stimulate the hair cells in the organ of Corti. Name the appropriate ear structures in their correct sequence and insert your responses in the answer blanks.

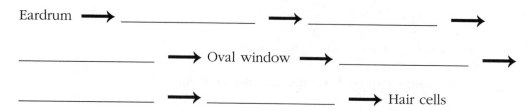

Eardrum ⟶ _____ ⟶ _____ ⟶

_____ ⟶ Oval window ⟶ _____ ⟶

_____ ⟶ _____ ⟶ Hair cells

18. Figure 8–4 is a view of the structures of the membranous labyrinth. Correctly identify the following major areas of the labyrinth on the figure: *membranous semicircular canals, saccule* and *utricle,* and the *cochlear duct.* Next, correctly identify each of the receptor types shown in enlarged views (organ of Corti, crista ampullaris, and macula). Finally, using terms from the key choices below, identify all receptor structures provided with leader lines. (Some of these terms may need to be used more than once.)

Figure 8–4

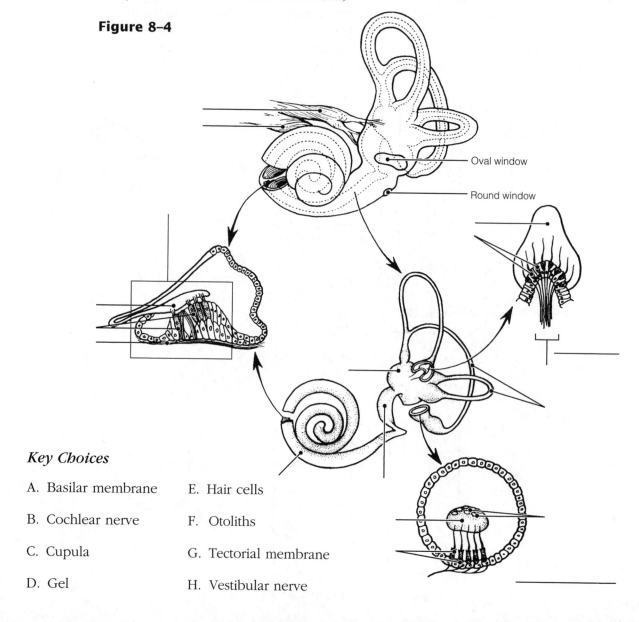

Oval window

Round window

Key Choices

A. Basilar membrane   E. Hair cells

B. Cochlear nerve   F. Otoliths

C. Cupula   G. Tectorial membrane

D. Gel   H. Vestibular nerve

19. Complete the following statements on the functioning of the static and dynamic equilibrium receptors by inserting the letter or term from the key choices in the answer blanks.

Key Choices

A. Angular/rotatory   E. Gravity   I. Semicircular canals

B. Cupula   F. Perilymph   J. Static

C. Dynamic   G. Proprioception   K. Utricle

D. Endolymph   H. Saccule   L. Vision

_____ 1.

_____ 2.

_____ 3.

_____ 4.

_____ 5.

_____ 6.

_____ 7.

_____ 8.

_____ 9.   _____ 10.   _____ 11.

The receptors for __(1)__ equilibrium are found in the crista ampullaris of the __(2)__. These receptors respond to changes in __(3)__ motion. When motion begins, the __(4)__ fluid lags behind and the __(5)__ is bent, which excites the hair cells. When the motion stops suddenly, the fluid flows in the opposite direction and again stimulates the hair cells. The receptors for __(6)__ equilibrium are found in the maculae of the __(7)__ and __(8)__. These receptors report the position of the head in space. Tiny stones found in a gel overlying the hair cells roll in response to the pull of __(9)__. As they roll, the gel moves and tugs on the hair cells, exciting them. Besides the equilibrium receptors of the inner ear, the senses of __(10)__ and __(11)__ are also important in helping to maintain equilibrium.

20. Indicate whether the following conditions relate to conduction deafness (*C*) or sensorineural (central) deafness (*S*). Place the correct letter choice in each answer blank.

_____ 1. Can result from a bug wedged in the external auditory meatus

_____ 2. Can result from damage to the cochlear nerve

_____ 3. Sound is heard in one ear but not in the other, during both bone and air conduction

_____ 4. Often improved by a hearing aid

_____ 5. Can result from otitis media

_____ 6. Can result from otosclerosis, excessive earwax, or a perforated eardrum

_____ 7. Can result from a blood clot in the auditory cortex of the brain

21. List three things about which a person with equilibrium problems might complain. Place your responses in the answer blanks.

_____ , _____ , and _____

22. Circle the term that does not belong in each of the following groupings.

1. Hammer   Anvil   Pinna   Stirrup

2. Tectorial membrane   Crista ampullaris   Semicircular canals   Cupula

3. Gravity   Angular motion   Sound waves   Rotation

4. Utricle   Saccule   Auditory tube   Vestibule

5. Vestibular nerve   Optic nerve   Cochlear nerve   Vestibulocochlear nerve

CHEMICAL SENSES: SMELL AND TASTE

23. Complete the following statements by inserting your responses in the answer blanks.

_____ 1.

_____ 2.

_____ 3.

_____ 4.

_____ 5.

_____ 6.

_____ 7.

_____ 8.

_____ 9.

_____ 10.

_____ 11.

_____ 12.

_____ 13.

_____ 14.

Three cranial nerves involved in transmitting impulses for the sense of taste are the __(1)__ , __(2)__ , and __(3)__ . Impulses for the sense of smell are transmitted by the __(4)__ nerve. The receptors for smell are located in the __(5)__ of the nasal passages; the act of __(6)__ increases the sensation because it brings more air into contact with the receptors. The receptors for taste are found in cluster-like areas called __(7)__ , most of which are located on the sides of __(8)__ or __(9)__ papillae. The five basic taste sensations are __(10)__ , __(11)__ , __(12)__ , __(13)__ , and __(14)__ . The most protective receptors are thought to be those that respond to __(15)__ substances. When nasal passages are congested, the sense of taste is decreased. This indicates that much of what is considered taste actually depends on the sense of __(16)__ . It is impossible to taste substances with a __(17)__ tongue because foods must be dissolved (or in solution) to excite the taste receptors. The sense of smell is closely tied to the emotional centers of the brain (limbic region), and many odors bring back __(18)__ .

_____ 15.   _____ 17.

_____ 16.   _____ 18.

24. On Figure 8–5A, label the two types of tongue papillae containing taste buds. On Figure 8–5B, color the taste buds green. On Figure 8–5C, color the gustatory cells red, the basal cells blue, and the cranial nerve fibers yellow. Add appropriate labels to the leader lines provided to identify the *taste pore* and *microvilli* of the gustatory cells.

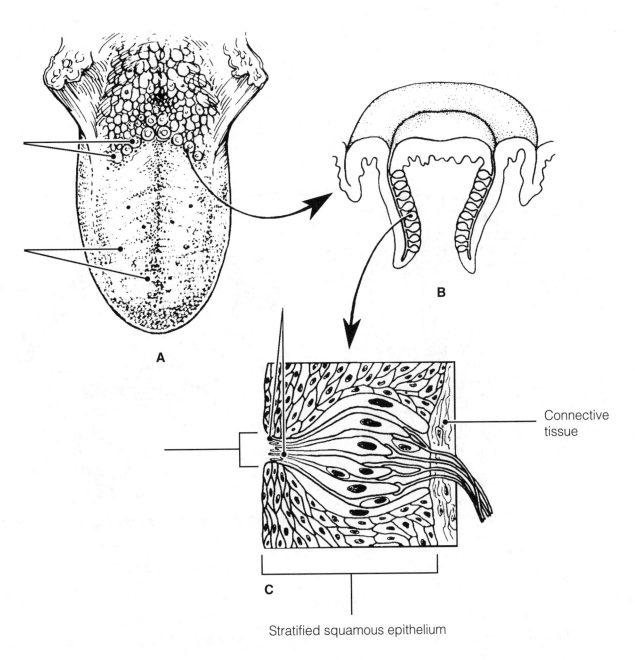

Connective tissue

Stratified squamous epithelium

Figure 8–5

25. Figure 8–6 illustrates the site of the olfactory epithelium in the nasal cavity (part A is an enlarged view of the olfactory receptor area). Select different colors to identify the structures listed below and use them to color the coding circles and corresponding structures in the illustration. Then add a label and leader line to identify the olfactory "hairs" and add arrows to indicate the direction of impulse transmission. Finally, respond to the questions following the diagram.

○ Olfactory neurons (receptor cells)   ○ Cribriform plate of the ethmoid bone

○ Olfactory bulb   ○ Olfactory nerve filaments

○ Fibers of the olfactory tract

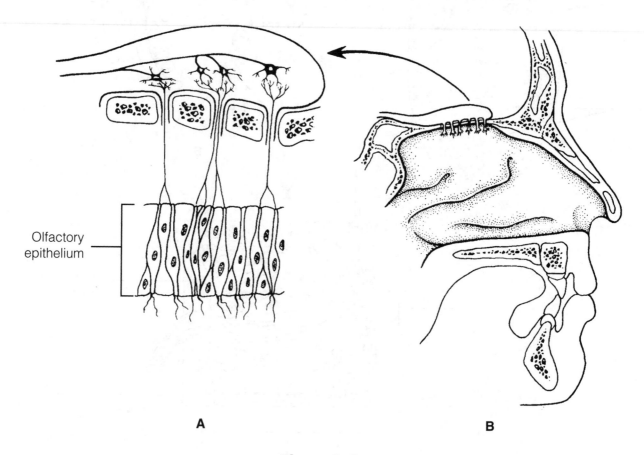

Olfactory
epithelium

A   B

Figure 8–6

1. What substance "captures" airborne odors (that is, acts as a solvent)?_____

2. How are olfactory neurons classified structurally?_____

26. Circle the term that does not belong in each of the following groupings.

1. Sweet   Musky   Sour   Bitter   Salty

2. Bipolar neuron   Epithelial cell   Olfactory receptor   Ciliated

3. Gustatory cell   Taste pore   Papillae   Neuron

4. Vagus nerve   Facial nerve   Glossopharyngeal nerve   Olfactory nerve

5. Olfactory receptor   High sensitivity   Variety of stimuli   Four receptor types

6. Sugars   Sweet   Saccharine   Metal ions   Amino acids

DEVELOPMENTAL ASPECTS OF THE SPECIAL SENSES

27. Complete the following statements by inserting your responses in the answer blanks.

_____ 1.

_____ 2.

_____ 3.

_____ 4.

_____ 5.

_____ 6.

_____ 7.

_____ 8.

_____ 9.

The special sense organs are actually part of the __(1)__ and are formed very early in the embryo. Maternal infections, particularly __(2)__, may cause both deafness and __(3)__ in the developing child. Of the special senses, the sense of __(4)__ requires the most learning or takes longest to mature. All infants are __(5)__, but generally by school age emmetropic vision has been established. Beginning sometime after the age of 40, the eye lenses start to become less __(6)__ and cannot bend properly to refract the light. As a result, a condition of farsightedness, called __(7)__, begins to occur. __(8)__, a condition in which the lens becomes hazy or discolored, is a frequent cause of blindness. In old age, a gradual hearing loss, called __(9)__, occurs. A declining efficiency of the chemical senses is also common in the elderly.

 INCREDIBLE JOURNEY

A Visualization Exercise for the Special Senses

You . . . see a discontinuous sea of glistening, white rock slabs. . . .

28. Where necessary, complete statements by inserting the missing words in the answer blanks.

_____ 1.

_____ 2.

_____ 3.

_____ 4.

_____ 5.

_____ 6.

_____ 7.

_____ 8.

_____ 9.

_____ 10.

_____ 11.

_____ 12.

1. Your present journey will take you through your host's inner ear to observe and document what you have learned about how hearing and equilibrium receptors work.

2. This is a very tightly planned excursion. Your host has been instructed to move his head at specific intervals and will be exposed to various sounds so that you can make certain observations. For this journey you are miniaturized and injected into the bony cavity of the inner ear, the __(1)__, and are to make your way through its various chambers in a limited amount of time.

Your first observation is that you are in a warm sea of __(2)__ in the vestibule. To your right are two large sacs, the __(3)__ and __(4)__. You swim over to one of these membranous sacs, cut a small semicircular opening in the wall, and wiggle through. Because you are able to see very little in the dim light, you set out to explore this area more fully. As you try to move, however, you find that your feet are embedded in a thick, gluelike substance. The best you can manage is slow-motion movements through this __(5)__.

It is now time for your host's first scheduled head movement. Suddenly your world tips sharply sideways. You hear a roar (rather like an avalanche) and look up to see a discontinuous sea of glistening, white rock slabs sliding toward you. You protect yourself from these __(6)__ by ducking down between the hair cells that are bending vigorously with the motion of the rocks. Now that you have seen and can document the operation of a __(7)__, a sense organ of __(8)__ equilibrium, you quickly back out through the hole you made.

Keeping in mind the schedule and the fact that it is nearly time for your host to be exposed to tuning forks, you swim quickly to the right, where you see what looks like the opening of a cave with tall seaweed waving gently in the current. Abruptly, as you enter the cave, you find that you are no longer in control of your movements but instead are swept along in a smooth undulating pattern through the winding passageway of the cave, which you now know is the cavity of the __(9)__. As you move up and down with the waves you see hair cells of the __(10)__, the sense organ for __(11)__, being vigorously disturbed below you. Flattening yourself against the chamber wall to prevent being carried further by the waves, you wait for the stimulus to stop. Meanwhile you are delighted by the electrical activity of the hair cells below you. As they depolarize to send impulses along the __(12)__, the landscape appears to be alive with fireflies.

_____13.

_____14.

_____15.

_____16.

Now that you have witnessed the events for this particular sense receptor, you swim back through the vestibule toward your final observation area at the other end of the bony chambers. You recognize that your host is being stimulated again because of the change in fluid currents, but since you are not close to any of the sensory receptors, you are not sure just what the stimulus is. Then, just before you, three dark openings appear, the __(13)__ . You swim into the middle opening and see a strange structure that looks like the brush end of an artist's paintbrush; you swim upward and establish yourself on the soft brushy top portion. This must be the __(14)__ of the __(15)__ , the sensory receptor for __(16)__ equilibrium. As you rock back and forth in the gentle currents, a sudden wave of fluid hits you. Clinging to the hairs as the fluid thunders past you, you realize that there will soon be another such wave in the opposite direction. You decide that you have seen enough of the special senses and head back for the vestibule to leave your host once again.

AT THE CLINIC

29. An infant girl with strabismus is brought to the clinic. Tests show that she can control both eyes independently. What therapy will be tried before surgery?

30. A man in his early 60s comes to the clinic complaining of fuzzy vision. An eye examination reveals clouding of his lenses. What is his problem and what factors might have contributed to it?

31. Albinism is a condition in which melanin pigment is not made. How do you think vision is affected by albinism?

32. A man claiming to have difficulty seeing at night seeks help at the clinic. What is the technical name for this disorder? What dietary supplement will be recommended? If the condition has progressed too far, what retinal structures will degenerate?

33. A child is brought to the speech therapist because she does not pronounce high-pitched sounds (like "s"). If it is determined that the spiral organ of Corti is the source of the problem, which region of the organ would be defective? Is this conduction or sensorineural deafness?

34. Little Biff's uncle tells the physician that 3-year-old Biff has frequent earaches and that a neighbor claims that Biff needs to have "ear tubes" put in. Upon questioning, the uncle reveals that Biff is taking swimming lessons and he can't remember the last time the boy had a sore throat. Does Biff have otitis media or otitis externa? Does he need ear tubes? Explain your reasoning.

35. Harry fell off a tall ladder and fractured the anterior cranial fossa of his skull. On arrival at the hospital, a watery, blood-tinged fluid was dripping from his right nostril. Several days later, Harry complained that he could no longer smell. What nerve was damaged in his fall?

36. Brian is brought to the clinic by his parents, who noticed that his right eye does not rotate laterally very well. The doctor explains that the nerve serving the lateral rectus muscle is not functioning properly. To what nerve is he referring?

37. When Mrs. Martinez visits her ophthalmologist, she complains of pain in her right eye. The intraocular pressure of that eye is found to be abnormally elevated. What is the name of Mrs. Martinez's probable condition? What causes it? What might be the outcome if the problem is not corrected?

38. Henri, a chef in a five-star French restaurant, has been diagnosed with leukemia. He is about to undergo chemotherapy, which will kill rapidly dividing cells in his body. He needs to continue working between bouts of chemotherapy. What consequences of chemotherapy would you predict that might affect his job as a chef?

✓ THE FINALE: MULTIPLE CHOICE

39. Select the best answer or answers from the choices given.

1. Gustatory cells are:
   A. bipolar neurons
   B. multipolar neurons
   C. unipolar neurons
   D. specialized epithelial cells

2. Alkaloids excite gustatory hairs mostly at the:
   A. tip of the tongue
   B. back of the tongue
   C. circumvallate papillae
   D. fungiform papillae

3. Cranial nerves that are part of the gustatory pathway include:
   A. trigeminal   C. hypoglossal
   B. facial   D. glossopharyngeal

4. The receptors for olfaction are:
   A. the ends of dendrites of bipolar neurons
   B. cilia
   C. specialized nonneural receptor cells
   D. olfactory hairs

5. Which cranial nerve controls contraction of the circular smooth muscle of the iris?
   A. Trigeminal   C. Oculomotor
   B. Facial   D. Abducens

6. Which of the following would be found in the fovea centralis?
   A. Ganglion neurons   C. Cones
   B. Bipolar neurons   D. Rhodopsin

7. The vitreous humor:
   A. helps support the lens
   B. holds the retina in place
   C. contributes to intraocular pressure
   D. is constantly replenished

8. Blockage of which of the following is suspected in glaucoma?
   A. Ciliary processes
   B. Retinal blood vessels
   C. Choroid vessels
   D. Scleral venous sinus

9. Refraction can be altered for near or far vision by the:
   A. cornea
   B. ciliary muscles
   C. vitreous humor
   D. neural layer of the retina

10. Convergence:
    A. requires contraction of the medial rectus muscles of both eyes
    B. is needed for near vision
    C. involves transmission of impulses along the abducens nerves
    D. can promote eye strain

11. Objects in the periphery of the visual field:
    A. stimulate cones
    B. cannot have their color determined
    C. can be seen in low light intensity
    D. appear fuzzy

12. Depth perception is caused by all of the following factors *except* which one(s)?
    A. The eyes are frontally located.
    B. There is total crossover of the optic nerve fibers at the optic chiasma.
    C. There is partial crossover of the optic nerve fibers at the optic chiasma.
    D. Each visual cortex receives input from both eyes.

13. Which structures are contained within the petrous portion of the temporal bone?

    A. Tympanic cavity

    B. Mastoid air cells

    C. External auditory meatus

    D. Stapedius muscle

14. Movement of the _____ membrane triggers bending of hairs of the hair cells in the spiral organ of Corti.

    A. tympanic   C. basilar

    B. tectorial   D. vestibular

15. Sounds entering the external auditory meatus are eventually converted to nerve impulses via a chain of events including:

    A. vibration of the eardrum

    B. vibratory motion of the ossicles against the round window

    C. stimulation of hair cells in the organ of Corti

    D. resonance of the basilar membrane

16. Which of the following structures is involved in static equilibrium?

    A. Maculae   C. Crista ampullaris

    B. Saccule   D. Otoliths

17. Which of the following are paired incorrectly?

    A. Cochlear duct—cupula

    B. Saccule—macula

    C. Ampulla—otoliths

    D. Semicircular duct—ampulla

18. Taste receptor cells are stimulated by:

    A. chemicals binding to the nerve fibers supplying them

    B. chemicals binding to their microvilli

    C. stretching of their microvilli

    D. impulses from the sensory nerves supplying them

9 THE ENDOCRINE SYSTEM

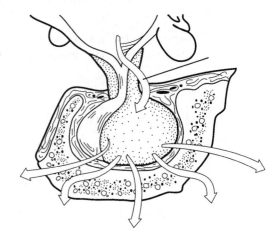

The endocrine system, vital to homeostasis, plays an important role in regulating the activity of body cells. By acting through bloodborne chemical messengers, called hormones, the endocrine system organs orchestrate cellular changes that lead to growth and development, reproductive capability, and the physiological homeostasis of many body systems.

This chapter covers the location of the various endocrine organs in the body, the general function of the various hormones, and the consequences of their hypersecretion or hyposecretion.

THE ENDOCRINE SYSTEM AND HORMONE FUNCTION—AN OVERVIEW

1. Complete the following statements by choosing answers from the key choices. Record the answers in the answer blanks.

Key Choices

A. Cardiovascular system   C. More rapid   E. Nervous system

B. Hormones   D. Nerve impulses   F. Slower and more prolonged

_____ 1.

_____ 2.

_____ 3.

_____ 4.

_____ 5.

The endocrine system is a major controlling system in the body. Its means of control, however, is much __(1)__ than that of the __(2)__, the other major body system that acts to maintain homeostasis. Perhaps the reason for this is that the endocrine system uses chemical messengers, called __(3)__, instead of __(4)__. These chemical messengers enter the blood and are carried throughout the body by the activity of the __(5)__.

2. Complete the following statements by choosing answers from the key choices. Record the answers in the answer blanks.

Key Choices

A. Altering activity
B. Anterior pituitary
C. Hormonal
D. Humoral
E. Hypothalamus

F. Negative feedback
G. Neural
H. Neuroendocrine
I. Receptors
J. Releasing hormones

K. Steroid or amino acid–based
L. Stimulating new or unusual activities
M. Sugar or protein
N. Target cell(s)

_____ 1.

_____ 2.

_____ 3.

_____ 4.

_____ 5.

_____ 6.

_____ 7.

_____ 8.

_____ 9.

_____10.

_____11.

All cells do not respond to endocrine system stimulation. Only those that have the proper __(1)__ on their cell membranes are activated by the chemical messengers. These responsive cells are called the __(2)__ of the various endocrine glands. Hormones promote homeostasis by __(3)__ of body cells rather than by __(4)__. Most hormones are __(5)__ molecules.

The various endocrine glands are prodded to release their hormones by nerve fibers (a __(6)__ stimulus), by other hormones (a __(7)__ stimulus), or by the presence of increased or decreased levels of various other substances in the blood (a __(8)__ stimulus). The secretion of most hormones is regulated by a __(9)__ system, in which increasing levels of that particular hormone "turn off" its stimulus. The __(10)__ is called the master endocrine gland because it regulates so many other endocrine organs. However, it is in turn controlled by __(11)__ secreted by the __(12)__. The structure identified as #12 is also part of the brain, so it is appropriately called a __(13)__ organ.

_____ 12. _____ 13.

3. For each key phrase, decide whether it better describes the mode of action of a steroid or amino acid–based hormone, and insert its key letter in the appropriate answer blank.

Key Choices

A. Binds to a plasma membrane receptor

B. Binds to a receptor in the cell's nucleus

C. Is lipid soluble

D. Activates a gene to transcribe messenger RNA

E. Acts through a second messenger such as cyclic AMP (cyclic adenine monophosphate)

Steroid hormones: _____   Amino acid–based hormones: _____

THE MAJOR ENDOCRINE ORGANS

4. Figure 9–1 depicts the anatomical relationships between the hypothalamus and the anterior and posterior lobes of the pituitary in a highly simplified way. First, identify each of the structures listed below by color coding and coloring them on the diagram. Then, on the appropriate lines write in the names of the hormones that influence each of the target organs shown at the bottom of the diagram. Color the target organ diagrams as you like.

◯ Hypothalamus ◯ Anterior pituitary

◯ Turk's saddle of the sphenoid bone ◯ Posterior pituitary

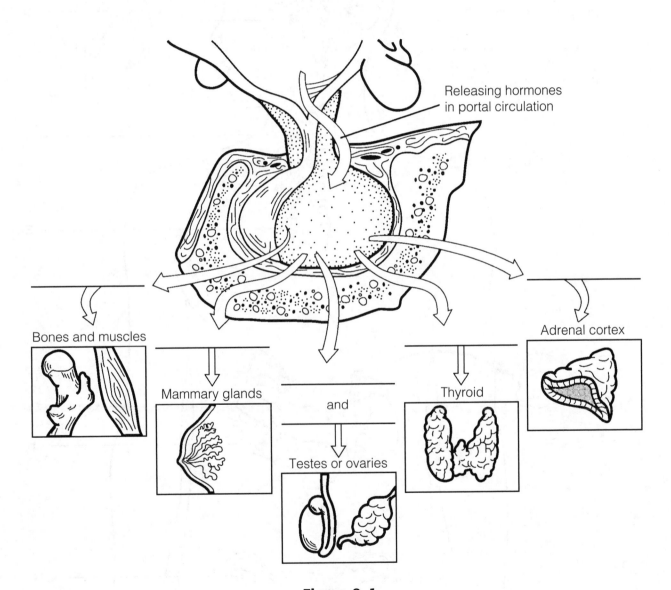

Figure 9–1

5. Figure 9–2 is a diagram of the various endocrine organs of the body. Next to each letter on the diagram, write the name of the endocrine-producing organ (or area). Then select different colors for each and color the corresponding organs in the illustration. To complete your identification of the hormone-producing organs, name the organs (not illustrated) described in items K and L.

K. Small glands that ride "horseback" on the thyroid

_____

L. Endocrine-producing organ present only in pregnant women

_____

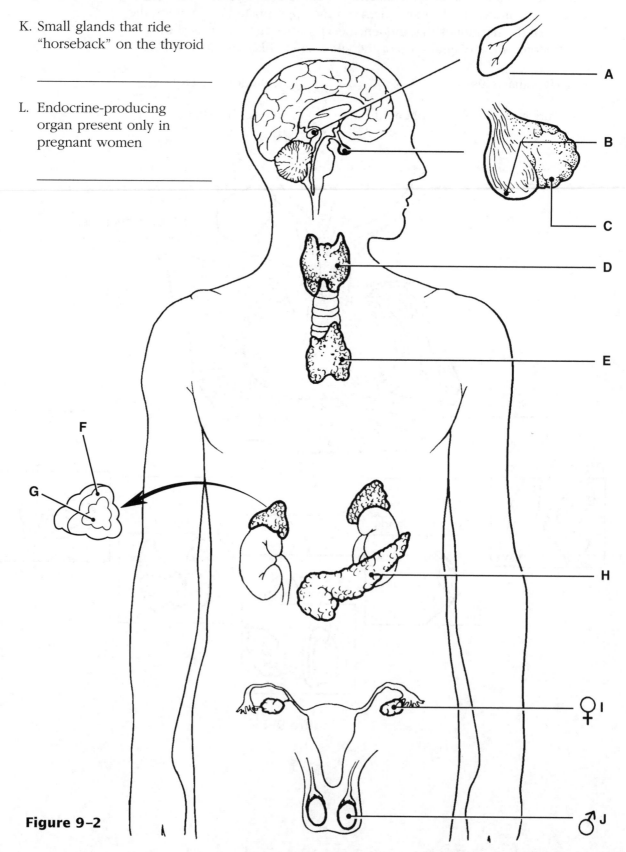

Figure 9–2

6. For each of the following hormones, indicate the organ (or organ part)
producing or releasing the hormone by inserting the appropriate letters
from Figure 9–2 in the answer blanks.

_____ 1. ACTH   _____ 8. Glucagon   _____ 15. PTH

_____ 2. ADH   _____ 9. Insulin   _____ 16. Growth hormone

_____ 3. Aldosterone   _____ 10. LH   _____ 17. Testosterone

_____ 4. Cortisol   _____ 11. Melatonin   _____ 18. Thymosins

_____ 5. Epinephrine   _____ 12. Oxytocin   _____ 19. Thyroxine

_____ 6. Estrogen   _____ 13. Progesterone   _____ 20. TSH

_____ 7. FSH   _____ 14. Prolactin

7. Name the hormone that best fits each of the following descriptions.
Insert your responses in the answer blanks.

_____ 1. Basal metabolic hormone

_____ 2. Program T lymphocytes

_____ 3. Most important hormone regulating the amount of calcium
circulating in the blood; released when blood calcium levels drop

_____ 4. Helps to protect the body during long-term stressful situations
such as extended illness and surgery

_____ 5. Short-term stress hormone; aids in the fight-or-flight response;
increases blood pressure and heart rate, for example

_____ 6. Necessary if glucose is to be taken up by body cells

_____ 7. _____ 8.

_____ 9. _____ 10. Four tropic hormones

_____ 11. Acts antagonistically to insulin; produced by the same
endocrine organ

_____ 12. Hypothalamic hormone important in regulating water balance

_____ 13. _____ 14. Anterior pituitary hormones that
regulate the ovarian cycle

_____ 15. _____ 16. Directly regulate the menstrual
uterine cycle

_____ 17. Adrenal cortex hormone involved in regulating salt levels
of body fluids

_____ 18. _____ 19. Necessary for milk production
and ejection

8. Name the hormone that would be produced in inadequate amounts in the following conditions. Place your responses in the answer blanks.

_____ 1. Sexual immaturity

_____ 2. Tetany

_____ 3. Excessive urination without high blood glucose levels; causes dehydration and tremendous thirst

_____ 4. Goiter

_____ 5. Cretinism; a type of dwarfism in which the individual retains childlike proportions and is mentally retarded

_____ 6. Excessive thirst, high blood glucose levels, acidosis

_____ 7. Abnormally small stature, normal proportions

_____ 8. Miscarriage

_____ 9. Lethargy, hair loss, low basal metabolic rate, obesity (myxedema in the adult)

9. Name the hormone that would be produced in excessive amounts in the following conditions. Place your responses in the answer blanks.

_____ 1. Lantern jaw; large hands and feet (acromegaly in the adult)

_____ 2. Bulging eyeballs, nervousness, increased pulse rate, weight loss (Graves' disease)

_____ 3. Demineralization of bones; spontaneous fractures

_____ 4. Cushing's syndrome—moon face, depression of the immune system

_____ 5. Abnormally large stature, relatively normal body proportions

_____ 6. Abnormal hairiness; masculinization

10. List the cardinal symptoms of diabetes mellitus, and provide the rationale for the occurrence of each symptom.

1. _____

_____

2. _____

_____

3. _____

_____

11. The activity of many end organs is regulated by negative feedback. Figure
9–3A shows the basic elements of a homeostatic control system. Figure 9–3B
shows a feedback loop with selected parts missing. Assume, for this system,
that the stimulus that initiates it is declining T_3 and T_4 levels in the blood,
which produces a drop in metabolic rate. Fill in the information missing in
the boxes to correctly complete this feedback loop. Also indicate whether it is
a negative or positive feedback loop.

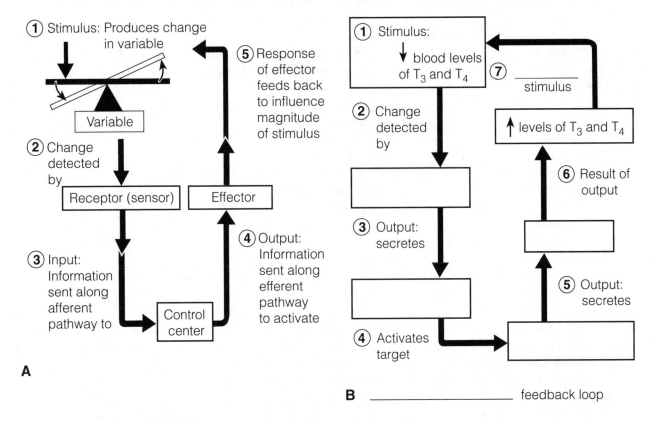

Figure 9–3

12. Circle the term that does not belong in each of the following groupings.

1. Posterior lobe   Hormone storage   Nervous tissue   Anterior lobe

2. Steroid hormone   Protein hormone   Second messenger   Membrane receptors

3. Catecholamines   Norepinephrine   Epinephrine   Cortisol

4. Calcitonin   Increases blood Ca^{2+} Thyroid gland   Enhances Ca^{2+} deposit

5. Glucocorticoids   Steroids   Aldosterone   Growth hormone

6. Thyroid follicles T_3 and T_4 Glucose metabolism   Parafollicular cells

OTHER HORMONE-PRODUCING TISSUES AND ORGANS

13. Besides the major endocrine organs, isolated clusters of cells produce hormones within body organs that are usually not associated with the endocrine system. A number of these hormones are listed in the table below. Fill in the missing information (blank spaces) on these hormones in the table.

Hormone	Chemical makeup	Source	Effects
Gastrin	Peptide		
Secretin		Duodenum	
Cholecystokinin	Peptide		
Erythropoietin		Kidney in response to hypoxia	
Active vitamin D_3		Skin; activated by kidneys	
Atrial natriuretic peptide (ANP)	Peptide		
Human chorionic gonadotropin (hCG)	Protein		
Leptin		Adipose tissue	

DEVELOPMENTAL ASPECTS OF THE ENDOCRINE SYSTEM

14. Complete the following statements by inserting your responses in the answer blanks.

_____ 1.

_____ 2.

_____ 3.

_____ 4.

_____ 5.

_____ 6.

_____ 7.

Under ordinary conditions, the endocrine organs operate smoothly until old age. However, a __(1)__ in an endocrine organ may lead to __(2)__ of its hormones. A lack of __(3)__ in the diet may result in undersecretion of thyroxine. Later in life, a woman experiences a number of symptoms such as hot flashes and mood changes, which result from decreasing levels of __(4)__ in her system. This period of a woman's life is referred to as __(5)__, and it results in a loss of her ability to __(6)__. Because __(7)__ tolerance tends to decrease in an aging person (due to declining sensitivity to insulin), adult-onset diabetes is common.

INCREDIBLE JOURNEY

A Visualization Exercise for the Endocrine System

. . . you notice charged particles shooting pell-mell out of the bone matrix. . . .

15. Where necessary, complete statements by inserting the missing words in the answer blanks.

_____ 1.

_____ 2.

_____ 3.

_____ 4.

_____ 5.

_____ 6.

_____ 7.

_____ 8.

_____ 9.

For this journey, you will be miniaturized and injected into a vein of your host. Throughout the journey, you will be traveling in the bloodstream. Your instructions are to record changes in blood composition as you float along and to form some conclusions as to why they are occurring (that is, which hormone is being released).

Bobbing gently along in the slowly moving blood, you realize that there is a sugary taste to your environment; however, the sweetness begins to decrease quite rapidly. As the glucose levels of the blood have just decreased, obviously __(1)__ has been released by the __(2)__ so that the cells can take up glucose.

A short while later you notice that the depth of the blood in the vein in which you are traveling has dropped substantially. To remedy this potentially serious situation, the __(3)__ will have to release more __(4)__ so the kidney tubules will reabsorb more water. Within a few minutes the blood becomes much deeper; you wonder if the body is psychic as well as wise.

As you circulate past the bones, you notice charged particles shooting pell-mell out of the bone matrix and jumping into the blood. You conclude that the __(5)__ glands have just released PTH because the __(6)__ levels have increased in the blood. As you continue to move in the bloodstream, the blood suddenly becomes sticky sweet, indicating that your host must be nervous about something. Obviously, his __(7)__ has released __(8)__ to cause this sudden increase in blood glucose.

Sometime later, you become conscious of a humming activity around you, and you sense that the cells are very busy. Your host's __(9)__ levels appear to be sufficient because his cells are certainly not sluggish in their metabolic activities. You record this observation and prepare to end this journey.

AT THE CLINIC

16. Pete is very short for his chronological age of 8 years. What physical features will allow you to determine quickly whether to check GH or thyroxine levels?

17. A young girl is brought to the clinic by her father. The girl fatigues easily and seems mentally sluggish. You notice a slight swelling in the anterior neck. What condition do you suspect? What are some possible causes and their treatments?

18. A 2-year-old boy is brought to the clinic by his anguished parents. He is developing sexually and shows an obsessive craving for salt. Blood tests reveal hyperglycemia. What endocrine gland is hypersecreting?

19. When the carnival came to a small town, the local health professionals and consumer groups joined forces to enforce truth-in-advertising laws to protect selected employees of the carnival. They demanded that the fat man, the dwarf, the giant, and the bearded lady be billed as "people with endocrine system problems" (which of course removed all the sensationalism usually associated with these attractions). Identify the endocrine disorder in each case and explain how (or why) the disorder produced the characteristic features of these four show people.

20. The brain is "informed" when we are stressed, and the hypothalamus responds by secreting a releasing hormone called corticotropin-releasing hormone (CRH) that helps the body deal with the stressors. Outline this entire sequence, starting with CRH and ending with the release of cortisol. (Be sure to trace the hormone through the hypophyseal portal system and out of the pituitary gland.)

21. Mrs. Jackson claims she is not menstruating and reports that her breasts are producing milk, although she has never been pregnant. What hormone is being hypersecreted?

THE FINALE: MULTIPLE CHOICE

22. Select the best answer or answers from the choices given.

1. The major endocrine organs of the body:

   A. tend to be very large organs

   B. are closely connected with each other

   C. all contribute to the same function (digestion)

   D. tend to lie near the midline of the body

2. Of the following endocrine structures, which develops from the brain?

   A. Posterior pituitary   C. Thyroid gland

   B. Anterior pituitary   D. Thymus gland

3. Which is generally true of hormones?

   A. Exocrine glands produce them.

   B. They travel throughout the body in the blood.

   C. They affect only non-hormone-producing organs.

   D. All steroid hormones produce very similar physiological effects in the body.

4. Which of the following are tropic hormones secreted by the anterior pituitary gland?

   A. LH   C. TSH

   B. ACTH   D. FSH

5. Smooth muscle contractions are stimulated by:

   A. testosterone   C. prolactin

   B. FSH   D. oxytocin

6. Relative to the cyclic AMP second-messenger system, which of the following is *not* accurate?

   A. The activating hormone interacts with a receptor site on the plasma membrane.

   B. Binding of the hormone directly produces the second messenger.

   C. Activated enzymes catalyze the transformation of AMP to cyclic AMP.

   D. Cyclic AMP acts within the cell to alter cell function as is characteristic for that specific hormone.

7. Nerve input regulates the release of:

   A. oxytocin   C. melatonin

   B. epinephrine   D. cortisol

8. ANP, the hormone secreted by the heart, has exactly the opposite function to this hormone secreted by the adrenal cortex:

   A. epinephrine   C. aldosterone

   B. cortisol   D. testosterone

9. Hormones that act directly or indirectly to elevate blood glucose include:

   A. GH   C. insulin

   B. cortisol   D. ACTH

10. Hormones secreted by females include:

    A. estrogens   C. prolactin

    B. progesterone   D. testosterone

11. Which of the following are direct or indirect effects of growth hormone?

    A. Stimulates cells to take in amino acids and form proteins

    B. Important in determining final body size

    C. Increases blood levels of fatty acids

    D. Decreases utilization of glucose by most body cells

12. Hypothyroidism can cause:

    A. myxedema   C. cretinism

    B. Cushing's syndrome   D. exophthalmos

13. Which of the following is given as a drug to reduce inflammation?

    A. Epinephrine   C. Aldosterone

    B. Cortisol   D. ADH

14. Which of the following hormones is (are) released by neurons?

    A. Oxytocin   C. ADH

    B. Insulin   D. Cortisol

15. The major stimulus for release of thyroid hormone is:

    A. hormonal

    B. humoral

    C. neural

16. A hormone *not* involved in glucose metabolism is:

    A. glucagon

    B. cortisone

    C. aldosterone

    D. insulin

17. Parathyroid hormone:

    A. increases bone formation and lowers blood calcium levels

    B. increases calcium excretion from the body

    C. decreases calcium absorption from the gut

    D. demineralizes bone and raises blood calcium levels

18. The word root referring to body fluids is:

    A. mell   C. humor

    B. hormon   D. gen

19. Most hormones are made and released as needed. The exception to this generalization is:

    A. catecholamines   C. insulin

    B. thyroxine   D. aldosterone

10 BLOOD

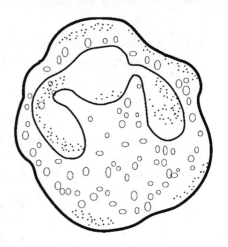

Blood, the "life fluid" that courses through the body's blood vessels, provides the means for the body's cells to receive vital nutrients and oxygen and dispose of their metabolic wastes. As blood flows past the tissue cells, exchanges continually occur between the blood and the tissue cells so that vital activities can go on continuously.

This chapter provides an opportunity to review the general characteristics of whole blood and plasma, to identify the various formed elements (blood cells), and to recall their functions. Blood groups, transfusion reactions, clotting, and various types of blood abnormalities are also considered.

COMPOSITION AND FUNCTIONS OF BLOOD

1. Complete the following description of the components of blood by writing the missing words in the answer blanks.

_____ 1.

_____ 2.

_____ 3.

_____ 4.

_____ 5.

_____ 6.

_____ 7.

_____ 8.

_____ 9.

_____ 10.

In terms of its tissue classification, blood is classified as a __(1)__ because it has living blood cells, called __(2)__, suspended in a nonliving fluid matrix called __(3)__. The "fibers" of blood only become visible during __(4)__.

If a blood sample is centrifuged, the heavier blood cells become packed at the bottom of the tube. Most of this compacted cell mass is composed of __(5)__, and the volume of blood accounted for by these cells is referred to as the __(6)__. The less dense __(7)__ rises to the top and constitutes about 45% of the blood volume. The so-called "buffy coat," composed of __(8)__ and __(9)__, is found at the junction between the other two blood elements. The buffy coat accounts for less than __(10)__ % of blood volume.

Blood is scarlet red in color when it is loaded with __(11)__; otherwise, it tends to be dark red.

_____ 11.

2. Using the key choices, identify the cell type(s) or blood elements that fit the following descriptions. Insert the correct term or letter response in the spaces provided.

Key Choices

A. Red blood

B. Megakaryocyte

C. Eosinophil

D. Basophil

E. Monocyte

F. Neutrophil

G. Lymphocyte

H. Formed elements

I. Plasma

_____ 1. Most numerous leukocyte

_____ 2. _____ 3. _____ 4. Granular leukocytes

_____ 5. Also called an erythrocyte; anucleate

_____ 6. _____ 7. Actively phagocytic leukocytes

_____ 8. _____ 9. Agranular leukocytes

_____ 10. Fragments to form platelets

_____ 11. (A) through (G) are examples of these

_____ 12. Increases during parasite attacks

_____ 13. Releases histamine during inflammatory reactions

_____ 14. After originating in bone marrow, may be formed in lymphoid tissue

_____ 15. Contains hemoglobin

_____ 16. Primarily water, noncellular; the fluid matrix of blood

_____ 17. Increases in number during prolonged infections

_____ 18. Least numerous leukocyte

_____ 19. _____ 20. Also called white blood cells (#19–23)

_____ 21. _____ 22. _____ 23.

3. Figure 10–1 depicts (in incomplete form) the erythropoietin mechanism for regulating the rate of erythropoiesis. Complete the statements that have answer blanks, and then choose colors (other than yellow) for the color-coding circles and corresponding structures on the diagram. Color all arrows on the diagram yellow. Finally, indicate the normal life span of erythrocytes.

◯ Kidney   ◯ Red bone marrow   ◯ Red blood cells (RBCs)

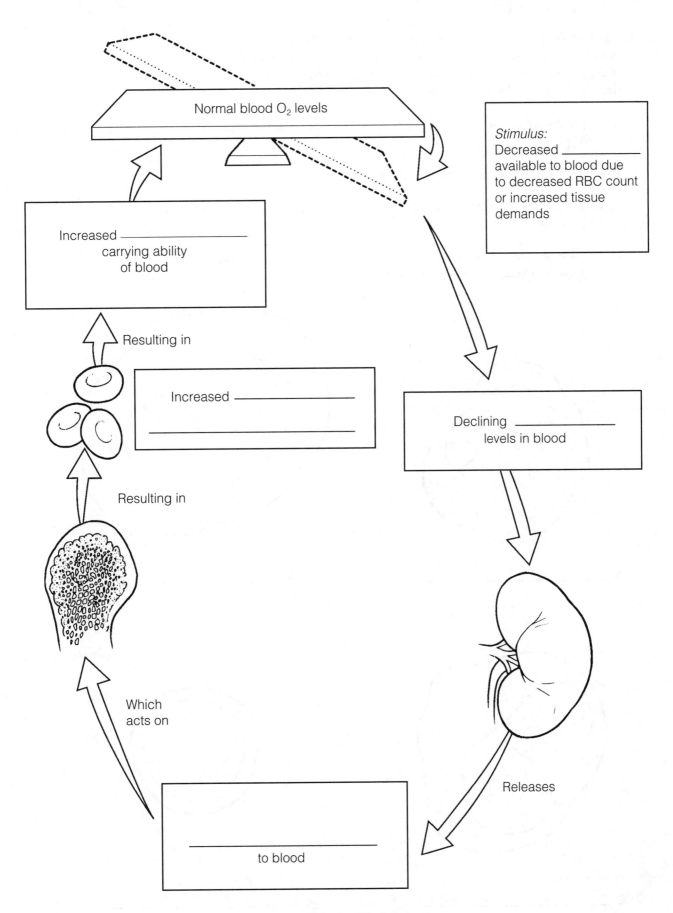

Normal blood O₂ levels

Stimulus:
Decreased _____ available to blood due to decreased RBC count or increased tissue demands

Increased _____ carrying ability of blood

Resulting in

Increased _____

Resulting in

Declining _____ levels in blood

Which acts on

Releases

_____ to blood

Figure 10–1

4. Four leukocytes are diagrammed in Figure 10–2. First, follow directions (given below) for coloring each leukocyte as it appears when stained with Wright's stain. Then, identify each leukocyte type by writing in the correct name in the blank below the illustration.

A. Color the granules pale violet, the cytoplasm pink, and the nucleus dark purple.

B. Color the nucleus deep blue and the cytoplasm pale blue.

C. Color the granules bright red, the cytoplasm pale pink, and the nucleus red/purple.

D. For this smallest white blood cell, color the nucleus deep purple/blue and the sparse cytoplasm pale blue.

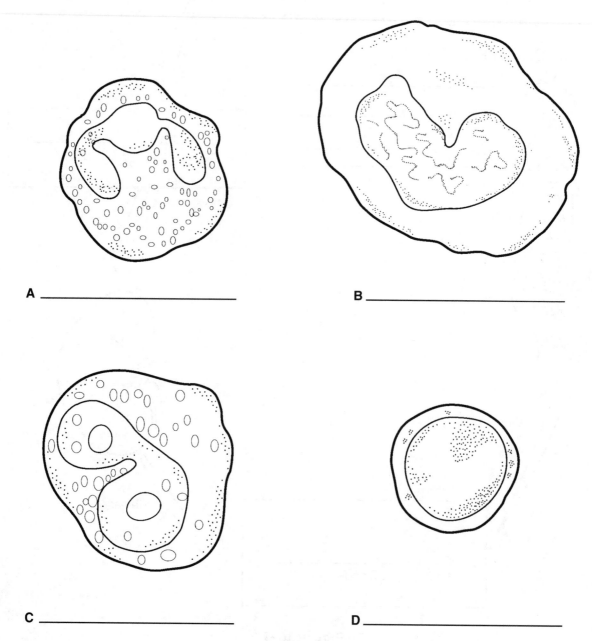

A _____

B _____

C _____

D _____

Figure 10–2

5. For each true statement, insert *T*. If any of the statements are false, correct the underlined term by inserting the correction in the answer blank.

_____ 1. White blood cells (WBCs) move into and out of blood vessels by the process of positive chemotaxis.

_____ 2. An abnormal decrease in the number of WBCs is leukopenia.

_____ 3. When blood becomes too acidic or too basic, both the respiratory system and the liver may be called into action to restore it to its normal pH range.

_____ 4. The normal pH range of blood is 7.00 to 7.45.

_____ 5. The cardiovascular system of an average adult contains approximately 4 liters of blood.

_____ 6. The only WBC type to arise from lymphoid stem cells is the lymphocyte.

_____ 7. An abnormal increase in the number of white blood cells is leukocytosis.

_____ 8. The normal RBC count is 3.5–4.5 million/mm³.

_____ 9. Normal hemoglobin values are in the area of 42%–47% of the volume of whole blood.

_____ 10. An anemia resulting from a decreased RBC number causes the blood to become more viscous.

_____ 11. Phagocytic agranular WBCs are eosinophils.

_____ 12. The leukocytes particularly important in the immune response are monocytes.

6. Circle the term that does not belong in each of the following groupings.

1. Erythrocytes   Lymphocytes   Monocytes   Eosinophils

2. Neutrophils   Monocytes   Basophils   Eosinophils

3. Hemoglobin   Lymphocyte   Oxygen transport   Erythrocytes

4. Platelets   Monocytes   Phagocytosis   Neutrophils

5. Thrombus   Aneurysm   Embolus   Clot

6. Plasma   Nutrients   Hemoglobin   Wastes

7. Myeloid stem cell   Lymphocyte   Monocyte   Basophil

7. Rank the following lymphocytes from 1 (most abundant) to 5 (least abundant) relative to their abundance in the blood of a healthy person.

_____ 1. Lymphocyte   _____ 3. Neutrophil   _____ 5. Monocyte

_____ 2. Basophil   _____ 4. Eosinophil

8. Check (✓) all the factors that would serve as stimuli for erythropoiesis.

_____ 1. Hemorrhage   _____ 3. Living at a high altitude

_____ 2. Aerobic exercise   _____ 4. Breathing pure oxygen

HEMOSTASIS

9. Using the key choices, correctly complete the following description of the blood-clotting process. Insert the key term or letter in the answer blanks.

Key Choices

A. Break   D. Fibrinogen   G. Prothrombin activator   J. Thrombin

B. Erythrocytes   E. Platelets   H. PF₃   K. Tissue factor

C. Fibrin   F. Prothrombin   I. Serotonin

_____ 1.

_____ 2.

_____ 3.

_____ 4.

_____ 5.

_____ 6.

_____ 7.

_____ 8.

_____ 9.   _____ 10.   _____ 11.

Clotting begins when a __(1)__ occurs in a blood vessel wall. Almost immediately, __(2)__ cling to the blood vessel wall and release __(3)__, which helps to decrease blood loss by helping to constrict the vessel. __(4)__, released by damaged cells in the area, interacts with __(5)__ on the platelet surfaces and other clotting factors to form __(6)__. This chemical substance causes __(7)__ to be converted to __(8)__. Once present, molecule #8 acts as an enzyme to attach __(9)__ molecules together to form long, threadlike strands of __(10)__, which then traps __(11)__ flowing by in the blood.

10. For each true statement, write _T_. If any statements are false, correct the underlined term by inserting the correction in the answer blank.

_____ 1. Normally, blood clots within <u>5–10</u> minutes.

_____ 2. The most important natural body anticoagulant is <u>histamine</u>.

_____ 3. <u>Hemostasis</u> means stoppage of blood flow.

BLOOD GROUPS AND TRANSFUSIONS

11. Correctly complete the following table concerning ABO blood groups.

Blood type	Agglutinogens or antigens	Agglutinins or antibodies in plasma	Can donate blood to type	Can receive blood from type
1. Type A	A			
2. Type B		anti-A		
3. Type AB			AB	
4. Type O	none			

12. What blood type is the *universal donor?* _____

The *universal recipient?* _____

13. When a person is given a transfusion of mismatched blood, a transfusion reaction occurs. Define the term "transfusion reaction" in the blanks provided here.

_____

_____

DEVELOPMENTAL ASPECTS OF BLOOD

14. Complete the following statements by inserting your responses in the answer blanks.

_____ 1.

_____ 2.

_____ 3.

_____ 4.

A fetus has a special type of hemoglobin, hemoglobin __(1)__, that has a particularly high affinity for oxygen. After birth, the infant's fetal RBCs are rapidly destroyed and replaced by hemoglobin A–containing RBCs. When the immature infant liver cannot keep pace with the demands to rid the body of hemoglobin breakdown products, the infant's tissues become yellowed, or __(2)__.

Genetic factors lead to several congenital diseases concerning the blood. An anemia in which RBCs become sharp and "logjam" in the blood vessels under conditions of low-oxygen tension in the blood is __(3)__ anemia. Bleeder's disease, or __(4)__, is a result of a deficiency of certain clotting factors.

→

_____ 5.

_____ 6.

_____ 7.

_____ 8.

_____ 9.

5. Diet is important to normal blood formation. Women are particularly prone to __(5)__ -deficiency anemia because of their monthly menses. A decreased efficiency of the gastric mucosa makes elderly individuals particularly susceptible to __(6)__ anemia as a result of a lack of intrinsic factor, which is necessary for vitamin __(7)__ absorption. An important problem in aged individuals is their tendency to form undesirable clots, or __(8)__. Both the young and the elderly are at risk for cancer of the blood, or __(9)__.

INCREDIBLE JOURNEY

A Visualization Exercise for the Blood

Once inside, you quickly make a slash in the vessel lining. . . .

15. Where necessary, complete statements by inserting the missing words in the answer blanks.

_____ 1.

_____ 2.

_____ 3.

_____ 4.

_____ 5.

_____ 6.

_____ 7.

_____ 8.

_____ 9.

For this journey, you will be miniaturized and injected into the external iliac artery and will be guided by a fluorescent monitor into the bone marrow of the iliac bone. You will observe and report events of blood cell formation, also called __(1)__, seen there; then you will move out of the bone into the circulation to initiate and observe the process of blood clotting, also called __(2)__. Once in the bone marrow, you watch as several large dark-nucleated stem cells, or __(3)__, begin to divide and produce daughter cells. To your right, the daughter cells eventually formed have tiny cytoplasmic granules and very peculiarly shaped nuclei that look like small masses of nuclear material connected by thin strands of nucleoplasm. You have just witnessed the formation of a type of white blood cell called the __(4)__. You describe its appearance and make a mental note to try to observe its activity later. Meanwhile you can tentatively report that this cell type functions as a __(5)__ to protect the body.

At another site, daughter cells arising from the division of a stem cell are difficult to identify initially. As you continue to observe the cells, you see that they, in turn, divide. Eventually some of their daughter cells eject their nuclei and flatten out to assume a disc shape. You assume that the kidneys must have released __(6)__ because those cells are __(7)__. That dark material filling their interior must be __(8)__ because those cells function to transport __(9)__ in the blood.

_____10.

_____11.

_____12.

_____13.

_____14.

_____15.

Now you turn your attention to the daughter cells being formed by the division of another stem cell. They are small round cells with relatively large round nuclei. In fact, their cytoplasm is very sparse. You record your observation of the formation of __(10)__ . They do not remain in the marrow very long after formation but seem to enter the circulation almost as soon as they are produced. Some of those cells produce __(11)__ or act in other ways in the immune response. At this point, although you have yet to see the formation of __(12)__ , __(13)__ , __(14)__ , or __(15)__ , you decide to proceed into the circulation to make the blood-clotting observations.

_____16.

_____17.

_____18.

_____19.

_____20.

_____21.

_____22.

_____23.

_____24.

_____25.

You maneuver yourself into a small venule to enter the general circulation. Once inside, you quickly make a slash in the vessel lining, or __(16)__ . Almost immediately, what appear to be hundreds of jagged cell fragments swoop into the area and plaster themselves over the freshly made incision. You record that __(17)__ have just adhered to the damaged site. As you are writing, your chemical monitor flashes the message, "vasoconstrictor substance released." You record that __(18)__ has been released based on your observation that the vessel wall seems to be closing in. Peering out at the damaged site, you see that long ropelike strands are being formed at a rapid rate and are clinging to the site. You report that the __(19)__ mesh is forming and is beginning to trap RBCs to form the basis of the __(20)__ . Even though you do not have the equipment to monitor the intermediate steps of this process, you know that the interaction of platelet PF_3 and other clotting factors must have generated __(21)__ , which then converted __(22)__ to __(23)__ . This second enzyme then joined the soluble __(24)__ molecules together to form the network of strands you can see.

You carefully back away from the newly formed clot. You do *not* want to disturb the area because you realize that if the clot detaches, it might become a life-threatening __(25)__ . Your mission here is completed, and you return to the entrance site.

AT THE CLINIC

16. Correctly respond to five questions (#1–5) referring to the following situation. Mrs. Carlyle is pregnant for the first time. Her blood type is Rh negative, her husband is Rh positive, and their first child has been determined to be Rh positive. Ordinarily, the first such pregnancy causes no major problems, but baby Carlyle is born blue and cyanotic.

1. What is this condition, a result of Rh incompatibility, called?

→

2. Why is the baby cyanotic?

3. Because this is Mrs. Carlyle's first pregnancy, how can you account for the baby's problem?

4. Assume that baby Carlyle was born pink and healthy. What measures should be taken to prevent the previously described situation from happening in a second pregnancy with an Rh-positive baby?

5. Mrs. Carlyle's sister has had two miscarriages before seeking medical help with her third pregnancy. Blood typing shows that she, like her sister, is Rh negative; her husband is Rh positive. What course of treatment will be followed?

17. Ms. Pratt is claiming that Mr. X is the father of her child. Ms. Pratt's blood type is O negative. Her baby boy has type A positive blood. Mr. X's blood is typed and found to be B positive. Could he be the father of her child? _____ If not, what blood type would the father be expected to have? _____

18. Cancer patients being treated with chemotherapy drugs designed to destroy rapidly dividing cells are monitored closely for changes in their RBC and WBC counts. Why?

19. A red marrow biopsy is ordered for two patients—a young child and an adult. The specimen is taken from the tibia of the child but from the iliac crest of the adult. Explain why different sites are used to obtain marrow samples in adults and children. (You might want to check Chapter 5 for this one.)

20. Mrs. Graves has just donated a pint of blood. Shortly thereafter, her bone marrow is gearing up to replace the loss. Which of the formed elements will be produced in the greatest quantities?

21. Mr. Rudd, who has just had surgery for stomach cancer, has been receiving weekly injections of vitamin B_{12}. Why is he receiving the vitamin injections? Why can't the vitamin be delivered in tablet form? What will be the result if he refuses to continue the injections?

THE FINALE: MULTIPLE CHOICE

22. Select the best answer or answers from the choices given.

1. Which of the following are true concerning erythrocytes?

   A. They rely on anaerobic respiration.

   B. A large part of their volume is hemoglobin.

   C. Their precursor is called a megakaryo-blast.

   D. Their shape increases membrane surface area.

2. A serious bacterial infection leads to more of these cells in the blood.

   A. Erythrocytes and platelets

   B. Neutrophils

   C. Erythrocytes and monocytes

   D. All formed elements

3. Sickling of RBCs can be induced in those with sickle cell anemia by:

   A. blood loss   C. stress

   B. vigorous exercise   D. fever

4. A child is diagnosed with sickle cell anemia. This means that:

   A. one parent had sickle cell anemia

   B. one parent carried the sickle cell gene

   C. both parents had sickle cell anemia

   D. both parents carried the sickle cell gene

5. Which would lead to increased erythropoiesis?

   A. Chronic bleeding ulcer

   B. Reduction in respiratory ventilation

   C. Decreased level of physical activity

   D. Reduced blood flow to the kidneys

6. Which of the following does *not* characterize leukocytes?

   A. Ameboid

   B. Phagocytic (some)

   C. Nucleated

   D. Cells found in largest numbers in the bloodstream

7. The blood cell that can attack a specific antigen is a(n):

   A. monocyte

   B. neutrophil

   C. lymphocyte

   D. eosinophil

8. The leukocyte that releases histamine and other inflammatory chemicals is the:

   A. basophil   C. monocyte

   B. eosinophil   D. neutrophil

9. Leukocytes share all of the following features *except*:

   A. diapedesis

   B. disease fighting

   C. distorted, lobed nuclei

   D. more active in connective tissues than in blood

10. In leukemia:

    A. the cancerous WBCs function normally

    B. the cancerous WBCs fail to specialize

    C. production of RBCs and platelets is decreased

    D. infection and bleeding can be life threatening

11. A condition resulting from thrombocytopenia is:

   A. thrombus formation

   B. embolus formation

   C. petechiae

   D. hemophilia

12. Which of the following can cause problems in a transfusion reaction?

   A. Donor antibodies attacking recipient RBCs

   B. Clogging of small vessels by agglutinated clumps of RBCs

   C. Lysis of donated RBCs

   D. Blockage of kidney tubules

13. If an Rh⁻ mother becomes pregnant, when can hemolytic disease of the newborn *not possibly* occur in the child?

   A. If the child is Rh⁻

   B. If the child is Rh⁺

   C. If the father is Rh⁺

   D. If the father is Rh⁻

14. What is the difference between a thrombus and an embolus?

   A. One occurs in the bloodstream, whereas the other occurs outside the bloodstream.

   B. One occurs in arteries, the other in veins.

   C. One is a blood clot, while the other is a parasitic worm.

   D. A thrombus must travel to become an embolus.

15. The plasma component that forms the fibrous skeleton of a clot consists of:

   A. platelets

   B. fibrinogen

   C. thromboplastin

   D. thrombin

16. The normal pH of blood is:

   A. 8.4   C. 7.4

   B. 7.8   D. 4.7

11 THE CARDIOVASCULAR SYSTEM

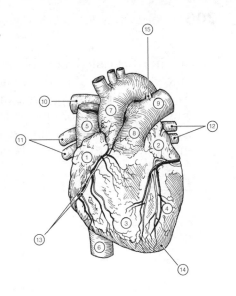

The major structures of the cardiovascular system, the heart and blood vessels, play a vital role in human physiology. The major function of the cardiovascular system is transportation. Using blood as the transport vehicle, the system carries nutrients, gases, wastes, antibodies, electrolytes, and many other substances to and from body cells. Its propulsive force is the contracting heart.

The anatomy and location of the heart and blood vessels and the important understandings of cardiovascular physiology (for example, cardiac cycle, electrocardiogram [ECG], and regulation of blood pressure) are the major topics of this chapter.

THE HEART
Anatomy of the Heart

1. Complete the following statements by inserting your answers in the answer blanks.

_____ 1.

_____ 2.

_____ 3.

_____ 4.

_____ 5.

_____ 6.

_____ 7.

_____ 8.

_____ 9.

The heart is a cone-shaped muscular organ located within the __(1)__ . Its apex rests on the __(2)__ , and its base is at the level of the __(3)__ rib. The coronary arteries that nourish the myocardium arise from the __(4)__ . The coronary sinus empties into the __(5)__ . Relative to the roles of the heart chambers, the __(6)__ are receiving chambers, whereas the __(7)__ are discharging chambers. The membrane that lines the heart and also forms the valve flaps is called the __(8)__ . The outermost layer of the heart is called the __(9)__ . The fluid that fills the pericardial sac acts to decrease __(10)__ during heart activity. The heart muscle, or myocardium, is composed of a specialized type of muscle tissue called __(11)__ .

_____ 10.

_____ 11.

2. The heart is called a double pump because it serves two circulations. Trace the flow of blood through the pulmonary and systemic circulations by writing the missing terms in the answer blanks. Then, color regions transporting O₂-poor blood blue and regions transporting O₂-rich blood red on Figure 11–1. Finally, identify the various regions of the circulation shown in Figure 11–1 by labeling them using the key choices.

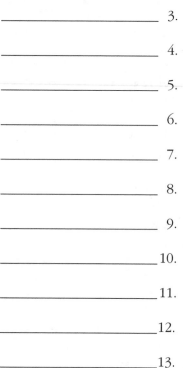

1. _____

2. _____

3. _____

4. _____

5. _____

6. _____

7. _____

8. _____

9. _____

10. _____

11. _____

12. _____

13. _____

From the right atrium through the tricuspid valve to the __(1)__, through the __(2)__ valve to the pulmonary trunk to the right and left __(3)__, to the capillary beds of the __(4)__, to the __(5)__, to the __(6)__ of the heart through the __(7)__ valve, to the __(8)__ through the __(9)__ semilunar valve, to the __(10)__, to the systemic arteries, to the __(11)__ of the body tissues, to the systemic veins, to the __(12)__ and __(13)__, which enter the right atrium of the heart.

Figure 11–1

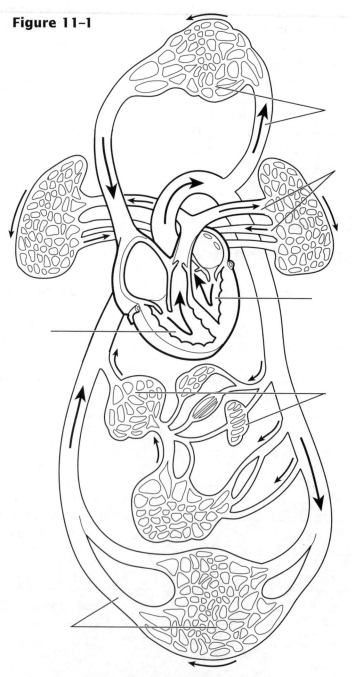

Key Choices

A. Vessels serving head and upper limbs

B. Vessels serving body trunk and lower limbs

C. Vessels serving the viscera

D. Pulmonary circulation

E. Pulmonary "pump"

F. Systemic "pump"

3. Figure 11–2 is an anterior view of the heart. Identify each numbered structure
and write its name in the corresponding numbered answer blank. Then,
select different colors for each structure provided with a color-coding circle,
and use them to color the coding circles and corresponding structures on the
figure.

○ _____ 1.   ○ _____ 6.   ○ _____ 11.

○ _____ 2.   ○ _____ 7.   _____ 12.

○ _____ 3.   ○ _____ 8.   _____ 13.

○ _____ 4.   _____ 9.   _____ 14.

_____ 5.   _____ 10.   ○ _____ 15.

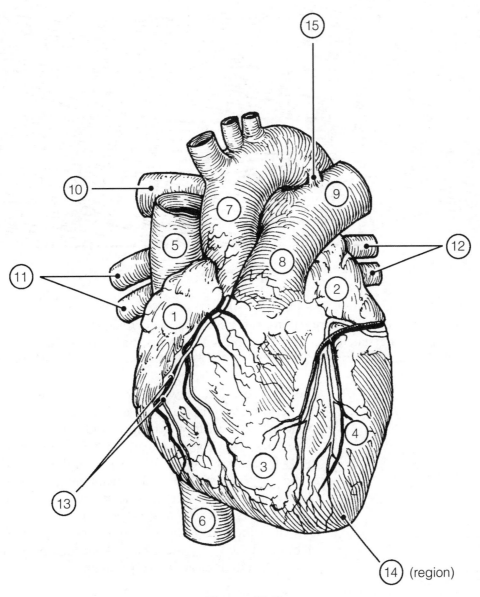

Figure 11–2

4. Figure 11–3 is a schematic drawing of the microscopic structure of cardiac muscle. Using different colors, color the coding circles of the structures listed below and the corresponding structures on the figure.

◯ Nuclei (with nucleoli)   ◯ Muscle fibers

◯ Intercalated discs   ◯ Striations

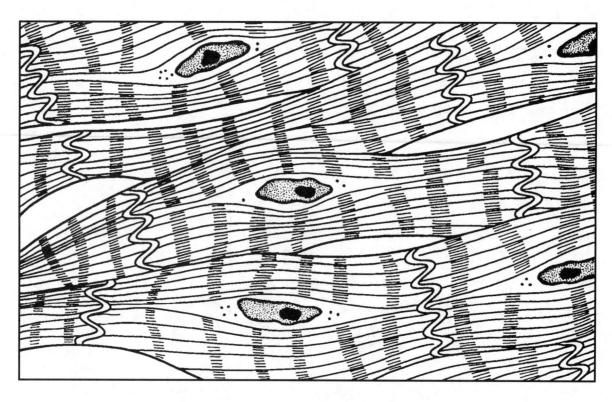

Figure 11–3

5. The events of one complete heartbeat are referred to as the cardiac cycle. Complete the following statements that describe these events. Insert your answers in the answer blanks.

_____ 1.

_____ 2.

_____ 3.

_____ 4.

_____ 5.

_____ 6.

_____ 7.

_____ 8.

The contraction of the ventricles is referred to as __(1)__, and the period of ventricular relaxation is called __(2)__. The monosyllables describing heart sounds during the cardiac cycle are __(3)__. The first heart sound is a result of closure of the __(4)__ valves; the second heart sound is caused by closure of the __(5)__ valves. The heart chambers that have just been filled when you hear the first heart sound are the __(6)__, and the chambers that have just emptied are the __(7)__. Immediately after the second heart sound, the __(8)__ are filling with blood, and the __(9)__ are empty. Abnormal heart sounds, or __(10)__, usually indicate valve problems.

_____ 9.

_____ 10.

6. Figure 11–4 is a diagram of the frontal section of the heart. Follow the instructions below to complete this exercise.

First, draw arrows to indicate the direction of blood flow through the heart. Draw the pathway of the oxygen-rich blood with red arrows, and trace the pathway of oxygen-poor blood with blue arrows.

Second, identify each of the elements of the intrinsic conduction system (numbers 1–5 on the figure) by inserting the appropriate terms in the blanks left of the figure. Then, indicate with green arrows the pathway that impulses take through this system.

Third, correctly identify each of the heart valves (numbers 6–9 on the figure) by inserting the appropriate terms in the blanks left of the figure, and draw in and identify by name the cordlike structures that anchor the flaps of the atrioventricular (AV) valves.

Fourth, use the numbers from the figure to identify the structures described below. Place the numbers in the lettered answer blanks.

_____ A. _____ B. Prevent backflow into the ventricles when the heart is relaxed

_____ C. _____ D. Prevent backflow into the atria when the ventricles are contracting

_____ E. AV valve with three flaps

_____ F. AV valve with two flaps

_____ G. The pacemaker of the intrinsic conduction system

_____ H. The point in the intrinsic conduction system where the impulse is temporarily delayed

_____ 1.

_____ 2.

_____ 3.

_____ 4.

_____ 5.

_____ 6.

_____ 7.

_____ 8.

_____ 9.

Figure 11–4

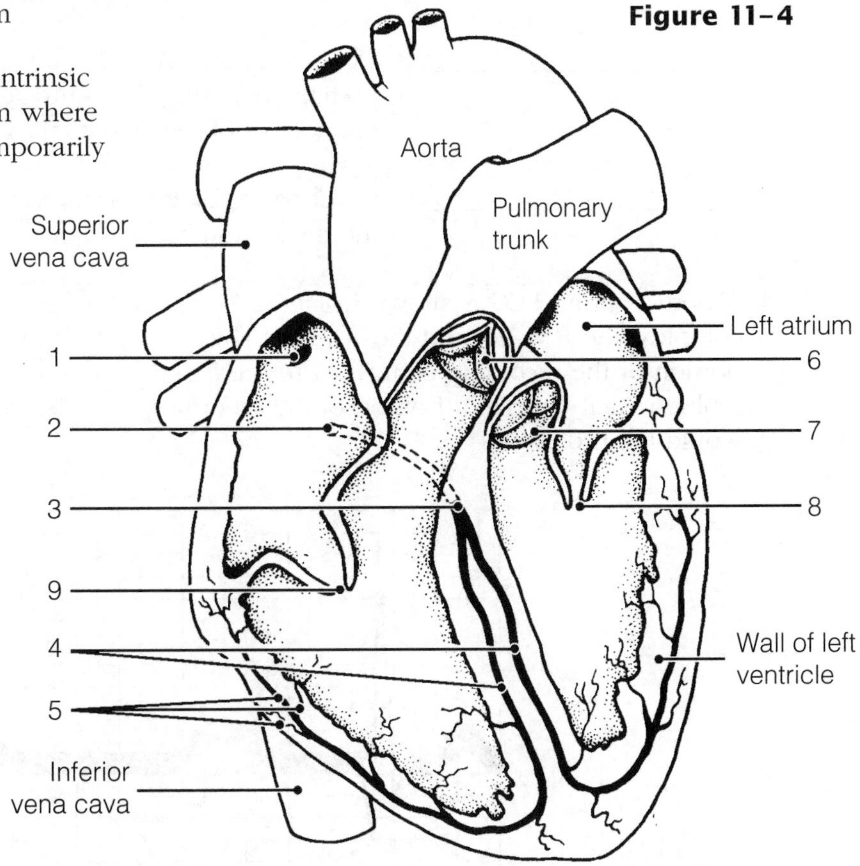

Physiology of the Heart

7. Match the terms provided in Column B with the statements given in Column A. Place the correct term or letter response in the answer blanks.

Column A

_____ 1. A recording of the electrical activity of the heart

_____ 2. The period when the atria are depolarizing

_____ 3. The period when the ventricles are repolarizing

_____ 4. The period during which the ventricles are depolarizing, which precedes their contraction

_____ 5. An abnormally slow heartbeat, that is, slower than 60 beats per minute

_____ 6. A condition in which the heart is uncoordinated and useless as a pump

_____ 7. An abnormally rapid heartbeat, that is, faster than 100 beats per minute

_____ 8. Damage to the AV node, totally or partially releasing the ventricles from the control of the sinoatrial (SA) node

_____ 9. Chest pain, resulting from ischemia of the myocardium

Column B

A. Angina pectoris

B. Bradycardia

C. Electrocardiogram

D. Fibrillation

E. Heart block

F. P wave

G. QRS wave

H. T wave

I. Tachycardia

8. A portion of an ECG is shown in Figure 11–5. On the figure identify the QRS complex, the P wave, and the T wave. Then, using a red pencil, bracket a portion of the recording equivalent to the length of one cardiac cycle. Using a blue pencil, bracket a portion of the recording in which the *ventricles* would be in diastole.

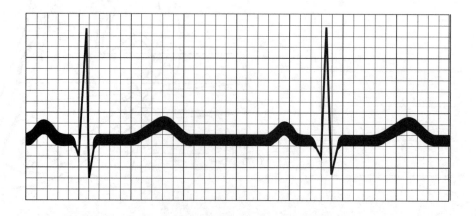

Figure 11–5

9. Complete the following statements relating to cardiac output by writing
the missing terms in the answer blanks.

_____ 1.

_____ 2.

_____ 3.

_____ 4.

_____ 5.

_____ 6.

_____ 7.

_____ 8.

In the relationship CO = HR × SV, CO stands for __(1)__, HR
stands for __(2)__, and SV stands for __(3)__. For the normal
resting heart, the value of HR is __(4)__, and the value of SV is
__(5)__. The normal average adult cardiac output, therefore, is
__(6)__. The time for the entire blood supply to pass through
the body is once each __(7)__.

According to Starling's law of the heart, the critical factor that
determines force of heartbeat, or __(8)__, is the degree of
__(9)__ of the cardiac muscle just before it contracts. Consequently, the force of heartbeat can be increased by increasing
the amount of __(10)__ returned to the heart.

_____ 9. _____10.

10. Check (✓) all factors that lead to an *increase* in cardiac output by influencing
either heart rate or stroke volume.

_____ 1. Epinephrine   _____ 6. Activation of the sympathetic nervous system

_____ 2. Thyroxine   _____ 7. Activation of the vagus nerves

_____ 3. Hemorrhage   _____ 8. Low blood pressure

_____ 4. Fear   _____ 9. High blood pressure

_____ 5. Exercise   _____ 10. Fever

11. For each of the following statements that is true, write *T* in the answer blank.
For any false statements, correct the underlined term by writing the correct
term in the answer blank.

_____ 1. The resting heart rate is fastest in <u>adult</u> life.

_____ 2. Because the heart of the highly trained athlete hypertrophies,
its <u>stroke volume</u> decreases.

_____ 3. If the <u>right</u> side of the heart fails, pulmonary congestion occurs.

_____ 4. In <u>peripheral</u> congestion, the feet, ankles, and fingers swell.

_____ 5. The pumping action of the healthy heart ordinarily maintains a
balance between cardiac output and <u>venous return</u>.

12. Circle the term that does not belong in each of the following groupings.

1. Pulmonary trunk   Vena cava   Right side of heart   Left side of heart

2. QRS wave   T wave   P wave   Electrical activity of the ventricles

3. AV valves closed   AV valves opened   Ventricular systole   Semilunar valves open

4. Papillary muscles   Aortic semilunar valve   Tricuspid valve   Chordae tendineae

5. Tricuspid valve   Mitral valve   Bicuspid valve   Left AV valve

6. Ischemia   Infarct   Scar tissue repair   Heart block

BLOOD VESSELS

Microscopic Anatomy of Blood Vessels

13. Complete the following statements concerning blood vessels.

_____ 1.   The central cavity of a blood vessel is called the __(1)__. Reduction of the diameter of this cavity is called __(2)__, and enlarge-

_____ 2.   ment of the vessel diameter is called __(3)__. Blood is carried to the heart by __(4)__ and away from the heart by __(5)__. Capillary

_____ 3.   beds are supplied by __(6)__ and drained by __(7)__.

_____ 4.   _____ 6.

_____ 5.   _____ 7.

14. Briefly explain in the space provided the need for valves in veins but not in arteries.

_____

_____

15. Name two events *occurring within the body* that aid venous return. Place your responses in the blanks that follow.

_____ and _____

16. First, select different colors for each of the three blood vessel tunics listed in the key choices and illustrated in Figure 11–6 on page 213. Color the color-coding circles and the corresponding structures in the three diagrams. In the blanks beneath the illustrations correctly identify each vessel type. In the additional spaces provided, list the structural details that allowed you to make the identifications. Then, using the key choices, identify the blood vessel tunics described in each of the following descriptions. Insert the term or letter of the key choice in the answer blanks.

Key Choices

A. ◯ Tunica intima   B. ◯ Tunica media   C. ◯ Tunica externa

_____ 1. Single thin layer of endothelium

_____ 2. Bulky middle coat, containing smooth muscle and elastin

_____ 3. Provides a smooth surface to decrease resistance to blood flow

_____ 4. The only tunic of capillaries

_____ 5. Also called the adventitia

_____ 6. The only tunic that plays an active role in blood pressure regulation

_____ 7. Supporting, protective coat

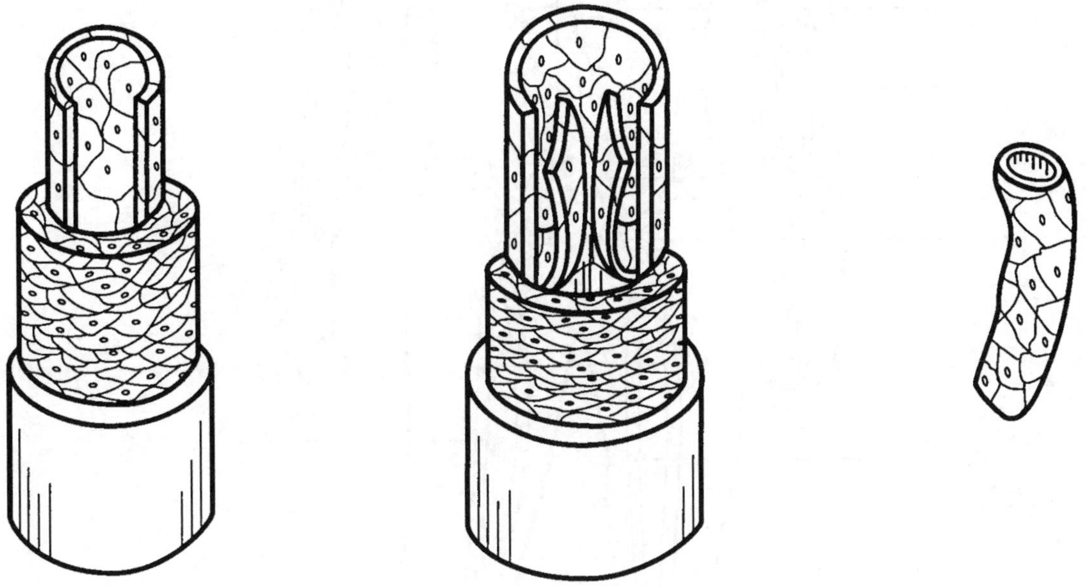

Figure 11–6

A. _____   B. _____   C. _____

_____   _____   _____

_____   _____   _____

Gross Anatomy of Blood Vessels

17. Figures 11–7 and 11–8 on pages 214 and 215 illustrate the location of the most important arteries and veins of the body. The veins are shown in Figure 11–7. Color the veins blue and then identify each vein provided with a leader line on the figure. The arteries are shown in Figure 11–8. Color them red and then identify those indicated by leader lines on the figure. NOTE: If desired, the vessels identified may be colored differently to aid you in their later identification.

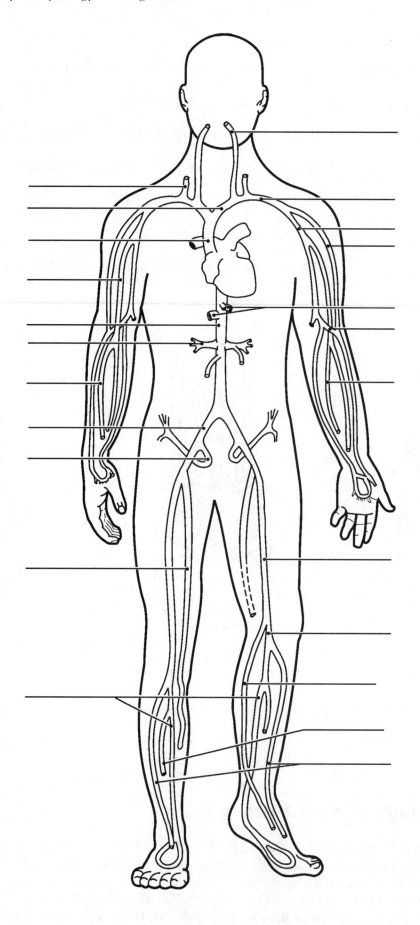

Figure 11-7 Veins

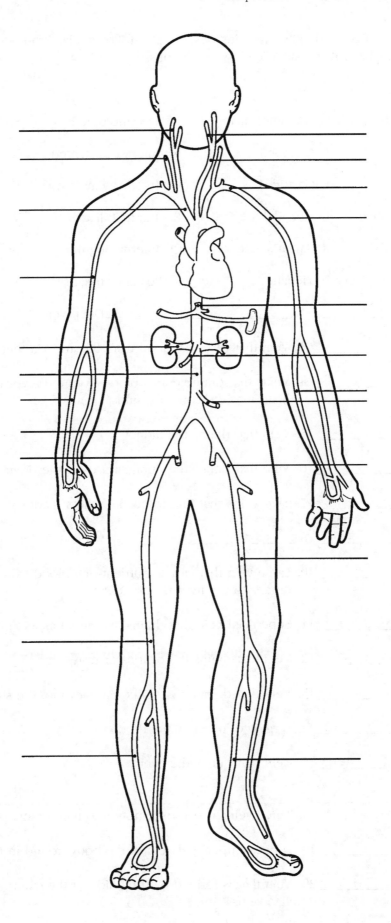

Figure 11–8   Arteries

18. Using the key choices, identify the veins described as follows. Place the correct term or letter response in the answer blanks.

Key Choices

A. Anterior tibial	G. Common iliac	M. Hepatic portal	S. Radial
B. Azygos	H. Femoral	N. Inferior mesenteric	T. Renal
C. Basilic	I. Gastric	O. Inferior vena cava	U. Subclavian
D. Brachiocephalic	J. Gonadal	P. Internal iliac	V. Superior mesenteric
E. Cardiac	K. Great saphenous	Q. Internal jugular	W. Superior vena cava
F. Cephalic	L. Hepatic	R. Posterior tibial	X. Ulnar

_____ 1. _____ 2. Deep veins, draining the forearm

_____ 3. Vein that receives blood from the arm via the axillary vein

_____ 4. Veins that drain venous blood from the myocardium of the heart into the coronary sinus

_____ 5. Vein that drains the kidney

_____ 6. Vein that drains the dural sinuses of the brain

_____ 7. Two veins that join to become the superior vena cava

_____ 8. _____ 9. Veins that drain the leg and foot

_____ 10. Large vein that carries nutrient-rich blood from the digestive organs to the liver for processing

_____ 11. Superficial vein that drains the lateral aspect of the arm

_____ 12. Vein that drains the ovaries or testes

_____ 13. Vein that drains the thorax, empties into the superior vena cava

_____ 14. Largest vein below the thorax

_____ 15. Vein that drains the liver

_____ 16. _____ 17. _____ 18. Three veins that form/empty into the hepatic portal vein

_____ 19. Longest superficial vein of the body; found in the leg

_____ 20. Vein that is formed by the union of the external and internal iliac veins

_____ 21. Deep vein of the thigh

19. Figure 11–9 shows the pulmonary circuit. Identify all vessels that have leader lines. Color the vessels (and heart chambers) transporting oxygen-rich blood *red*; color those transporting carbon dioxide-rich blood *blue*.

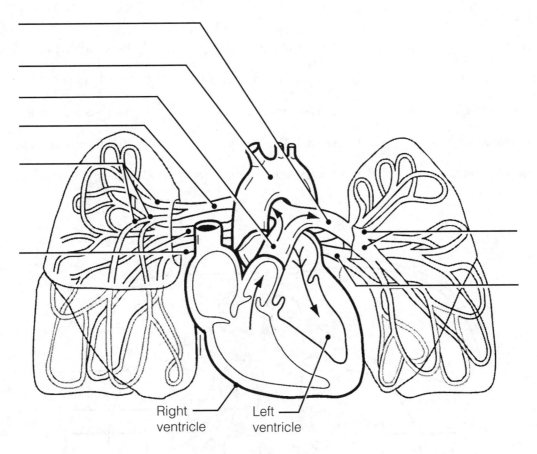

Right ventricle

Left ventricle

Figure 11–9

20. Using the key choices, identify the special circulations described below.

Key Choices

A. Cerebral   C. Hepatic   E. Skeletal muscle

B. Coronary   D. Pulmonary   F. Skin

_____ 1. Blood flow increases markedly when the body temperature rises

_____ 2. Arteries characteristically have thin walls and large lumens

_____ 3. Vessels do not constrict but are compressed during systole

_____ 4. Receives constant blood flow whether the body is at rest or strenuously exercising

_____ 5. Much lower arterial pressure than that in systemic circulation

_____ 6. Impermeable tight junctions in capillary endothelium

_____ 7. During vigorous physical activity, receives up to two-thirds of blood flow

21. The abdominal vasculature is depicted in Figure 11–10. Using the key choices, identify the following vessels by selecting the correct letters. Color the diagram as you wish.

Key Choices

A. Aorta

B. Celiac trunk

C. Common iliac arteries

D. Gonadal arteries

E. Hepatic veins

F. Inferior mesenteric artery

G. Inferior vena cava

H. Lumbar arteries

I. External iliac artery

J. Superior mesenteric artery

K. Renal arteries

L. Renal veins

M. Left gonadal vein

N. Right gonadal vein

O. Internal iliac artery

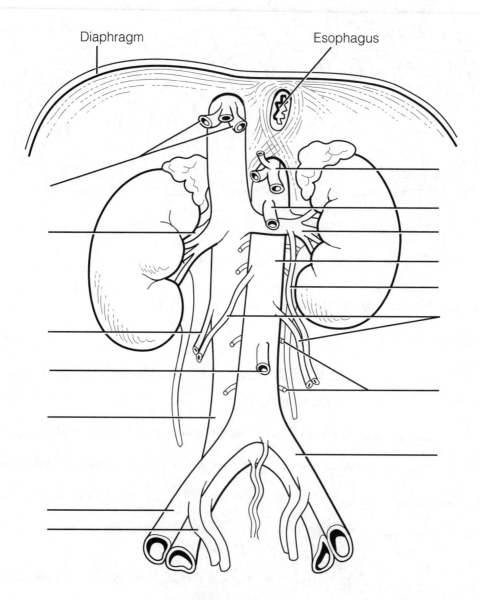

Diaphragm

Esophagus

Figure 11–10

22. Figure 11–11 is a diagram of the hepatic portal circulation. Select different colors for the structures listed below and use them to color the color-coding circles and corresponding structures on the illustration.

◯ Inferior mesenteric vein   ◯ Splenic vein   ◯ Hepatic portal vein

◯ Superior mesenteric vein   ◯ Gastric vein

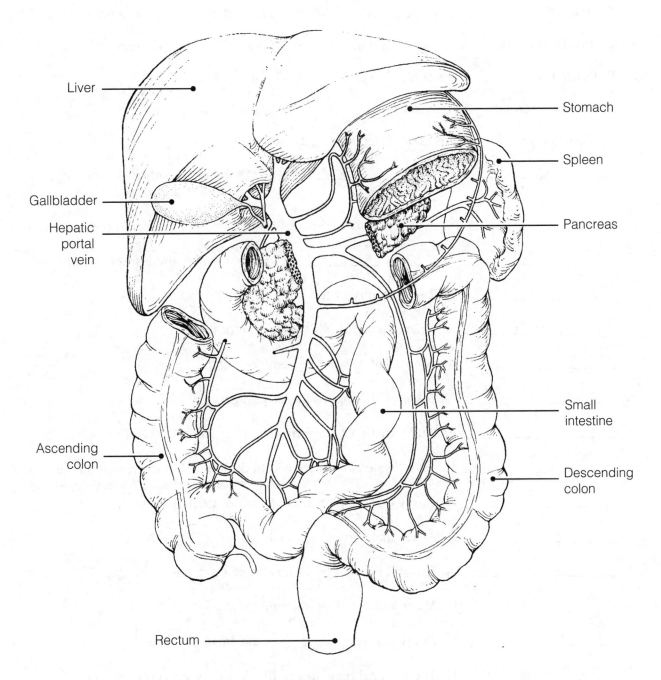

Figure 11–11

23. Using the key choices, identify the *arteries* described as follows. Place the correct term or letter response in the spaces provided.

Key Choices

A. Anterior tibial	H. Coronary	O. Intercostals	V. Renal
B. Aorta	I. Deep artery of thigh	P. Internal carotid	W. Subclavian
C. Brachial	J. Dorsalis pedis	Q. Internal iliac	X. Superior mesenteric
D. Brachiocephalic	K. External carotid	R. Peroneal	Y. Vertebral
E. Celiac trunk	L. Femoral	S. Phrenic	Z. Ulnar
F. Common carotid	M. Hepatic	T. Posterior tibial	
G. Common iliac	N. Inferior mesenteric	U. Radial	

_____ 1. _____ 2. Two arteries formed by the division of the brachiocephalic trunk

_____ 3. First artery that branches off the ascending aorta; serves the heart

_____ 4. _____ 5. Two paired arteries, serving the brain

_____ 6. Largest artery of the body

_____ 7. Arterial network on the dorsum of the foot

_____ 8. Artery that serves the posterior thigh

_____ 9. Artery that supplies the diaphragm

_____ 10. Artery that splits to form the radial and ulnar arteries

_____ 11. Artery generally auscultated to determine blood pressure in the arm

_____ 12. Artery that supplies the last half of the large intestine

_____ 13. Artery that serves the pelvis

_____ 14. External iliac becomes this artery on entering the thigh

_____ 15. Major artery serving the arm

_____ 16. Artery that supplies most of the small intestine

_____ 17. The terminal branches of the dorsal, or descending, aorta

_____ 18. Arterial trunk that has three major branches, which serve the liver, spleen, and stomach

_____ 19. Major artery, serving the tissues external to the skull

_____ 20. _____ 21. _____ 22.
Three arteries, serving the leg inferior to the knee

_____ 23. Artery generally used to feel the pulse at the wrist

_____ 24. Damage to the left semilunar valve would interfere with blood
flow into this vessel

24. Figure 11–12 illustrates the arterial circulation of the brain. Select different
colors for the following structures and use them to color the coding circles
and corresponding structures in the diagram.

◯ Basilar artery   ◯ Communicating branches

◯ Anterior cerebral arteries   ◯ Middle cerebral arteries

◯ Posterior cerebral arteries

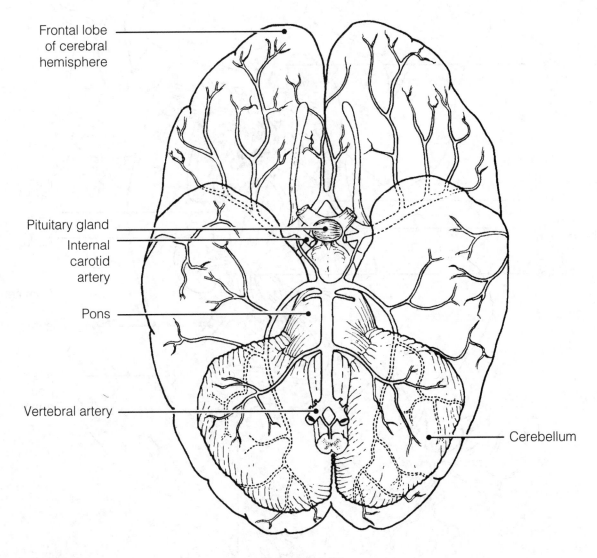

Frontal lobe
of cerebral
hemisphere

Pituitary gland

Internal
carotid
artery

Pons

Vertebral artery

Cerebellum

Figure 11–12

25. Figure 11–13 illustrates the special fetal structures listed below. Select different colors for each and use them to color coding circles and corresponding structures in the diagram.

◯ Foramen ovale   ◯ Ductus arteriosus   ◯ Ductus venosus

◯ Umbilical arteries   ◯ Umbilical cord   ◯ Umbilical vein

Figure 11–13

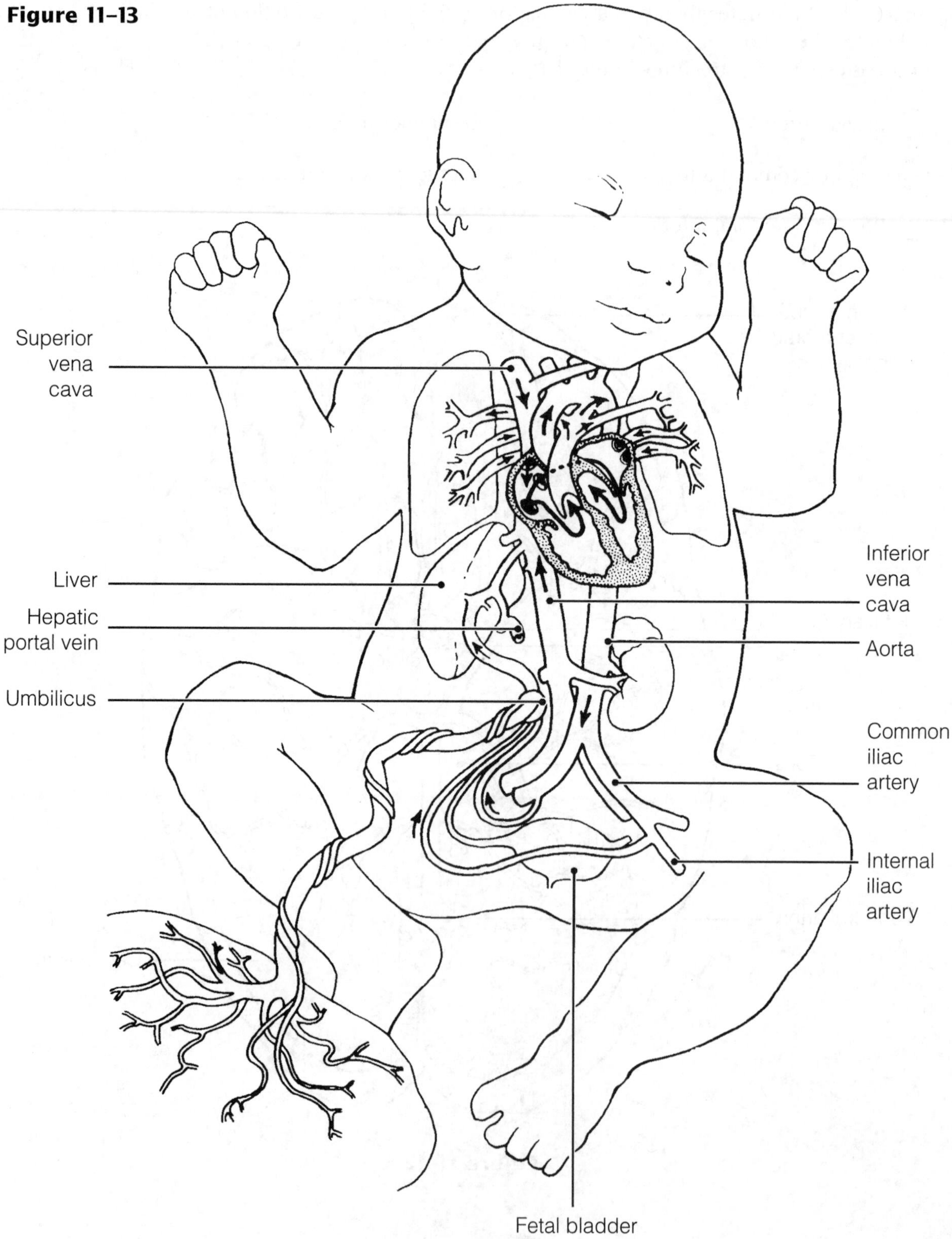

Superior vena cava

Liver

Hepatic portal vein

Umbilicus

Inferior vena cava

Aorta

Common iliac artery

Internal iliac artery

Fetal bladder

26. Eight structures unique to the special circulations of the body are described here. Identify each, using the key choices. Place the correct terms or letters in the answer blanks.

Key Choices

A. Anterior cerebral artery   E. Ductus venosus   H. Posterior cerebral artery

B. Basilar artery   F. Foramen ovale   I. Umbilical artery

C. Circle of Willis   G. Middle cerebral artery   J. Umbilical vein

D. Ductus arteriosus

_____ 1. An anastomosis that allows communication between the posterior and anterior blood supplies of the brain

_____ 2. The vessel carrying oxygen and nutrient-rich blood to the fetus from the placenta

_____ 3. The shunt that allows most fetal blood to bypass the liver

_____ 4. Two pairs of arteries, arising from the internal carotid artery

_____ 5. The posterior cerebral arteries, serving the brain, arise from here

_____ 6. Fetal shunt between the aorta and pulmonary trunk that allows the lungs to be bypassed by the blood

_____ 7. Opening in the interatrial septum that shunts fetal blood from the right to the left atrium, thus bypassing the fetal lungs

27. Briefly explain in the space provided why the lungs are largely bypassed by the circulating blood in the fetus.

_____

_____

Physiology of Circulation

28. Circle the term that does not belong in each of the following groupings.

1. High pressure   Vein   Artery   Spurting blood

2. Carotid artery   Cardiac vein   Coronary sinus   Coronary artery

3. Increased venous return   Respiratory pump   Vasodilation   Milking action of skeletal muscles

4. High blood pressure   Hemorrhage   Weak pulse   Low cardiac output

5. Resistance   Friction   Vasodilation   Vasoconstriction

29. The following section relates to understandings concerning blood pressure and pulse. Match the items given in Column B with the appropriate descriptions provided in Column A. Place the correct term or letter response in the answer blanks.

Column A

Column B

_____ 1. Expansion and recoil of an artery during heart activity

A. Over arteries

_____ 2. Pressure exerted by the blood against the blood vessel walls

B. Blood pressure

C. Cardiac output

_____ 3. _____ 4. Factors related to blood pressure

D. Constriction of arterioles

_____ 5. Event primarily responsible for peripheral resistance

E. Diastolic blood pressure

_____ 6. Blood pressure during heart contraction

F. Peripheral resistance

G. Pressure points

_____ 7. Blood pressure during heart relaxation

H. Pulse

I. Sounds of Korotkoff

_____ 8. Site where blood pressure determinations are normally made

J. Systolic blood pressure

_____ 9. Points at the body surface where the pulse may be felt

K. Over veins

_____ 10. Sounds heard over a blood vessel when the vessel is partially compressed

30. Complete the following statements about capillary functions by placing answers from the key in the answer blanks. Use terms or letters from the key.

Key Choices

A. Blood   E. Fat soluble   H. Osmotic pressure

B. Capillary clefts   F. Hydrostatic pressure   I. Vesicles

C. Diffusion   G. Interstitial fluid   J. Water soluble

D. Fenestrations

_____ 1.

_____ 2.

_____ 3.

All exchanges to and from the blood and tissue cells occur through the __(1)__. Generally speaking, substances tend to move according to their concentration gradients by the process of __(2)__. Substances that are __(3)__ pass directly through the plasma membranes of the capillary endothelial cells; other

_____ 4.   substances pass by means of or via __(4)__ , __(5)__ , or __(6)__ .
The most permeable capillaries are those exhibiting __(7)__ .

_____ 5.   Capillaries that have __(8)__ and __(9)__ tend to be leaky, and
forces acting at capillary beds cause fluid flows.

_____ 6.

_____ 7.

_____ 8.   _____ 9.

31. Indicate what effect the following factors have on blood pressure. Indicate
an increase in pressure by *I* and a decrease in pressure by *D*. Place the
correct letter response in the answer blanks.

_____ 1. Increased diameter of the arterioles   _____ 8. Physical exercise

_____ 2. Increased blood viscosity   _____ 9. Physical training

_____ 3. Increased cardiac output   _____ 10. Alcohol

_____ 4. Increased pulse rate   _____ 11. Hemorrhage

_____ 5. Anxiety, fear   _____ 12. Nicotine

_____ 6. Increased urine output   _____ 13. Arteriosclerosis

_____ 7. Sudden change in position from
reclining to standing

32. For each of the following statements that is true, insert *T* in the answer
blank. If any of the statements are false, correct the underlined term by
inserting the correct word in the answer blank.

_____ 1. Renin, released by the kidneys, causes a <u>decrease</u> in blood
pressure.

_____ 2. The decreasing efficiency of the sympathetic nervous system
vasoconstrictor functioning, due to aging, leads to a type of
hypotension called <u>sympathetic</u> hypotension.

_____ 3. Two body organs in which vasoconstriction rarely occurs are
the heart and the <u>kidneys</u>.

_____ 4. A <u>sphygmomanometer</u> is used to take the apical pulse.

_____ 5. The pulmonary circulation is a <u>high</u>-pressure circulation.

_____ 6. The fetal equivalent of (functional) lungs and liver is the <u>placenta</u>.

_____ 7. Cold has a <u>vasodilating</u> effect.

_____ 8. <u>Thrombophlebitis</u> is called the silent killer.

33. Figure 11–14 is a diagram of a capillary bed. Arrows indicate the direction of blood flow. Select five different colors and color the coding circles and their structures on the figure. Then answer the questions that follow by referring to Figure 11–14. Notice that questions 1–9 concern fluid flows at capillary beds and the forces (hydrostatic and osmotic pressures) that promote such fluid shifts.

○ Arteriole   ○ Vascular shunt   ○ Postcapillary venule

○ Precapillary sphincters   ○ True capillaries

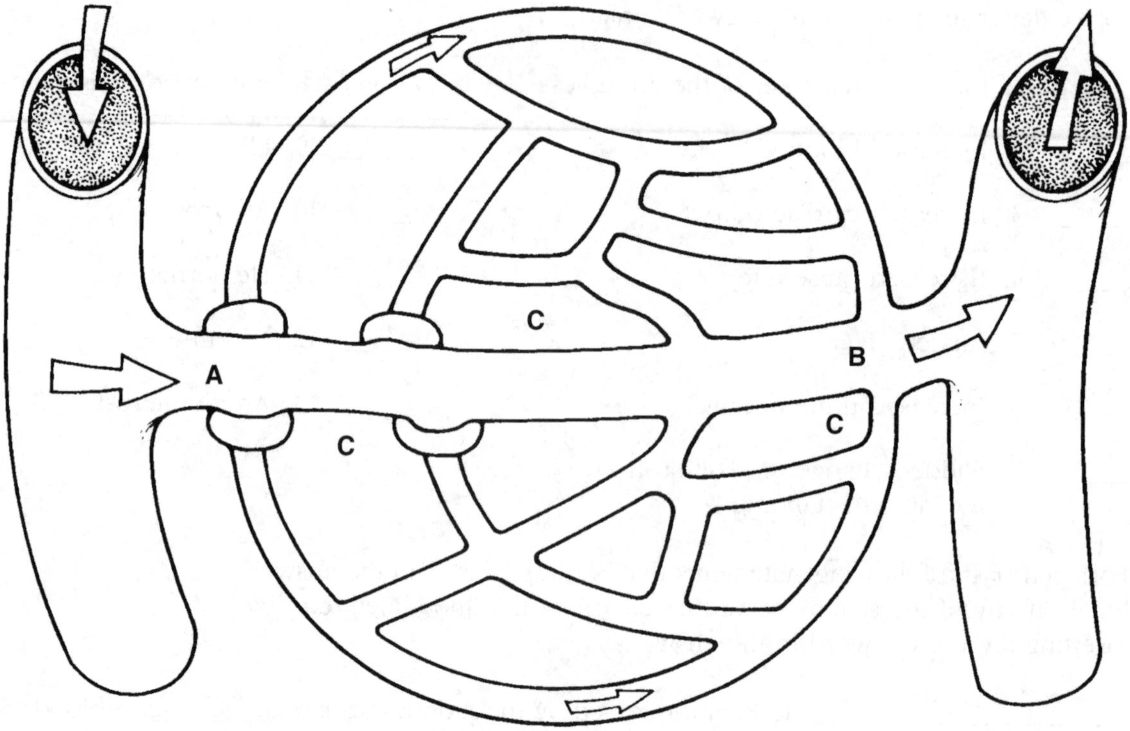

Figure 11–14

1. If the precapillary sphincters are contracted, by which route will the blood flow?

_____

2. Under normal conditions, in which area does hydrostatic pressure predominate: A, B, or C?

_____

3. Which area has the highest osmotic pressure? _____

4. Which pressure is in excess and causes fluids to move from A to C? (Be specific as to whether the force exists in the capillary or the interstitial space.)

_____

5. Which pressure causes fluid to move from A to B? _____

6. Which pressure causes fluid to move from C to B? _____

7. Which blood protein is most responsible for osmotic pressure? _____

8. Where does the greater net flow of water out of the capillary occur? _____

9. If excess fluid does not return to the capillary, where does it go? _____

_____

34. Respond to the following exercise by placing brief answers in the spaces
provided. Assume someone has been injured in an automobile accident and
is bleeding profusely. What pressure point could you compress to help stop
the bleeding from the following areas?

_____ 1. Thigh   _____ 4. Lower jaw

_____ 2. Forearm   _____ 5. Thumb

_____ 3. Calf   _____ 6. Temple

DEVELOPMENTAL ASPECTS
OF THE CARDIOVASCULAR SYSTEM

35. Complete the following statements by inserting your responses in the
answer blanks.

_____ 1.

_____ 2.

_____ 3.

_____ 4.

_____ 5.

_____ 6.

_____ 7.

_____ 8.

_____ 9.

_____ 10.

_____ 11.

The cardiovascular system forms early, and the heart
is acting as a functional pump by the __(1)__ week of
development. The ductus arteriosus and foramen ovale allow
the blood to bypass the nonfunctioning fetal __(2)__. Another
fetal structure, the __(3)__, allows most of the blood to bypass
the liver. The fetus is supplied with oxygen and nutrients via
the __(4)__, which carries blood from the __(5)__ to the
__(6)__. Metabolic wastes and carbon dioxide are removed
from the fetus in blood carried by the __(7)__. These special
bypass structures that exist to bypass the fetal lungs and liver
become __(8)__ shortly after birth. Congenital heart defects
(some resulting from the failure of the bypass structures to
close) account for half of all infant __(9)__ resulting from
congenital defects.

__(10)__ is a degenerative process that begins in youth
but may take its toll in later life by promoting a myocardial
infarct or stroke. Generally women have less of this degenerative process than men until after __(11)__, when estrogen
production ends.

→

_____ 12.

_____ 13.

_____ 14.

_____ 15.

_____ 16.

Regular __(12)__ increases the efficiency of the cardiovascular system and helps to slow the progress of __(13)__ . A vascular problem that affects many in "standing professions" is __(14)__ . In this condition, the valves become incompetent, and the veins become twisted and enlarged, particularly in the __(15)__ and __(16)__ .

INCREDIBLE JOURNEY

A Visualization Exercise for the Cardiovascular System

All about you are huge white cords, hanging limply from two flaps of endothelial tissue. . . .

36. Where necessary, complete the statements by inserting the missing word(s) in the answer blanks.

_____ 1.

_____ 2.

_____ 3.

_____ 4.

_____ 5.

_____ 6.

_____ 7.

Your journey starts in the pulmonary vein and includes a trip to part of the systemic circulation and a special circulation. You ready your equipment and prepare to be miniaturized and injected into your host.

Almost immediately after injection, you find yourself swept into a good-sized chamber, the __(1)__ . However, you do not stop in this chamber but continue to plunge downward into a larger chamber below. You land with a big splash and examine your surroundings. All about you are huge white cords, hanging limply from two flaps of endothelial tissue far above you. You report that you are sitting in the __(2)__ chamber of the heart, seeing the flaps of the __(3)__ valve above you. The valve is open, and its anchoring cords, the __(4)__ , are lax. Because this valve is open, you conclude that the heart is in the __(5)__ phase of the cardiac cycle.

Gradually you notice that the chamber walls seem to be closing in. You hear a thundering boom, and the whole chamber vibrates as the valve slams shut above you. The cords, now rigid and strained, form a cage about you, and you feel extreme external pressure. Obviously, the heart is in a full-fledged __(6)__ . Then, high above on the right, the "roof" opens, and you are forced through this __(7)__ valve. A fraction of a second later, you hear another tremendous boom that sends shock waves through the whole area. Out of the corner of your eye, you see that the valve below you is closed, and it looks rather like a pie cut into three wedges.

_____ 8.

_____ 9.

_____10.

_____11.

_____12.

_____13.

_____14.

_____15.

_____16.

_____17.

_____18.

_____19.

_____20.

_____21.

8. As you are swept along in this huge artery, the __(8)__, you pass several branch-off points but continue to careen along, straight down at a dizzying speed until you approach the __(9)__ artery, feeding the small intestine. After entering this artery and passing through successively smaller and smaller subdivisions of it, you finally reach the capillary bed of the small intestine. You watch with fascination as nutrient molecules move into the blood through the single layer of __(10)__ cells forming the capillary wall. As you move to the opposite shore of the capillary bed, you enter a venule and begin to move superiorly once again. The venules draining the small intestine combine to form the __(11)__ vein, which in turn combines with the __(12)__ vein to form the hepatic portal vein that carries you into the liver. As you enter the liver, you are amazed at the activity there. Six-sided hepatic cells, responsible for storing glucose and making blood proteins, are literally grabbing __(13)__ out of the blood as it percolates slowly past them. Protective __(14)__ cells are removing bacteria from the slowly moving blood. Leaving the liver through the __(15)__ vein, you almost immediately enter the huge __(16)__, which returns blood from the lower part of the body to the __(17)__ of the heart. From here, you move consecutively through the right chambers of the heart into the __(18)__. Soon that vessel splits and you are carried into a __(19)__ artery, which carries you to the capillary beds of the __(20)__ and then back to the left side of the heart once again. After traveling through the left side of the heart again, you leave your host when you are aspirated out of the __(21)__ artery, which extends from the aorta to the axillary artery of the armpit.

AT THE CLINIC

37. A man, en route to the hospital emergency room by ambulance, is in fibrillation. What is his cardiac output likely to be? He arrives at the emergency entrance DOA (dead on arrival). His autopsy reveals a blockage of the posterior interventricular artery. What is the cause of death?

38. Excessive vagal stimulation can be caused by severe depression. How would this be reflected in a routine physical examination?

39. Mrs. Suffriti has swollen ankles and signs of degenerating organ functions. What is a likely diagnosis?

40. A routine scan of an elderly man reveals partial occlusion of the right internal carotid artery, yet blood supply to his cerebrum is unimpaired. What are two possible causes of the occlusion? What anastomosis is maintaining blood supply to the brain and by what (probable) route(s)?

41. Mr. Abdul, a patient with a bone marrow cancer, is polycythemic. Will his blood pressure be high or low? Why?

42. After a bout with bacterial endocarditis, scar tissue often stiffens the edges of the heart valves. How would this be picked up in a routine examination?

43. Len, an elderly man, is bedridden after a hip fracture. He complains of pain in his legs, and thrombophlebitis is diagnosed. What is thrombophlebitis, and what life-threatening complication can develop?

44. Mr. Langley is telling his friend about his recent visit to his doctor for a checkup. During his story, he mentions that the ECG revealed that he had a defective mitral valve and a heart murmur. Mr. Langley apparently misunderstood some of what the doctor explained to him about the diagnostic process. What has he misunderstood?

45. A less-than-respectable news tabloid announced that "Doctors show that exercise shortens life. Life expectancy is programmed into a set number of heartbeats; the faster your heart beats, the sooner you die." Even if this "theory" were true, what is wrong with the conclusion concerning exercise?

46. Mrs. Tuney says that when she stands up after lying down in the afternoon she gets very dizzy. Her husband grumbles, "It's because she keeps the danged house too warm." He's right (in this particular case). Explain how this might cause her dizziness.

47. Mary Anne is taking a calcium channel blocking drug. What effect on her stroke volume (SV) would you expect this medication to have?

48. You are conducting animal research at Hampshire University. You have just chemically stimulated the ACh receptors on the rat's heart. How would you expect this to affect that heart's stroke volume?

49. How does the pulsating blood pressure in the largest arteries relate to their structures?

☑ THE FINALE: MULTIPLE CHOICE

50. Select the best answer or answers from the choices given.

1. The innermost layer of the pericardial sac is the:

   A. epicardium

   B. fibrous pericardium

   C. parietal layer of the serous pericardium

   D. visceral layer of the serous pericardium

2. The thickest layer of the heart wall is:

   A. endocardium   C. epicardium

   B. myocardium   D. fibrous pericardium

3. Atrioventricular valves are held closed by:

   A. papillary muscles

   B. trabeculae carneae

   C. pectinate muscles

   D. chordae tendineae

4. The fibrous skeleton of the heart:

   A. supports valves

   B. anchors vessels

   C. provides electrical insulation to separate the atrial mass from the ventricular mass

   D. anchors cardiac muscle fibers

5. Freshly oxygenated blood is first received by the:

   A. right ventricle   C. right atrium

   B. left ventricle   D. left atrium

6. Atrial repolarization coincides in time with the:

   A. P wave   C. QRS wave

   B. T wave   D. P-Q interval

7. Soon after the onset of ventricular systole the:

   A. AV valves close

   B. semilunar valves open

   C. first heart sound is heard

   D. aortic pressure increases

8. Which of the following depolarizes next after the AV node?

   A. Atrial myocardium

   B. Ventricular myocardium

   C. Bundle branches

   D. Purkinje fibers

9. Which of the regulatory chemicals listed involve or target the kidneys?

   A. Angiotensin   C. ADH

   B. Aldosterone   D. ANP

10. Cardiovascular conditioning results in:

    A. ventricular hypertrophy

    B. bradycardia

    C. increase in SV

    D. increase in CO

11. Which of the following is (are) part of the tunica intima?

    A. Simple squamous epithelium

    B. Basement membrane

    C. Loose connective tissue

    D. Smooth muscle

12. In comparing a parallel artery and vein, you would find that:

    A. the artery wall is thicker

    B. the artery diameter is greater

    C. the artery lumen is smaller

    D. the artery endothelium is thicker

13. Fenestrated capillaries occur in the:

    A. liver

    B. kidney

    C. cerebrum

    D. intestinal mucosa

14. Which of the following is (are) part of a capillary bed?

    A. Precapillary sphincter

    B. Vascular shunt

    C. True capillaries

    D. Terminal arteriole

15. Which of the following can function as a blood reservoir?

    A. Brachiocephalic artery

    B. Cerebral capillaries

    C. Dural sinuses

    D. Inferior vena cava

16. An increase in which of the following results in increased filtration from capillaries to the interstitial space?

    A. Capillary hydrostatic pressure

    B. Interstitial fluid hydrostatic pressure

    C. Capillary osmotic pressure

    D. Duration of precapillary sphincter contraction

17. Vessels involved in the circulatory pathway to and from the brain are the:

    A. brachiocephalic artery

    B. subclavian artery

    C. internal jugular vein

    D. internal carotid artery

18. Which of the following are associated with aging?

    A. Increasing blood pressure

    B. Weakening of venous valves

    C. Arteriosclerosis

    D. Stenosis of the ductus arteriosus

19. Which layer of the artery wall thickens most in atherosclerosis?

    A. Tunica media

    B. Tunica intima

    C. Tunica adventitia

    D. Tunica externa

20. Based on the vessels named pulmonary trunk, thyrocervical trunk, and celiac trunk, the term *trunk* must refer to:

    A. a vessel in the heart wall

    B. a vein

    C. a capillary

    D. a large artery from which other arteries branch

21. Which of these vessels is bilaterally symmetrical (i.e., one vessel of the pair occurs on each side of the body)?

    A. Internal carotid artery

    B. Brachiocephalic artery

    C. Azygos vein

    D. Superior mesenteric vein

22. A stroke that occludes a posterior cerebral artery will most likely affect:

    A. hearing

    B. vision

    C. smell

    D. higher thought processes

23. Tracing the drainage of the *superficial* venous blood from the leg, we find that blood enters the great saphenous vein, femoral vein, inferior vena cava, and right atrium. Which veins are missing from that sequence?

    A. Coronary sinus and superior vena cava

    B. Posterior tibial and popliteal

    C. Fibular (peroneal) and popliteal

    D. External and common iliacs

24. Tracing the drainage of venous blood from the small intestine, we find that blood enters the superior mesenteric vein, hepatic vein, inferior vena cava, and right atrium. Which vessels are missing from that sequence?

    A. Coronary sinus and left atrium

    B. Celiac and common hepatic veins

    C. Internal and common iliac veins

    D. Hepatic portal vein and liver sinusoids

12 THE LYMPHATIC SYSTEM AND BODY DEFENSES

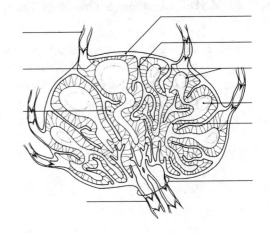

The lymphatic system, with its many lymphoid organs and vessels derived from veins of the cardiovascular system, is a rather strange system. Although both types of organs help to maintain homeostasis, these two elements of the lymphatic system have substantially different roles. The lymphatic vessels help keep the cardiovascular system functional by maintaining blood volume. The lymphoid organs help defend the body from pathogens by providing operating sites for phagocytes and cells of the immune system.

The immune system, which serves as the body's *specific defense system*, is a unique functional system made up of billions of individual cells, most of which are lymphocytes. The sole function of this defensive system is to protect the body against an incredible array of pathogens. In general, these "enemies" fall into three major camps: (1) microorganisms (bacteria, viruses, and fungi) that have gained entry into the body, (2) foreign tissue cells that have been transplanted (or, in the case of red blood cells, infused) into the body, and (3) the body's own cells that have become cancerous. The result of the immune system's activities is immunity, or specific resistance to disease.

The body is also protected by a number of nonspecific defenses provided by intact surface membranes such as skin and mucosae, and by a variety of cells and chemicals that can quickly mount an attack against foreign substances. The specific and nonspecific defenses enhance each other's effectiveness.

Chapter 12 tests your understanding of the functional roles of the various lymphatic system elements and both the nonspecific and specific body defenses.

THE LYMPHATIC SYSTEM

Lymphatic Vessels

1. Complete the following statements by writing the missing terms in the answer blanks.

_____ 1.

_____ 2.

_____ 3.

_____ 4.

_____ 5.

_____ 6.

Together the cardiovascular and lymphatic systems make up the circulatory system. Although the cardiovascular system has a pump (the heart) and arteries, veins, and capillaries, the lymphatic system lacks two of these structures: the __(1)__ and __(2)__. Like the __(3)__ of the cardiovascular system, the vessels of the lymphatic system are equipped with __(4)__ to prevent backflow. The lymphatic vessels act primarily to pick up leaked fluid, now called __(5)__, and return it to the bloodstream. About __(6)__ of fluid is returned every 24 hours.

2. Figure 12–1 provides an overview of the lymphatic vessels. In part A, the relationship between lymphatic vessels and the blood vessels of the cardiovascular system is depicted schematically. Part B shows the different types of lymphatic vessels in a simple way. First, color-code and color the following structures in Figure 12–1.

◯ Heart   ◯ Veins   ◯ Lymphatic vessels/lymph node

◯ Arteries   ◯ Blood capillaries   ◯ Loose connective tissue around blood and lymph capillaries

Then, identify by labeling these specific structures in part B:

A. Lymph capillaries   C. Lymphatic collecting vessels   E. Valves

B. Lymph duct   D. Lymph node   F. Vein

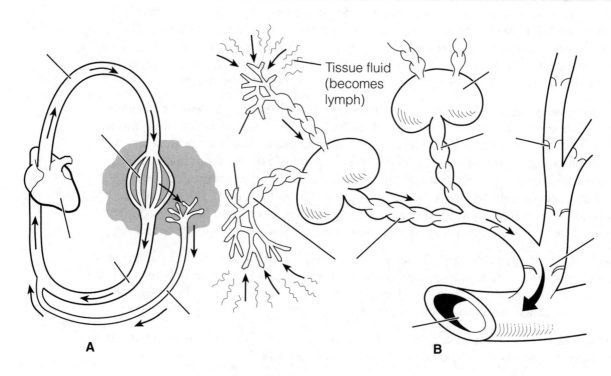

Figure 12–1

3. Circle the term that does not belong in each of the following groupings.

1. Blood capillary   Lymph capillary   Blind-ended   Permeable to proteins

2. Edema   Blockage of lymphatics   Elephantiasis   Inflammation   Abundant supply of lymphatics

3. Skeletal muscle pump   Flow of lymph   Respiratory pump   High-pressure gradient   Action of smooth muscle cells in walls of lymph vessels

4. Minivalves   Endothelial cell overlap   Impermeable   Lymphatic capillaries

Lymph Nodes and Other Lymphoid Organs

4. Match the terms in Column B with the appropriate descriptions in Column A. More than one choice may apply in some cases.

Column A

_____ 1. A blood reservoir

_____ 2. Monitor composition of lymph

_____ 3. Located between the lungs at the base of the throat

_____ 4. Collectively called MALT

_____ 5. Prevents bacteria from breaching the intestinal wall

Column B

A. Lymph nodes

B. Peyer's patches

C. Spleen

D. Thymus

E. Tonsils

5. Figure 12–2 depicts several different lymphoid organs. Label all lymphoid organs indicated by a leader line and add labels as necessary to identify the sites where the axillary, cervical, and inguinal lymph nodes would be located. Color the lymphoid organs as you like, and then shade in light green the portion of the body that is drained by the right lymphatic duct.

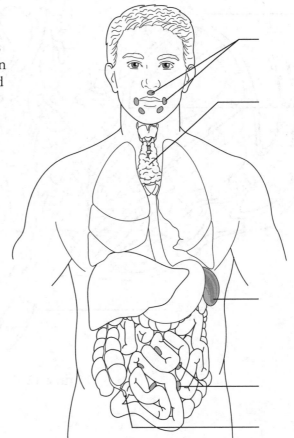

Figure 12–2

6. Figure 12–3 is a diagram of a lymph node. First, using the terms with color-coding circles, label all structures on the diagram that have leader lines. Color those structures as well. Then, add arrows to the diagram to show the direction of lymph flow through the organ. Circle the region that would approximately correspond to the medulla of the organ. Finally, answer the questions that follow.

◯ Germinal centers of follicles

◯ Cortex (other than germinal centers)

◯ Medullary cords

◯ Capsule and trabeculae

◯ Hilum

◯ Afferent lymphatics

◯ Efferent lymphatics

◯ Sinuses (subcapsular and medullary)

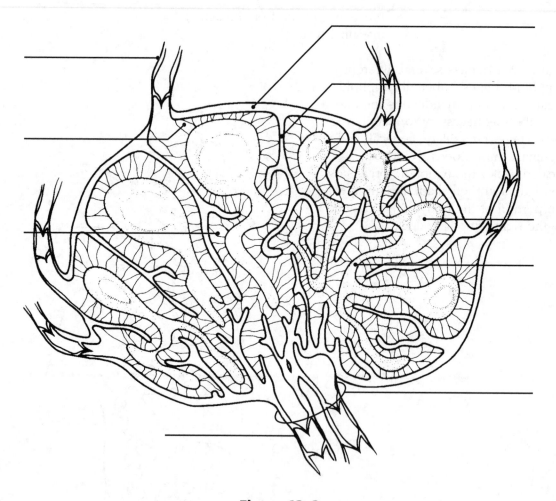

Figure 12–3

1. Which cell type is found in greatest abundance in the germinal centers?

   _____

2. What is the function of their daughter cells, the plasma cells?

   _____

3. What is the major cell type in cortical areas other than the germinal centers?

   _____

4. The third important cell type in lymph nodes (usually found clustered around

   the medullary sinuses) are the _____ .

   These cells act as _____ .

5. Of what importance is the fact that there are fewer efferent than afferent
   lymphatics associated with lymph nodes?

   _____

   _____

   _____

6. What structures ensure the one-way flow of lymph through the node?

   _____

7. The largest collections of lymph nodes are found in what three body regions?

   _____

7. Match the terms in Column B with the appropriate descriptions in Column A.
   More than one choice may apply in some cases.

Column A	Column B
_____ 1. The largest lymphatic organ	A. Lymph nodes
_____ 2. Filter lymph	B. Peyer's patches
_____ 3. Particularly large and important during youth; helps to program T cells of the immune system	C. Spleen
	D. Thymus
_____ 4. Found in the wall of the GI tract	E. Tonsils
_____ 5. Removes aged and defective red blood cells	

BODY DEFENSES
Nonspecific (Innate) Body Defenses

8. The three major elements of the body's nonspecific defense system are: the

(1) _____ , consisting of the skin and _____ ;

defensive cells, such as (2) _____ and phagocytes; and a whole

deluge of (3) _____ .

9. Indicate the sites of activity or the secretions of the nonspecific defenses by writing the correct terms in the answer blanks.

1. Lysozyme is found in the body secretions called _____ and _____ .

2. Fluids with an acid pH are found in the _____ and _____ .

3. Sebum is a product of the _____ glands and acts at the surface

   of the _____ .

4. Mucus is produced by mucus-secreting glands found in the respiratory and

   _____ system mucosae.

10. Figure 12–4 diagrams the events involved in the inflammatory response. Assume the following events have already occurred: tissue injury and invasion of microbes, and release of inflammatory chemicals by mast cells. Each subsequent event is represented by a square with one or more arrows. From the list below, write the correct number in each event square in the figure. Then, color-code and color the structures that appear below the numbered list.

1. WBCs are drawn to the injured area by the release of inflammatory chemicals.

2. Tissue repair occurs.

3. Local blood vessels dilate, and the capillaries become engorged with blood.

4. Phagocytosis of microbes occurs.

5. Fluid containing clotting proteins is lost from the bloodstream and enters the injured tissue area.

6. Diapedesis occurs.

◯ Monocyte   ◯ Neutrophil(s)   ◯ Endothelium of capillary

◯ Epithelium   ◯ Macrophage   ◯ Microorganisms

◯ Erythrocyte(s)   ◯ Subcutaneous tissue   ◯ Fibrous repair tissue

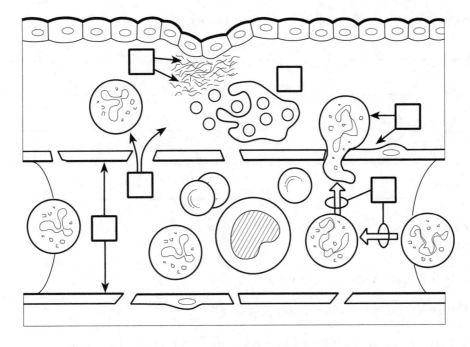

Figure 12–4

11. Circle the term that does not belong in each of the following groupings.

1. Redness   Pain   Swelling   Itching   Heat

2. Neutrophils   Macrophages   Phagocytes   Natural killer cells

3. Inflammatory chemicals   Histamine   Kinins   Interferon

4. Intact skin   Intact mucosae   Inflammation   First line of defense

5. Interferons   Antiviral   Antibacterial   Proteins

12. Match the terms in Column B with the descriptions of the nonspecific defenses of the body in Column A. More than one choice may apply.

Column A	Column B
_____ 1. Have antimicrobial activity	A. Acids
_____ 2. Provide mechanical barriers	B. Lysozyme
_____ 3. Provide chemical barriers	C. Mucosae
_____ 4. Entraps microorganisms entering the respiratory passages	D. Mucus
	E. Protein-digesting enzymes
_____ 5. Part of the first line of defense	F. Sebum
	G. Skin

13. Describe the protective role of cilia in the respiratory tract. _____

_____

_____

14. Define *phagocytosis*. _____

_____

_____

15. Check (✓) all phrases that correctly describe the role of fever in body protection.

_____ 1. Is a normal response to pyrogens

_____ 2. Protects by denaturing tissue proteins

_____ 3. Reduces the availability of iron and zinc required for bacterial proliferation

_____ 4. Increases metabolic rate

16. Match the terms in Column B with the descriptions in Column A concerning events of the inflammatory response.

Column A	Column B
_____ 1. Accounts for redness and heat in an inflamed area	A. Chemotaxis
_____ 2. Inflammatory chemical released by injured cells	B. Diapedesis
_____ 3. Promote release of white blood cells from the bone marrow	C. Edema
_____ 4. Cellular migration directed by a chemical gradient	D. Fibrin mesh
_____ 5. Results from accumulation of fluid leaked from the bloodstream	E. Histamine
_____ 6. Phagocytic offspring of monocytes	F. Increased blood flow to an area
_____ 7. Leukocytes pass through the wall of a capillary	G. Inflammatory chemicals
_____ 8. First phagocytes to migrate into the injured area	H. Macrophages
_____ 9. Walls off the area of injury	I. Neutrophils

17. Complete the following description of the activation and activity of complement by writing the missing terms in the answer blanks.

_____ 1.

_____ 2.

_____ 3.

_____ 4.

_____ 5.

_____ 6.

Complement is a system of plasma __(1)__ that circulate in the blood in an inactive form. Complement is __(2)__ when it becomes attached to the surface of foreign cells (bacteria, fungi, red blood cells). One result of this complement fixation is that __(3)__ appear in the membrane of the foreign cell. This allows __(4)__ to rush in, which causes __(5)__ of the foreign cell. Some of the chemicals released during complement fixation enhance phagocytosis. This is called __(6)__. Others amplify the inflammatory response.

18. Describe the event that leads to the synthesis of interferon and the result of its synthesis.

_____

_____

Specific (Adaptive) Body Defenses: The Immune System

Antigens

19. What are three important characteristics of the adaptive immune

response? _____, _____,

and _____.

20. Complete the following statements relating to antigens by writing the missing terms in the answer blanks.

_____ 1.

_____ 2.

_____ 3.

_____ 4.

Antigens are substances capable of mobilizing the __(1)__. Of all the foreign molecules that act as complete antigens, __(2)__ are the most potent. Small molecules are not usually antigenic, but when they bind to self-cell surface proteins they may act as __(3)__, and then the complex is recognized as foreign, or __(4)__.

Cells of the Immune System: An Overview

21. Using the key choices, select the term that correctly completes each statement. Insert the appropriate term or letter in the answer blanks.

Key Choices

A. Antigen(s)   D. Cellular immunity   G. Lymph nodes

B. B cells   E. Humoral immunity   H. Macrophages

C. Blood   F. Lymph   I. T cells

_____ 1.

_____ 2.

_____ 3.

_____ 4.

_____ 5.

_____ 6.

_____ 7.

_____ 8.

_____ 9.

Immunity is resistance to disease resulting from the presence of foreign substances or __(1)__ in the body. When this resistance is provided by antibodies released to body fluids, the immunity is called __(2)__. When living cells provide the protection, the immunity is referred to as __(3)__. The major actors in the immune response are two lymphocyte populations, the __(4)__ and the __(5)__. Phagocytic cells that act as accessory cells in the immune response are the __(6)__. Because pathogens are likely to use both __(7)__ and __(8)__ as a means of getting around the body, __(9)__ and other lymphatic tissues (which house the immune cells) are in an excellent position to detect their presence.

22. A schematic of the life cycle of the lymphocytes involved in immunity is shown in Figure 12–5. First, select different colors for the areas listed below and use them to color the coding circles and the corresponding regions in the figure. If there is overlap, use stripes of a second color to indicate the second identification. Then respond to the statements following the figure, which relate to the two-phase differentiation process of B and T cells.

◯ Area where immature lymphocytes arise

◯ Area seeded by immunocompetent B and T cells

◯ Area where T cells become immunocompetent

◯ Area where the antigen challenge and clonal selection are likely to occur

◯ Area where B cells become immunocompetent

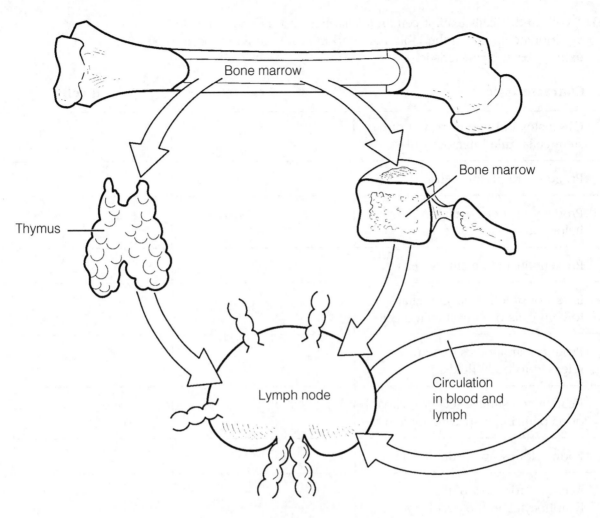

Figure 12–5

1. What signifies that a lymphocyte has become immunocompetent?

   _____

2. During what period of life does immunocompetence develop?

   _____

3. What determines which antigen a particular T or B cell will be able to recognize?
   A. its genes or B. "its" antigen

4. What triggers the process of clonal selection in a T or B cell?
   A. its genes or B. binding to "its" antigen

5. During development of immunocompetence, the ability to tolerate _____
   must also occur if the immune system is to function normally.

23. T cells and B cells exhibit certain similarities and differences. Check (✓) the appropriate spaces in the table below to indicate the lymphocyte type that exhibits each characteristic.

Characteristic	T cell	B cell
Originates in bone marrow from stem cells called hemocytoblasts		
Progeny are plasma cells		
Progeny include regulatory, helper, and cytotoxic cells		
Progeny include memory cells		
Is responsible for directly attacking foreign cells or virus-infected cells		
Produces antibodies that are released to body fluids		
Bears a cell-surface receptor capable of recognizing a specific antigen		
Forms clones upon stimulation		
Accounts for most of the lymphocytes in the circulation		

24. Circle the term that does not belong in each of the following groupings.

1. Antibodies   Gamma globulin   Cytokines   Immunoglobulins

2. Protein   Complete antigen   Nucleic acid   Hapten

3. Lymph nodes   Liver   Spleen   Thymus   Bone marrow

Humoral (Antibody-Mediated) Immune Response

25. The basic structure of an antibody molecule is diagrammed in Figure 12–6. Select different colors, and color in the coding circles below and the corresponding areas on the diagram.

◯ heavy chains   ◯ light chains

Add labels to the diagram to correctly identify the type of bonds holding the polypeptide chains together. Also label the constant (C) and variable (V) regions of the antibody, and add "polka dots" to the variable portions. Then, answer the two questions following the figure.

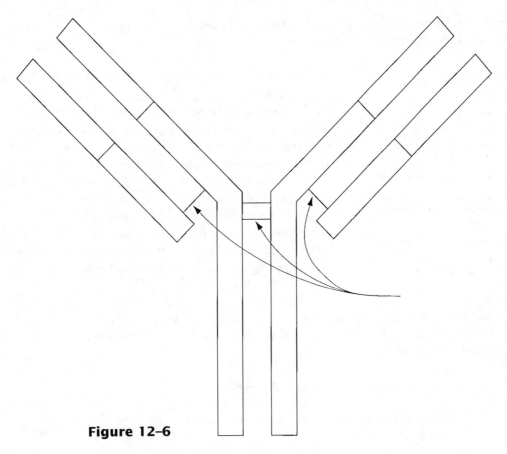

Figure 12–6

1. Which portion of the antibody—V or C—is its antigen-binding site?

_____

2. Which portion acts to determine antibody class and specific function?

_____

26. Match the antibody classes in Column B to their descriptions in Column A. Place the correct term(s) or letter response(s) in the answer blanks.

Column A	Column B
_____ 1. Bound to the surface of a B cell	A. IgA
_____ 2. Crosses the placenta	B. IgD
_____ 3. The first antibody released during the primary response	C. IgE
_____ 4. Fixes complement (two classes)	D. IgG
_____ 5. Is a pentamer	E. IgM
_____ 6. The most abundant antibody found in blood plasma and the chief antibody released during secondary responses	
_____ 7. Binds to the surface of mast cells and mediates an allergic response	
_____ 8. Predominant antibody found in mucus, saliva, and tears	

27. Complete the following descriptions of antibody function by writing the missing terms in the answer blanks.

_____ 1.

_____ 2.

_____ 3.

_____ 4.

_____ 5.

_____ 6.

_____ 7.

Antibodies can inactivate antigens in various ways, depending on the nature of the __(1)__. __(2)__ is the chief ammunition used against cellular antigens such as bacteria and mismatched red blood cells. The binding of antibodies to sites on bacterial exotoxins or viruses that can cause cell injury is called __(3)__. The cross-linking of cellular antigens into large lattices by antibodies is called __(4)__; Ig __(5)__, with its 10 antigen binding sites, is particularly efficient in this mechanism. When molecules are cross-linked into lattices by antibodies, the mechanism is more properly called __(6)__. In virtually all these cases, the protective mechanism mounted by the antibodies serves to disarm and/or immobilize the antigens until they can be disposed of by __(7)__.

28. Determine whether each of the following situations provides, or is an example of, active or passive immunity. If passive, write *P* in the blank; if active, write *A* in the blank.

_____ 1. An individual receives Sabin polio vaccine

_____ 2. Antibodies migrate through a pregnant woman's placenta into the vascular system of her fetus

_____ 3. A student nurse receives an injection of gamma globulin (containing antibodies to the hepatitis virus) after she has been exposed to viral hepatitis

_____ 4. "Borrowed" immunity

_____ 5. Immunological memory is provided

_____ 6. An individual suffers through chickenpox

29. There are several important differences between primary and secondary immune response(s). If the following statements best describe a primary response, write *P* in the blank; if a secondary response, write *S* in the blank.

_____ 1. The initial response to an antigen; gearing-up stage

_____ 2. A lag period of several days occurs before antibodies specific to the antigen appear in the bloodstream

_____ 3. Antibody levels increase rapidly and remain high for an extended period

_____ 4. Immunological memory is established

_____ 5. The second, third, and subsequent responses to the same antigen

Cellular (Cell-Mediated) Immune Response

30. Several populations of T cells exist. Match the terms in Column B to the descriptions in Column A. Place the correct term or letter response in the answer blanks.

Column A

_____ 1. Binds with and releases chemicals that activate B cells, T cells, and macrophages

_____ 2. Activated by recognizing both its antigen and a self-protein presented on the surface of a macrophage

_____ 3. Turns off the immune response when the "enemy" has been routed

_____ 4. Directly attacks and lyses cellular pathogens

_____ 5. Initiates secondary response to a recognized antigen

Column B

A. Helper T cell

B. Cytotoxic T cell

C. Regulatory T cell

D. Memory T cell

31. Using the key choices, select the terms that correspond to the descriptions of substances or events by inserting the appropriate term or letter in the answer blanks.

Key Choices

A. Anaphylactic shock   D. Complement   F. Inflammation

B. Antibodies   E. Cytokines   G. Interferon

C. Chemotaxis factors

_____ 1. A protein released by macrophages and activated T cells that helps to protect other body cells from viral multiplication

_____ 2. Any types of molecules that attract neutrophils and other protective cells into a region where an immune response is ongoing

_____ 3. Proteins released by plasma cells that mark antigens for destruction by phagocytes or complement

_____ 4. A consequence of the release of histamine and of complement activation

_____ 5. C and G are examples of this class of molecules

_____ 6. A group of plasma proteins that amplifies the immune response by causing lysis of cellular pathogens once it has been "fixed" to their surface

_____ 7. Class of chemicals released by macrophages

32. Organ transplants are often unsuccessful because self-proteins vary in different individuals. However, chances of success increase if certain important procedures are followed. The following questions refer to this important area of clinical medicine.

1. Assuming that autografts and isografts are not possible, what is the next most successful

   graft type and what is its source? _____

2. What two cell types are important in rejection phenomena?

   _____

3. Why are immunosuppressive drugs (or therapy) provided after transplant surgery, and what is the major shortcoming of this therapy?

   _____

   _____

   _____

33. Figure 12–7 is a flowchart of the immune response that tests your understanding of the interrelationships of that process. Several terms have been omitted from this schematic. First, complete the figure by inserting appropriate terms from the key choices below. (Note that oval blanks indicate that the required term identifies a cell type, and rectangular blanks represent the names of chemical molecules. Also note that solid lines represent stimulatory or enhancing effects, whereas broken lines indicate inhibition.) Then color the coding circles and the corresponding ovals, indicating the cell types identified.

Key Choices

Cell types:

◯ B cell

◯ Helper T cell

◯ Cytotoxic T cell

◯ Macrophage

◯ Memory B cell

◯ Memory T cell

◯ Neutrophils

◯ Plasma cell

Molecules:

Antibodies

Chemotactic factors

Complement

Cytokines

Interferon

Perforin

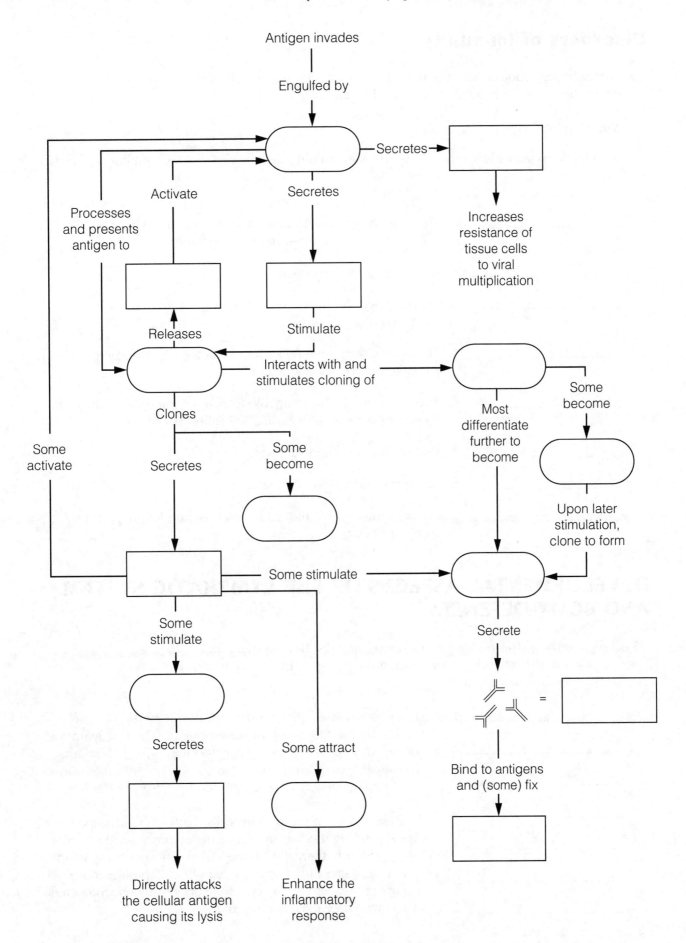

Figure 12–7

Disorders of Immunity

34. Using the key choices, identify the type of immunity disorder described. Insert the appropriate term or letter in the answer blank.

Key Choices

A. Allergy/Hypersensitivity   B. Autoimmune disease   C. Immunodeficiency

_____ 1. AIDS and SCID

_____ 2. The immune system mounts an extraordinarily vigorous response to an otherwise harmless antigen

_____ 3. A hypersensitivity reaction

_____ 4. Occurs when the production or activity of immune cells or complement is abnormal

_____ 5. The body's own immune system produces the disorder; a breakdown of self-tolerance

_____ 6. Affected individuals are unable to combat infections that would present no problem for normally healthy people

_____ 7. Multiple sclerosis and rheumatic fever

_____ 8. Hay fever and contact dermatitis

_____ 9. Typical symptoms of the acute response are tearing, a runny nose, and itching skin

DEVELOPMENTAL ASPECTS OF THE LYMPHATIC SYSTEM AND BODY DEFENSES

35. Complete the following statements concerning the development and operation of the immune system during the life span by inserting your answers in the answer blanks.

_____ 1.

_____ 2.

_____ 3.

_____ 4.

_____ 5.

_____ 6.

_____ 7.

Lymphatic vessels that "bud" from developing __(1)__ are visible by the fifth week of development. The first lymphoid organs to appear are the __(2)__ and the __(3)__. Most other lymphoid organs are poorly formed before birth; their development is believed to be controlled by __(4)__ hormones.

The earliest lymphocyte stem cells that can be identified appear during the first month of development in the fetal __(5)__. Shortly thereafter, bone marrow becomes the lymphocyte origin site; but after birth, lymphocyte proliferation occurs in the __(6)__. The development of immunocompetence has usually been accomplished by __(7)__.

_____ 8.   During old age, the effectiveness of the immune system __(8)__,
                              and elders are more at risk for __(9)__, __(10)__, and __(11)__. Part
_____ 9.   of the declining defenses may reflect the fact that __(12)__ anti-
                              bodies are unable to get to the mucosal surfaces where they
_____10.   carry out their normal protective role.

_____11.

_____12.

INCREDIBLE JOURNEY

A Visualization Exercise for the Immune System

*Something quite enormous and looking much like an octopus
is nearly blocking the narrow tunnel just ahead.*

36. Where necessary, complete statements by inserting the missing word(s) in
the answer blanks.

_____ 1.   For this journey, you are equipped with scuba gear before
                              you are miniaturized and injected into one of your host's lym-
_____ 2.   phatic vessels. He has been suffering with a red, raw "strep
                              throat" and has swollen cervical lymph nodes. Your assign-
_____ 3.   ment is to travel into a cervical lymph node and observe the
                              activities going on there that reveal that your host's immune
_____ 4.   system is doing its best to combat the infection.

_____ 5.   On injection, you enter the lymph with a "WHOOSH" and
                              then bob gently in the warm yellow fluid. As you travel
along, you see what seem to be thousands of spherical bacteria and a few large globular __(1)__
molecules that, no doubt, have been picked up by the tiny lymphatic capillaries. Shortly thereafter,
a large dark mass, shaped like a kidney bean, looms just ahead. This has to be a __(2)__, you conclude, and you dig in your wet suit pocket to find the waterproof pen and recording tablet.

As you enter the gloomy mass, the lymphatic stream becomes shallow and begins to flow
sluggishly. So that you can explore this little organ fully, you haul yourself to your feet and begin
to wade through the slowly moving stream. On each bank you see a huge ball of cells that have
large nuclei and such a scant amount of cytoplasm that you can barely make it out. You write,
"Sighted the spherical germinal centers composed of __(3)__." As you again study one of the cell
masses, you spot one cell that looks quite different and reminds you of a nest of angry hornets
because it is furiously spewing out what seems to be a horde of tiny Y-shaped "bees." "Ah ha,"
you think, "another valuable piece of information." You record, "Spotted a __(4)__ making and
releasing __(5)__."

→

_____ 6.

_____ 7.

_____ 8.

_____ 9.

_____ 10.

_____ 11.

That done, you turn your attention to scanning the rest of the landscape. Suddenly you let out an involuntary yelp. Something quite enormous and looking much like an octopus is nearly blocking the narrow tunnel just ahead. Your mind whirls as it tries to figure out the nature of this cellular "beast" that appears to be guarding the channel. Then it hits you—this has to be a __(6)__ on the alert for foreign invaders (more properly called __(7)__), which it "eats" when it catches them. The giant cell roars, "Halt, stranger, and be recognized," and you dig frantically in your pocket for your identification pass. As you drift toward the huge cell, you hold the pass in front of you, hands trembling because you know this cell could liquefy you as quick as the blink of an eye. Again the cell bellows at you, "Is this some kind of a security check? I'm on the job, as you can see!" Frantically you shake your head "NO," and the cell lifts one long tentacle and allows you to pass. As you squeeze by, the cell says, "Being inside, I've never seen my body's outside. I must say, humans are a rather strange-looking lot!" Still shaking, you decide that you are in no mood for a chat and hurry along to put some distance between yourself and this guard cell.

Immediately ahead are what appear to be hundreds of the same type of cell sitting on every ledge and in every nook and cranny. Some are busily snagging and engulfing unfortunate strep bacteria that float too close. The slurping sound is nearly deafening. Then something grabs your attention: The surface of one of these cells is becoming dotted with some of the same donut-shaped chemicals that you see on the strep bacteria membranes; a round cell, similar, but not identical, to those you earlier saw in the germinal centers, is starting to bind to one of these "doorknobs." You smile smugly because you know you have properly identified the octopus-like cells. You then record your observations as follows: "Cells like the giant cell just identified act as __(8)__. I have just observed one in this role during its interaction with a helper __(9)__ cell."

You decide to linger a bit to see if the round cell becomes activated. You lean against the tunnel walls and watch quietly, but your wait is brief. Within minutes, the cell that was binding to the octopus-like cell begins to divide, and then its daughter cells divide again and again at a head-spinning pace. You write, "I have just witnessed the formation of a __(10)__ of like cells." Most of the daughter cells enter the lymph stream, but a few of them settle back and seem to go into a light sleep. You decide that the "napping cells" don't have any role to play in helping get rid of your host's present strep infection but instead will provide for __(11)__ and become active at a later date.

You glance at your watch and wince as you realize that it is already 5 minutes past the time for your retrieval. You have already concluded that this is a dangerous place for those who don't "belong" and are far from sure about how long your pass is good, so you swim hurriedly from the organ into the lymphatic stream to reach your pickup spot.

AT THE CLINIC

37. A young man is rushed to the emergency room after fainting. His blood pressure is alarmingly low, and his companion reports the man collapsed shortly after being stung by a wasp. What has caused his hypotension? What treatment will be given immediately?

38. Patty Hourihan is a strict environmentalist and a new mother. Although she is very much against using disposable diapers, she is frustrated by the fact that her infant breaks out in a diaper rash when she uses cloth diapers. Considering that new cloth diapers do not cause the rash, but washed ones do, what do you think the problem is?

39. James, a 36-year-old engineer, appeared at the clinic in an extremely debilitated condition. He had purple-brown lesions on his skin and a persistent cough. A physical examination revealed swollen lymph nodes. Laboratory tests revealed a low lymphocyte count. Information taken during the personal history revealed that James is homosexual. The skin lesions proved to be evidence of Kaposi's sarcoma. What is James's problem?

40. About 6 months after an automobile accident in which her neck was severely lacerated, a young woman comes to the clinic for a routine checkup. Visual examination shows a slight swelling just inferior to her larynx; her skin is dry and her face is puffy. When questioned, the woman reports that she fatigues easily, has been gaining weight, and her hair is falling out. What do you think is wrong?

41. Young Joe Chang went sledding, and the runner of a sled hit him in the left side and ruptured his spleen. Joe almost died because he did not get to the hospital fast enough. Upon arrival, a splenectomy was performed. What, would you guess, is the immediate danger of spleen rupture? Will Joe require a transplant for spleen replacement?

42. Use of birth control pills decreases the acidity of the vaginal tract. Why might this increase the incidence of vaginal infection (vaginitis)?

43. After surgery to remove lymphatic vessels associated with the removal of a melanoma, what condition can be expected relative to lymph drainage? Is this a permanent problem? Why or why not?

44. David's lymphatic stream contains a high number of plasma cells. Has the relative number of antibodies in his bloodstream increased or decreased at this time? What is the basis of your response?

45. Is the allergen in poison ivy sap a water-soluble or lipid-soluble molecule? Explain your reasoning.

✓ THE FINALE: MULTIPLE CHOICE

46. Select the best answer or answers from the choices given.

1. Statements that apply to lymphatic capillaries include the following:

   A. The endothelial cells have continuous tight junctions.

   B. They are open ended like straws.

   C. Minivalves prevent the backflow of fluid into the interstitial spaces.

   D. The endothelial cells are anchored by filaments to the surrounding structures.

2. Chyle flows into the:

   A. lacteals

   B. intestinal lymph nodes

   C. intestinal trunk

   D. cisterna chyli

3. Which parts of the lymph node show increased activity when antibody production is high?

   A. Germinal centers

   B. Outer follicle

   C. Medullary cords

   D. Sinuses

4. The classification *lymphoid tissues* includes:

   A. the adenoids

   B. the spleen

   C. bone marrow

   D. the thyroid gland

5. The spleen functions to:

   A. remove aged red blood cells (RBCs)

   B. house lymphocytes

   C. filter lymph

   D. store some blood components

6. Which characteristics are associated with the thymus?

   A. Providing immunocompetence

   B. Hormone secretion

   C. Hypertrophy in later life

   D. Atrophy in later life

7. The tonsils:

   A. have a complete epithelial capsule

   B. have crypts to trap bacteria

   C. filter lymph

   D. contain germinal centers

8. Possible antigen-presenting cells (APCs) include:

   A. dendritic cells

   B. Langerhans' cells

   C. macrophages

   D. neutrophils

9. Effector T cells secrete:

   A. tumor necrosis factor

   B. histamine

   C. perforin

   D. interleukin 2

10. Neutrophils die in the line of duty because:

    A. they ingest infectious organisms

    B. their membranes become sticky and they are attacked by macrophages

    C. they secrete cellular toxins, which affect them in the same way they affect pathogens

    D. the buildup of tissue fluid pressure causes them to lyse

11. Macrophages:

    A. form exudate

    B. present antigens

    C. secrete interleukin 1

    D. activate helper T cells

12. Antibodies secreted in mother's milk:

    A. are IgG antibodies

    B. are IgA antibodies

    C. provide natural active immunity

    D. provide natural passive immunity

13. Conditions for which passive artificial immunity is the treatment of choice include:

    A. measles

    B. botulism

    C. rabies

    D. venomous snakebite

14. Which of these antibody classes is often arranged as a dimer?

    A. IgG

    B. IgM

    C. IgA

    D. IgD

15. Which of the following antibody capabilities causes a transfusion reaction with A or B blood cell antigens?

    A. Neutralization

    B. Precipitation

    C. Complement fixation

    D. Agglutination

16. Which of the following terms is applicable to the use of part of the patient's great saphenous vein in coronary bypass surgery?

    A. Isograft

    B. Xenograft

    C. Allograft

    D. Autograft

17. "Who" or "what" does the selecting that initiates clonal selection?

    A. Antigen   C. T cell

    B. Antibody   D. B cell

18. The cell type most often invaded by the HIV virus is:

    A. helper T cell

    B. plasma cell

    C. cytotoxic T cell

    D. B cell

13 THE RESPIRATORY SYSTEM

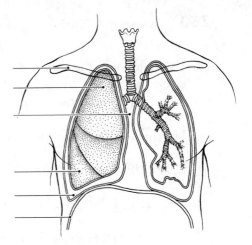

Body cells require an abundant and continuous supply of oxygen to carry out their activities. As cells use oxygen, they release carbon dioxide, a waste product that must be eliminated from the body. The circulatory and respiratory systems are intimately involved in obtaining and delivering oxygen to body cells and in eliminating carbon dioxide from the body. The respiratory system is responsible for gas exchange between the pulmonary blood and the external environment (that is, external respiration). The respiratory system also plays an important role in maintaining the acid-base balance of the blood.

Questions and activities in this chapter consider both the anatomy and physiology of the respiratory system structures.

FUNCTIONAL ANATOMY OF THE RESPIRATORY SYSTEM

1. The respiratory system is divisible into conducting zone and respiratory zone structures.

1. Name the conducting zone structures._____

_____

2. What is their common function?_____

_____

3. Name the respiratory zone structures._____

_____

2. The following questions refer to the main bronchi. In the spaces provided, insert the letter *R* to indicate the right main bronchus and the letter *L* to indicate the left main bronchus.

1. Which of the main bronchi is larger in diameter? _____

2. Which of the main bronchi is more horizontal? _____

3. Which of the main bronchi is the most common site for lodging of a foreign object

   that has entered the respiratory passageways? _____

3. Complete the following statements by inserting your answers in the answer blanks.

_____ 1.

_____ 2.

_____ 3.

_____ 4.

_____ 5.

_____ 6.

_____ 7.

_____ 8.

_____ 9.

_____ 10.

_____ 11.

_____ 12.

_____ 13.

_____ 14.

_____ 15.

_____ 16.

Air enters the nasal cavity of the respiratory system through the __(1)__. The nasal cavity is divided by the midline __(2)__. The nasal cavity mucosa has several functions. Its major functions are to __(3)__, __(4)__, and __(5)__ the incoming air. Mucous membrane–lined cavities called __(6)__ are found in several bones surrounding the nasal cavities. They make the skull less heavy and probably act as resonance chambers for __(7)__. The passageway common to the digestive and respiratory systems, the __(8)__, is often referred to as the throat; it connects the nasal cavity with the __(9)__ below. Clusters of lymphatic tissue, __(10)__, are part of the defensive system of the body. Reinforcement of the trachea with __(11)__ rings prevents its collapse during __(12)__ changes that occur during breathing. The fact that the rings are incomplete posteriorly allows a food bolus to bulge __(13)__ during its transport to the stomach. The larynx or voice box is built from many cartilages, but the largest is the __(14)__ cartilage. Within the larynx are the __(15)__, which vibrate with exhaled air and allow an individual to __(16)__.

4. Circle the term that does not belong in each of the following groupings.

1. Sphenoidal   Maxillary   Mandibular   Ethmoidal   Frontal

2. Nasal cavity   Trachea   Alveolus   Larynx   Bronchus

3. Apex   Base   Hilum   Larynx   Pleura

4. Sinusitis   Peritonitis   Pleurisy   Tonsillitis   Laryngitis

5. Laryngopharynx   Oropharynx   Transports air and food   Nasopharynx

6. Alveoli   Respiratory zone   Alveolar sac   Main bronchus

5. Figure 13–1 is a sagittal view of the upper respiratory structures. First, correctly identify all structures provided with leader lines on the figure. Then select different colors for the structures listed below and use them to color in the coding circles and the corresponding structures on the figure.

◯ Nasal cavity   ◯ Larynx   ◯ Thyroid cartilage

◯ Pharynx   ◯ Paranasal sinuses   ◯ Cricoid cartilage

◯ Trachea

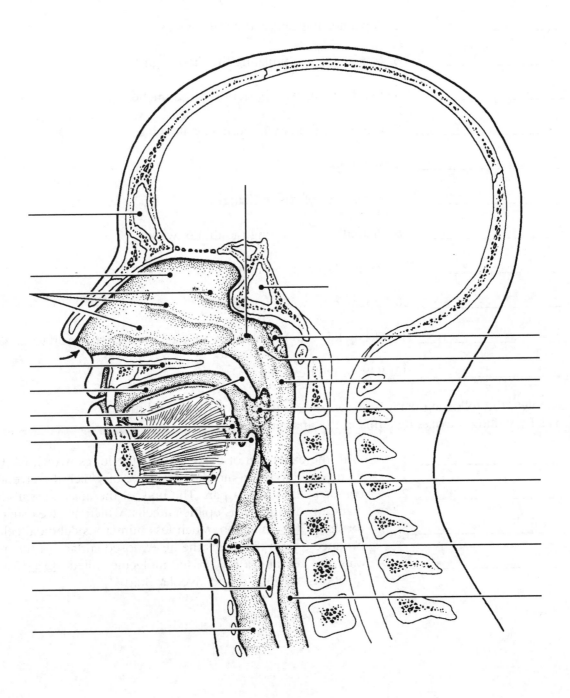

Figure 13–1

6. Using the key choices, select the terms identified in the following descriptions by inserting the appropriate term or letter in the answer blanks.

Key Choices

A. Alveoli   D. Epiglottis   G. Palate   J. Main bronchi   M. Vocal cords

B. Bronchioles   E. Esophagus   H. Parietal pleura   K. Trachea

C. Conchae   F. Glottis   I. Phrenic   L. Visceral pleura

_____ 1. Smallest conducting respiratory passageways

_____ 2. Separates the oral and nasal cavities

_____ 3. Major nerve, stimulating the diaphragm

_____ 4. Food passageway posterior to the trachea

_____ 5. Closes off the larynx during swallowing

_____ 6. Windpipe

_____ 7. Actual site of gas exchanges

_____ 8. Pleural layer covering the thorax walls

_____ 9. Pleural layer covering the lungs

_____ 10. Opening between vocal folds

_____ 11. Fleshy lobes in the nasal cavity which increase its surface area

_____ 12. Vibrate with expired air

7. Complete the following paragraph concerning the alveolar cells and their roles by writing the missing terms in the answer blanks.

_____ 1.

_____ 2.

_____ 3.

_____ 4.

With the exception of the stroma of the lungs, which is __(1)__ tissue, the lungs are mostly air spaces, of which the alveoli compose the greatest part. The bulk of the alveolar walls are made up of squamous epithelial cells, which are well suited for their __(2)__ function. Much less numerous cuboidal cells produce a fluid that coats the air-exposed surface of the alveolus and contains a lipid-based molecule called __(3)__ that functions to __(4)__ of the alveolar fluid.

8. Figure 13–2 is a diagram of the larynx and associated structures. On the
figure, identify each of the structures listed below. Select a different color
for each and use it to color in the coding circles and the corresponding
structures on the figure. Then answer the questions following the diagram.

◯ Hyoid bone   ◯ Tracheal cartilages   ◯ Cricoid cartilage

◯ Thyroid cartilage   ◯ Epiglottis

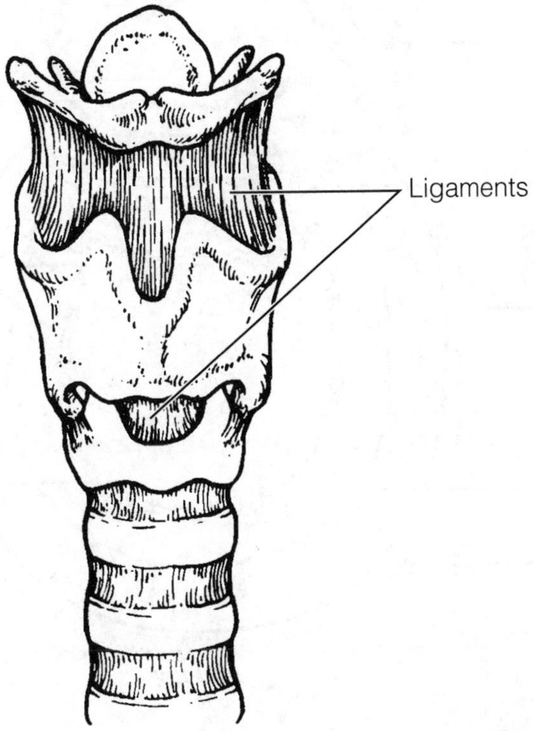

Ligaments

Figure 13–2

1. What are three functions of the larynx? _____

   _____

   _____

2. What type of cartilage forms the epiglottis? _____

3. What type of cartilage forms the other eight laryngeal cartilages? _____

4. Explain this difference. _____

   _____

   _____

5. What is the common name for the thyroid cartilage?_____

9. Figure 13–3 shows a cross section through the trachea. First, label the layers indicated by the leader lines. Next, color the following: mucosa (including the cilia, epithelium, lamina propria)—light pink; area containing the submucosal seromucous glands—purple; hyaline cartilage ring—blue; trachealis muscle—orange; and adventitia—yellow. Then, respond to the questions following the figure.

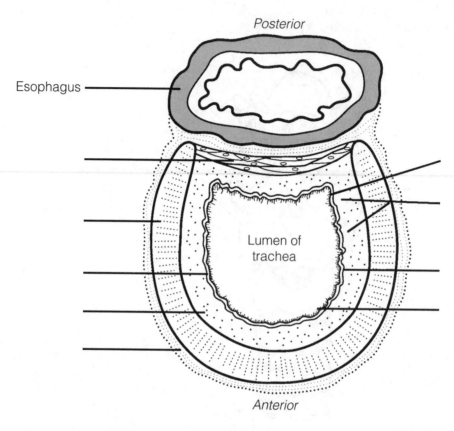

Posterior

Esophagus —

Lumen of
trachea

Anterior

Figure 13–3

1. What important role is played by the cartilage rings that reinforce the trachea?

   _____

2. Of what importance is the fact that the cartilage rings are incomplete posteriorly?

   _____

3. What occurs when the trachealis muscle contracts, and in what activities might this action

   be very helpful? _____

   _____

10. Figure 13–4 illustrates the gross anatomy of the lower respiratory system. Intact structures are shown on the left; respiratory passages are shown on the right. Select a different color for each of the structures listed below and use it to color in the coding circles and the corresponding structures on the figure. Then complete the figure by labeling the areas/structures that are provided with leader lines on the figure. Be sure to include the following: pleural space, mediastinum, apex of right lung, diaphragm, clavicle, and the base of the right lung.

○ Trachea   ○ Main (primary) bronchi   ○ Visceral pleura

○ Larynx   ○ Secondary bronchi   ○ Parietal pleura

○ Intact lung   ○ Tertiary bronchi

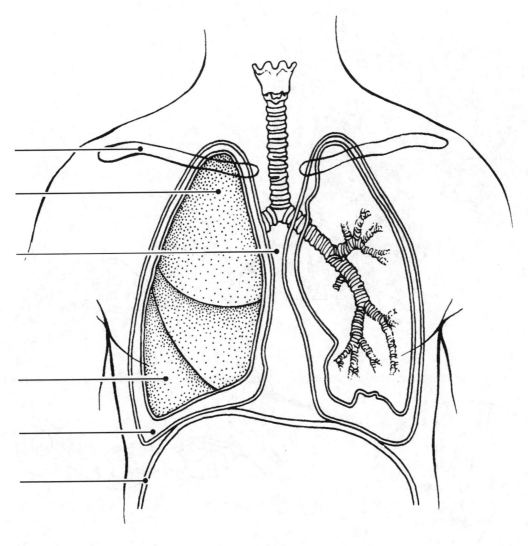

Figure 13–4

11. Figure 13–5 illustrates the microscopic structure of the respiratory unit of lung tissue. The external anatomy is shown in Figure 13–5A. Color the intact alveoli yellow, the pulmonary capillaries red, and the respiratory bronchioles green.

A cross section through an alveolus is shown on Figure 13–5B, and a blowup of the respiratory membrane is shown in Figure 13–5C. On these illustrations, color the alveolar epithelium yellow, the capillary endothelium pink, and the red blood cells in the capillary red. Also, label the alveolar chamber and color it pale blue. Finally, in Figure 13–5C label the region of the fused basement membranes. Then add the symbols for oxygen gas (O_2) and carbon dioxide gas (CO_2) in the sites where they would be in higher concentration and arrows correctly showing their direction of movement through the respiratory membrane.

Figure 13–5

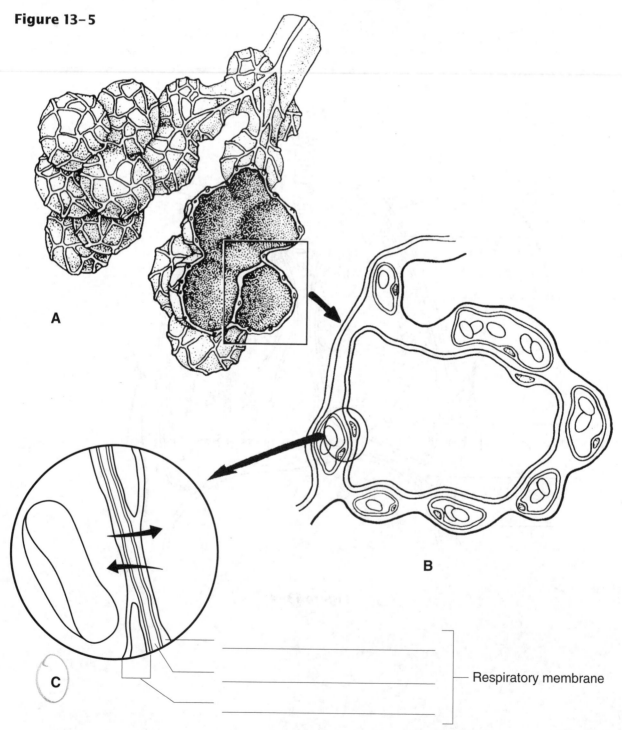

A

B

C

Respiratory membrane

RESPIRATORY PHYSIOLOGY

12. Using the key choices, select the terms identified in the following descriptions by inserting the appropriate term or letter in the answer blanks.

Key Choices

A. Atmospheric pressure   B. Intrapulmonary pressure   C. Intrapleural pressure

_____ 1. In healthy lungs, it is always lower than atmospheric pressure (that is, it is negative pressure)

_____ 2. Pressure of air outside the body

_____ 3. As it decreases, air flows into the passageways of the lungs

_____ 4. As it increases over atmospheric pressure, air flows out of the lungs

_____ 5. If this pressure becomes equal to the atmospheric pressure, the lungs collapse

_____ 6. Rises well over atmospheric pressure during a forceful cough

_____ 7. Also known as intra-alveolar pressure

13. Many changes occur within the lungs as the diaphragm (and external intercostal muscles) contract and then relax. These changes lead to the flow of air into and out of the lungs. The activity of the diaphragm is given in the left column of the following table. Several changes in condition are listed in the column heads to the right. Complete the table by checking (✓) the appropriate column to correctly identify the change that would be occurring relative to the diaphragm's activity in each case.

Activity of diaphragm	**Changes in**							
	Internal volume of thorax		Internal pressure in thorax		Size of lungs		Direction of airflow	
(↑ = increased) (↓ = decreased)	↑	↓	↑	↓	↑	↓	Into lung	Out of lung
Contracted, moves downward								
Relaxed, moves superiorly								

14. Use the key choices to respond to the following descriptions. Insert the correct term or letter in the answer blanks.

Key Choices

A. External respiration   C. Inspiration   E. Ventilation (breathing)

B. Expiration   D. Internal respiration

_____ 1. Period of breathing when air enters the lungs

_____ 2. Exchange of gases between the systemic capillary blood and body cells

_____ 3. Alternate flushing of air into and out of the lungs

_____ 4. Exchange of gases between alveolar air and pulmonary capillary blood

_____ 5. Period of breathing when air leaves the lungs

15. Although normal quiet expiration is largely passive because of lung recoil, when expiration must be more forceful (or the lungs are diseased), muscles that increase the abdominal pressure or depress the rib cage are enlisted.

1. Provide two examples of muscles that cause abdominal pressure to rise.

_____ and _____

2. Provide two examples of muscles that depress the rib cage.

_____ and _____

16. Four nonrespiratory movements are described here. Identify each by inserting your answers in the spaces provided.

1. Sudden inspiration, resulting from spasms of the diaphragm. _____

2. A deep breath is taken, the glottis is closed, and air is forced out of the lungs against

the glottis; clears the lower respiratory passageways. _____

3. As just described, but it clears the upper respiratory passageways. _____

4. Increases ventilation of the lungs; may be initiated by a need to increase oxygen levels

in the blood. _____

17. The following section concerns respiratory volume measurements. Using the key choices, select the terms identified in the following descriptions by inserting the appropriate term or letter in the answer blanks.

Key Choices

A. Dead space volume   C. Inspiratory reserve volume (IRV)   E. Tidal volume (TV)

B. Expiratory reserve   D. Residual volume (RV)   F. Vital capacity (VC)
   volume (ERV)

_____ 1. Respiratory volume inhaled or exhaled during normal breathing

_____ 2. Air in respiratory passages that does not contribute to gas exchange

_____ 3. Total amount of exchangeable air

_____ 4. Gas volume that allows gas exchange to go on continuously

_____ 5. Amount of air that can still be exhaled (forcibly) after a normal exhalation

18. Figure 13–6 is a diagram showing respiratory volumes. Complete the figure by making the following additions.

1. Bracket the volume representing the vital capacity and color the area yellow; label it VC.

2. Add green stripes to the area representing the inspiratory reserve volume and label it IRV.

3. Add red stripes to the area representing the expiratory reserve volume and label it ERV.

4. Identify and label the respiratory volume, which is *now just yellow*. Color the residual volume (RV) blue and label it appropriately on the figure.

5. Bracket and label the inspiratory capacity (IC).

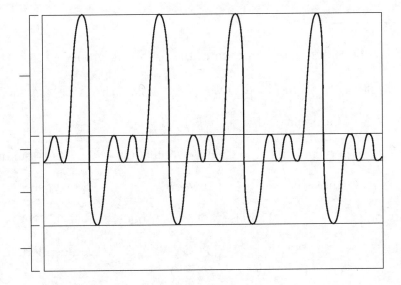

Figure 13–6

19. Use the key choices to correctly complete the following statements, which refer to gas exchanges in the body. Insert the correct letter response in the answer blanks.

Key Choices

A. Active transport

B. Air of alveoli to capillary blood

C. Carbon dioxide–poor and oxygen-rich

D. Capillary blood to alveolar air

E. Capillary blood to tissue cells

F. Diffusion

G. Higher concentration

H. Lower concentration

I. Oxygen-poor and carbon dioxide–rich

J. Tissue cells to capillary blood

_____ 1.

_____ 2.

_____ 3.

_____ 4.

_____ 5.

_____ 6.   _____ 7.   _____ 8.   _____ 9.

All gas exchanges are made by __(1)__. When substances pass in this manner, they move from areas of their __(2)__ to areas of their __(3)__. Thus oxygen continually passes from the __(4)__ and then from the __(5)__. Conversely, carbon dioxide moves from the __(6)__ and from __(7)__. From there it passes out of the body during expiration. As a result of such exchanges, arterial blood tends to be __(8)__ while venous blood is __(9)__.

20. Complete the following statements by inserting your answers in the answer blanks.

_____ 1.

_____ 2.

_____ 3.

_____ 4.

Most oxygen is transported bound to __(1)__ inside the red blood cells. Conversely, *most* carbon dioxide is carried in the form of __(2)__ in the __(3)__. Carbon monoxide poisoning is lethal because carbon monoxide competes with __(4)__ for binding sites.

21. Circle the term that does not belong in each of the following groupings.

1. ↑Respiratory rate   ↓In blood CO_2 Alkalosis   Acidosis

2. Acidosis   ↑Carbonic acid   ↓pH   ↑pH

3. Acidosis   Hyperventilation   Hypoventilation CO_2 buildup

4. Apnea   Cyanosis   ↑Oxygen   ↓Oxygen

5. ↑Respiratory rate   ↑Exercise   Anger   ↑CO_2 in blood

6. High altitude   ↓PO_2 ↑PCO_2 ↓Atmospheric pressure

RESPIRATORY DISORDERS

22. Match the terms in Column B with the pathological conditions described in Column A.

	Column A	Column B
_____	1. Lack or cessation of breathing	A. Apnea
_____	2. Normal breathing in terms of rate and depth	B. Asthma
_____	3. Labored breathing, or "air hunger"	C. Chronic bronchitis
_____	4. Chronic oxygen deficiency	D. Dyspnea
_____	5. Condition characterized by loss of lung elasticity and an increase in size of the alveolar chambers	E. Emphysema
		F. Eupnea
_____	6. Condition characterized by increased mucus production, which clogs respiratory passageways and promotes coughing	G. Hypoxia
		H. Lung cancer
_____	7. Respiratory passageways narrowed by bronchiolar spasms	I. Tuberculosis
_____	8. Together called COPD (chronic obstructive pulmonary disease)	
_____	9. Incidence strongly associated with cigarette smoking; outlook is poor	
_____	10. Infection spread by airborne bacteria; a recent alarming increase in drug-resistant cases	

DEVELOPMENTAL ASPECTS OF THE RESPIRATORY SYSTEM

23. Mrs. Jones gave birth prematurely to her first child. At birth, the baby weighed 2 lb 8 oz. Within a few hours, the baby had developed severe dyspnea and was becoming cyanotic. Therapy with a positive pressure ventilator was prescribed. Answer the following questions related to the situation just described. Place your responses in the answer blanks.

1. The infant's condition is referred to as _____

2. It occurs because of a relative lack of _____

3. The function of the deficient substance is to _____

_____

4. Explain what the positive pressure apparatus accomplishes. _____

24. Complete the following statements by inserting your answers in the answer blanks.

_____ 1.

_____ 2.

_____ 3.

_____ 4.

_____ 5.

_____ 6.

_____ 7.

_____ 8.

The respiratory rate of a newborn baby is approximately __(1)__ respirations per minute. In a healthy adult, the respiratory range is __(2)__ respirations per minute. Most problems that interfere with the operation of the respiratory system fall into one of the following categories: infections such as pneumonia, obstructive conditions such as __(3)__ and __(4)__, and/or conditions that destroy lung tissue, such as __(5)__. With age, the lungs lose their __(6)__, and the __(7)__ of the lungs decreases. Protective mechanisms also become less efficient, causing elderly individuals to be more susceptible to __(8)__.

INCREDIBLE JOURNEY

A Visualization Exercise for the Respiratory System

You carefully begin to pick your way down, using cartilages as steps.

25. Where necessary, complete statements by inserting the missing word(s) in the answer blanks.

_____ 1.

_____ 2.

_____ 3.

_____ 4.

_____ 5.

_____ 6.

_____ 7.

Your journey through the respiratory system is to be on foot. To begin, you simply will walk into your host's external nares. You are miniaturized, and your host is sedated lightly to prevent sneezing during your initial observations in the nasal cavity and subsequent descent.

You begin your exploration of the nasal cavity in the right nostril. One of the first things you notice is that the chamber is very warm and humid. High above, you see three large, round lobes, the __(1)__, which provide a large mucosal surface area for warming and moistening the entering air. As you walk toward the rear of this chamber, you see a large lumpy mass of lymphatic tissue, the __(2)__ in the __(3)__, or first portion of the pharynx. As you peer down the pharynx, you realize that it will be next to impossible to maintain your footing during the next part of your journey. It is nearly straight down, and the __(4)__ secretions are like grease. You sit down and dig your heels in to get started. After a quick slide, you land abruptly on one of a pair of flat, sheetlike structures that begin to vibrate rapidly, bouncing you up and down helplessly. You are also conscious of a rhythmic hum during this jostling, and you realize that you have landed on a __(5)__. You pick yourself up and look over the superior edge of the __(6)__, down into the seemingly endless esophagus behind. You chastise yourself for not remembering that the __(7)__ and respiratory

_____ 8.

_____ 9.

_____10.

_____11.

_____12.

_____13.

_____14.

_____15.

_____16.

_____17.

_____18.

_____19.

_____20.

pathways separate at this point. Hanging directly over your head is the leaflike __(8)__ cartilage. Normally, you would not have been able to get this far because it would have closed off this portion of the respiratory tract. With your host sedated, however, that protective reflex does not work.

You carefully begin to pick your way down, using the cartilages as steps. When you reach the next respiratory organ, the __(9)__, your descent becomes much easier, because the structure's C-shaped cartilages form a ladder-like supporting structure. As you climb down the cartilages, your face is stroked rhythmically by soft cellular extensions, or __(10)__. You remember that their function is to move mucus laden with bacteria or dust and other debris toward the __(11)__.

You finally reach a point where the descending passageway splits into two __(12)__, and because you want to control your progress (rather than slide downward), you choose the more horizontal __(13)__ branch. If you remain in the superior portion of the lungs, your return trip will be less difficult because the passageways will be more horizontal than steeply vertical. The passageways get smaller and smaller, slowing your progress. As you are squeezing into one of the smallest of the respiratory passageways, a __(14)__, you see a bright spherical chamber ahead. You scramble into this __(15)__, pick yourself up, and survey the area. Scattered here and there are lumps of a substance that look suspiciously like coal,

reminding you that your host is a smoker. As you stand there, a soft rustling wind seems to flow in and out of the chamber. You press your face against the transparent chamber wall and see disclike cells, __(16)__, passing by in the capillaries on the other side. As you watch, they change from a somewhat bluish color to a bright __(17)__ color as they pick up __(18)__ and unload __(19)__.

You record your observations and then contact headquarters to let them know you are ready to begin your ascent. You begin your return trek, slipping and sliding as you travel. By the time you reach the inferior edge of the trachea, you are ready for a short break. As you rest on the mucosa, you begin to notice that the air is becoming close and very heavy. You pick yourself up quickly and begin to scramble up the trachea. Suddenly and without warning, you are hit by a huge wad of mucus and catapulted upward and out onto your host's freshly pressed handkerchief! Your host has assisted your exit with a __(20)__.

 AT THE CLINIC

26. After a long bout of bronchitis, Ms. Dupee complains of a stabbing pain in her side with each breath. What is her probable condition?

27. The Kozloski family is taking a long auto trip. Michael, who has been riding in the back of the station wagon, complains of a throbbing headache. A little later, he seems confused and his face is flushed. What is your diagnosis of Michael's problem?

28. A new mother checks on her sleeping infant son, only to find that he has stopped breathing and is turning blue. The mother quickly picks up the baby and pats his back until he starts to breathe. What tragedy has been averted?

29. Joanne Willis, a long-time smoker, is complaining that she has developed a persistent cough. What is your first guess as to her condition? What has happened to her bronchial cilia?

30. Barbara is rushed to the emergency room after an auto accident. The 8th through 10th ribs on her left side have been fractured and have punctured the lung. What term is used to indicate lung collapse? Will both lungs collapse? Why or why not?

31. A young boy is diagnosed with cystic fibrosis. What effect will this have on his respiratory system?

32. Mr. and Ms. Rao took their sick 5-year-old daughter to the doctor. The girl was breathing entirely through her mouth, her voice sounded odd and whiny, and a puslike fluid was dripping from her nose. Which one of the tonsils was most likely infected in this child?

33. Assume you are a second-year nursing student. As your assignment, you are asked to explain how a history of heavy smoking might interfere with a patient's gas exchange.

34. Why does an EMT administering a Breathalyzer test for alcohol ask the person being tested to expel one deep breath instead of several shallow ones?

35. The cilia lining the respiratory passageways superior to the larynx beat inferiorly while those lining the larynx and below beat superiorly. What is the functional "reason" for this difference?

✓ THE FINALE: MULTIPLE CHOICE

36. Select the best answer or answers from the choices given.

1. Structures that are part of the respiratory zone include:

   A. terminal bronchioles

   B. respiratory bronchioles

   C. tertiary bronchi

   D. alveolar ducts

2. Which structures are associated with the production of speech?

   A. Cricoid cartilage

   C. Arytenoid cartilage

   B. Glottis

   D. Pharynx

3. The skeleton of the external nose consists of:

   A. cartilage and bone

   B. bone only

   C. hyaline cartilage only

   D. elastic cartilage only

4. Which of the following is *not* part of the conducting zone of the respiratory system?

   A. Pharynx   D. Lobar bronchi

   B. Alveolar sac   E. Larynx

   C. Trachea

5. Select the single false statement about the true vocal cords:

   A. They are the same as the vocal folds.

   B. They attach to the arytenoid cartilages via the vocal ligaments.

   C. Exhaled air flowing through the glottis vibrates them to produce sound.

   D. They are also called the vestibular folds.

6. The function of the cuboid cells of the alveolar walls is:

   A. to produce surfactant

   B. to propel mucous sheets

   C. phagocytosis of dust particles

   D. to allow rapid diffusion of respiratory gases

7. An examination of a lobe of the lung reveals many branches off the main passageway. These branches are:

   A. main bronchi   C. tertiary bronchi

   B. lobar bronchi   D. segmental bronchi

8. An alveolar sac:

   A. is an alveolus

   B. relates to an alveolus as a bunch of grapes relates to one grape

   C. is a huge, saclike alveolus in an emphysema patient

   D. is the same as an alveolar duct

9. The respiratory membrane (air-blood barrier) consists of:

   A. squamous cells, basal membranes, endothelial cells

   B. air, connective tissue, lung

   C. squamous and cuboidal epithelial cells and macrophages

   D. pseudostratified epithelium, lamina propria, capillaries

10. Oxygen and carbon dioxide are exchanged in the lungs and through all cell membranes by:

   A. active transportation

   B. diffusion

   C. filtration

   D. osmosis

11. Which of the following are characteristic of a bronchopulmonary segment?

   A. Removal causes collapse of adjacent segments

   B. Fed by a tertiary bronchus

   C. Supplied by its own branches of the pulmonary artery and vein

   D. Separated from other segments by its septum

12. During inspiration, intrapulmonary pressure is:

   A. greater than atmospheric pressure

   B. less than atmospheric pressure

   C. greater than intrapleural pressure

   D. less than intrapleural pressure

13. When the inspiratory muscles contract,

   A. the size of the thoracic cavity increases in diameter

   B. the size of the thoracic cavity increases in length

   C. the volume of the thoracic cavity decreases

   D. the size of the thoracic cavity increases in both length and diameter

14. Lung collapse is prevented by:

   A. high surface tension of alveolar fluid

   B. high surface tension of pleural fluid

   C. high pressure in the pleural cavities

   D. high elasticity of lung tissue

15. Resistance is increased by:

   A. epinephrine

   B. parasympathetic stimulation

   C. inflammatory chemicals

   D. contraction of the trachealis muscle

16. Which of the following changes accompanies the loss of elasticity associated with aging?

   A. Increase in tidal volume

   B. Increase in inspiratory reserve volume

   C. Increase in residual volume

   D. Increase in vital capacity

14 THE DIGESTIVE SYSTEM AND BODY METABOLISM

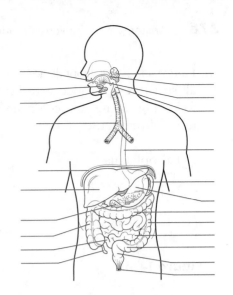

The digestive system processes food so that it can be absorbed and used by the body's cells. The digestive organs are responsible for food ingestion, digestion, absorption, and elimination of undigested remains from the body. In one sense, the digestive tract can be viewed as a disassembly line in which food is carried from one stage of its breakdown process to the next by muscular activity, and its nutrients are made available en route to the cells of the body. In addition, the digestive system provides for one of life's greatest pleasures—eating.

The anatomy of both alimentary canal and accessory digestive organs, mechanical and enzymatic breakdown, and absorption mechanisms are covered in this chapter. An introduction to nutrition and some important understandings about cellular metabolism (utilization of foodstuffs by body cells) are also considered in this chapter review.

ANATOMY OF THE DIGESTIVE SYSTEM

1. Complete the following statements by inserting your answers in the answer blanks.

_____ 1.

_____ 2.

_____ 3.

_____ 4.

_____ 5.

_____ 6.

_____ 7.

The digestive system is responsible for many body processes. Its functions begin when food is taken into the mouth, or (1) . The process called (2) occurs as food is broken down both chemically and mechanically. For the broken-down foods to be made available to the body cells, they must be absorbed through the digestive system walls into the (3) . Undigestible food remains are removed, or (4) , from the body in (5) . The organs forming a continuous tube from the mouth to the anus are collectively called the (6) . Organs located outside the digestive tract proper, which secrete their products into the digestive tract, are referred to as (7) digestive system organs.

2. Figure 14–1 is a frontal view of the digestive system. First, correctly identify all structures provided with leader lines. Then select different colors for the following organs and color the coding circles and the corresponding structures of the figure.

◯ Esophagus   ◯ Pancreas   ◯ Tongue

◯ Liver   ◯ Salivary glands   ◯ Uvula

◯ Large intestine ◯ Small intestine ◯ Stomach

Figure 14–1

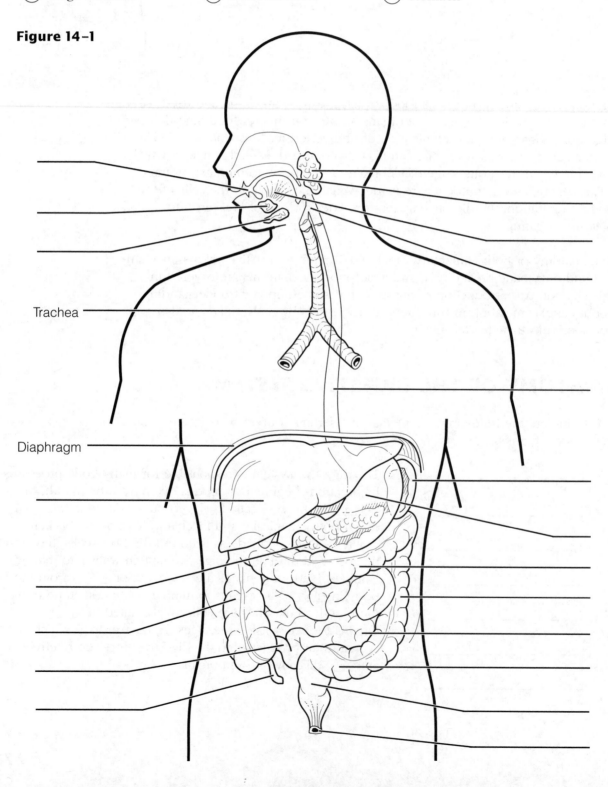

Trachea

Diaphragm

3. Figure 14–2 illustrates oral cavity structures. First, correctly identify all structures provided with leader lines. Then color the structure that attaches the tongue to the floor of the mouth red; color the portions of the roof of the mouth unsupported by bone blue; color the structures that are essentially masses of lymphatic tissue yellow; and color the structure that contains the bulk of the taste buds pink.

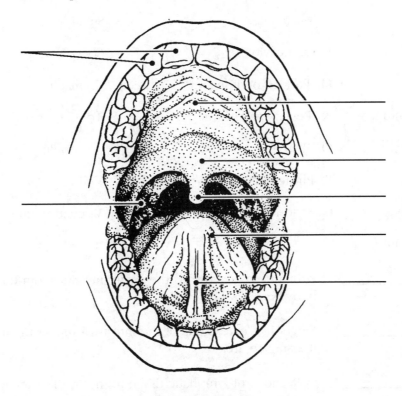

Figure 14–2

4. Various types of glands secrete substances into the alimentary tube. Match the glands listed in Column B to the functions/locations described in Column A. Place the correct term or letter response in the answer blanks.

Column A

_____ 1. Produce an enzyme-poor "juice" containing mucus; found in the submucosa of the small intestine

_____ 2. Secretion includes amylase, which begins starch digestion in the mouth

_____ 3. Ducts a variety of enzymes in an alkaline fluid into the duodenum

_____ 4. Produces bile, which is transported to the duodenum via the bile duct

_____ 5. Produce hydrochloric acid and pepsinogen

Column B

A. Gastric glands

B. Intestinal glands

C. Liver

D. Pancreas

E. Salivary glands

5. Using the key choices, select the terms identified in the following descriptions by inserting the appropriate term or letter in the answer blanks.

Key Choices

A. Anal canal   J. Mesentery   R. Rugae

B. Appendix   K. Microvilli   S. Small intestine

C. Colon   L. Oral cavity   T. Soft palate

D. Esophagus   M. Parietal peritoneum   U. Stomach

E. Greater omentum   N. Peyer's patches   V. Tongue

F. Hard palate   O. Pharynx   W. Vestibule

G. Haustra   P. Plicae circulares   X. Villi

H. Ileocecal valve   Q. Pyloric sphincter (valve)   Y. Visceral peritoneum

I. Lesser omentum

_____ 1. Structure that suspends the small intestine from the posterior body wall

_____ 2. Finger-like extensions of the intestinal mucosa that increase the surface area

_____ 3. Collections of lymphatic tissue found in the submucosa of the small intestine

_____ 4. Folds of the small intestine wall

_____ 5. Two anatomical regions involved in the mechanical breakdown of food

_____ 6. Organ that mixes food in the mouth

_____ 7. Common passage for food and air

_____, 8. Three extensions/modifications of the peritoneum

_____, and _____

_____ 9. Literally a food chute; has no digestive or absorptive role

_____ 10. Folds of the stomach mucosa

_____ 11. Saclike outpocketings of the large intestine wall

_____ 12. Projections of the plasma membrane of a cell that increase the cell's surface area

_____ 13. Prevents food from moving back into the small intestine once it has entered the large intestine

_____ 14. Organ responsible for most food and water absorption

_____ 15. Organ primarily involved in water absorption and feces formation

_____ 16. Area between the teeth and lips/cheeks

_____ 17. Blind sac hanging from the initial part of the colon

_____ 18. Organ in which protein digestion begins

_____ 19. Membrane attached to the lesser curvature of the stomach

_____ 20. Organ into which the stomach empties

_____ 21. Sphincter controlling the movement of food from the stomach into the duodenum

_____ 22. Uvula hangs from its posterior edge

_____ 23. Organ that receives pancreatic juice and bile

_____ 24. Serosa of the abdominal cavity wall

_____ 25. Region, containing two sphincters, through which feces are expelled from the body

_____ 26. Anterosuperior boundary of the oral cavity; supported by bone

_____ 27. Serous membrane forming part of the wall of the small intestine

6. Circle the term that does not belong in each of the following groupings.

1. Nasopharynx   Esophagus   Laryngopharynx   Oropharynx

2. Villi   Plicae circulares   Rugae   Microvilli

3. Salivary glands   Pancreas   Liver   Gallbladder

4. Duodenum   Cecum   Jejunum   Ileum

5. Ascending colon   Haustra   Circular folds   Cecum

6. Mesentery   Frenulum   Greater omentum   Parietal peritoneum

7. Parotid   Sublingual   Submandibular   Palatine

8. Protein-digesting enzymes   Saliva   Intrinsic factor   HCl

9. Colon   Water absorption   Protein absorption   Vitamin B absorption

7. Figure 14–3A is a longitudinal section of the stomach. First, use the following terms to identify the regions provided with leader lines on the figure.

Body   Pyloric region   Greater curvature   Cardioesophageal sphincter

Fundus   Pyloric valve   Lesser curvature

Then select different colors for each of the following structures/areas and use them to color the coding circles and corresponding structures/areas on the figure.

◯ Oblique muscle layer   ◯ Longitudinal muscle layer   ◯ Circular muscle layer

◯ Area where rugae are visible   ◯ Serosa

Figure 14–3B shows two types of secretory cells found in gastric glands. Identify the third type called *chief cells* by choosing a few cells deep in the glands and labeling them. Then, color the hydrochloric acid–secreting cells red, color the mucus-secreting cells yellow, and color the cells that produce protein-digesting enzymes blue.

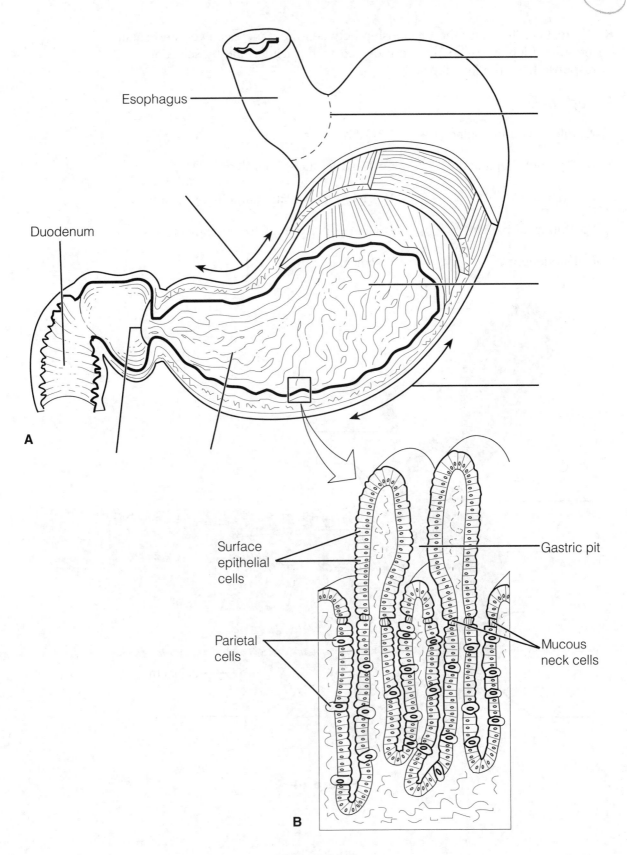

Esophagus

Duodenum

A

Surface
epithelial
cells

Gastric pit

Parietal
cells

Mucous
neck cells

B

Figure 14–3

8. Figure 14–4 illustrates the relationship between the pancreas, liver, and small intestine. Identify each structure provided with a leader line by selecting a response from the key choices.

Key Choices

A. Bile duct and sphincter

B. Common hepatic duct

C. Cystic duct

D. Duodenal papilla

E. Duodenum

F. Gallbladder

G. Hepatic ducts from liver

H. Hepatopancreatic ampulla and sphincter

I. Main pancreatic duct and sphincter

J. Pancreas

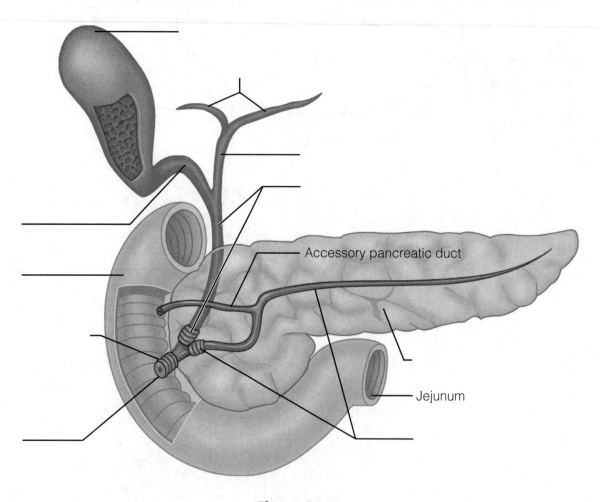

Accessory pancreatic duct

Jejunum

Figure 14–4

9. The walls of the alimentary canal have four typical layers, as illustrated in Figure 14–5. Identify each layer by placing its correct name in the space before the appropriate description. Then select different colors for each layer and use them to color the coding circles and corresponding structures on the figure. Finally, assume the figure shows a cross-sectional view of the small intestine and label the three structures provided with leader lines.

_____ ◯ 1. The secretory and absorptive layer

_____ ◯ 2. Layer composed of at least two muscle layers

_____ ◯ 3. Connective tissue layer, containing blood, lymph vessels, and nerves

_____ ◯ 4. Outermost layer of the wall; visceral peritoneum

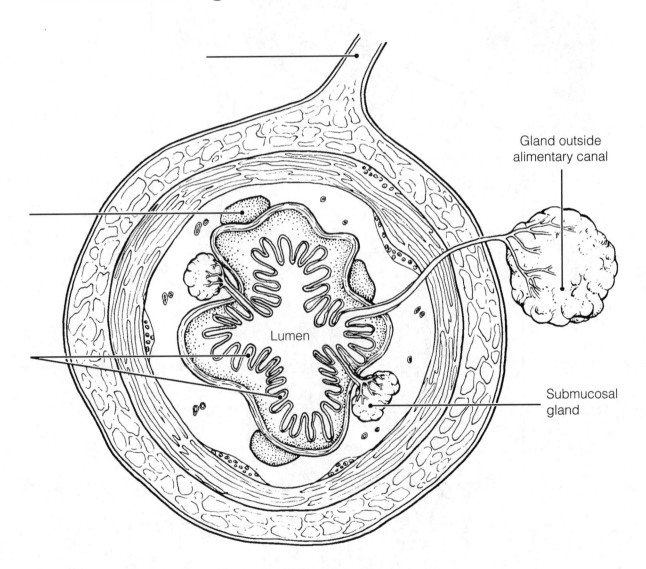

Figure 14–5

10. Figure 14–6 shows three views of the small intestine. First, label the villi in views B and C and the plicae circulares in views A and B. Then select different colors for each term listed below and use them to color in the coding circles and corresponding structures in view C.

◯ Surface epithelium   ◯ Lacteal   ◯ Capillary network

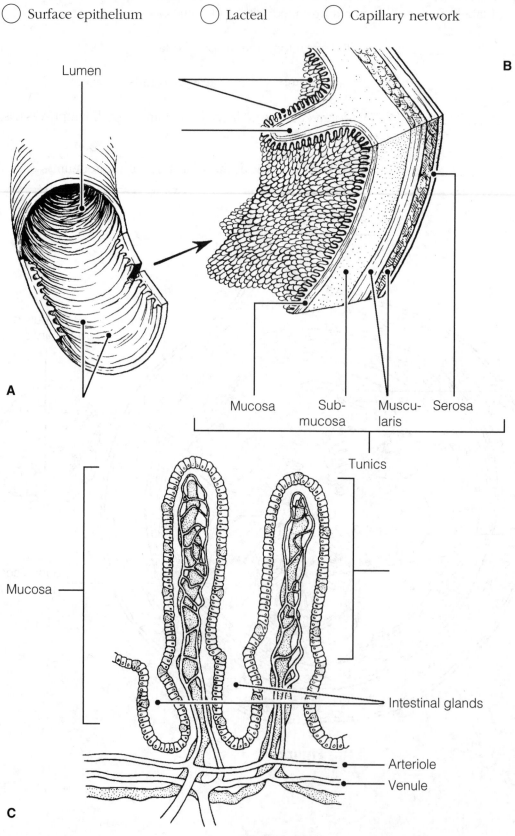

Figure 14–6

11. Three accessory organs are illustrated in Figure 14–7. Identify each of the three organs and the ligament provided with leader lines on the figure. Then select different colors for the following structures and use them to color the coding circles and the corresponding structures on the figure.

○ Common hepatic duct ○ Bile duct ○ Cystic duct ○ Pancreatic duct

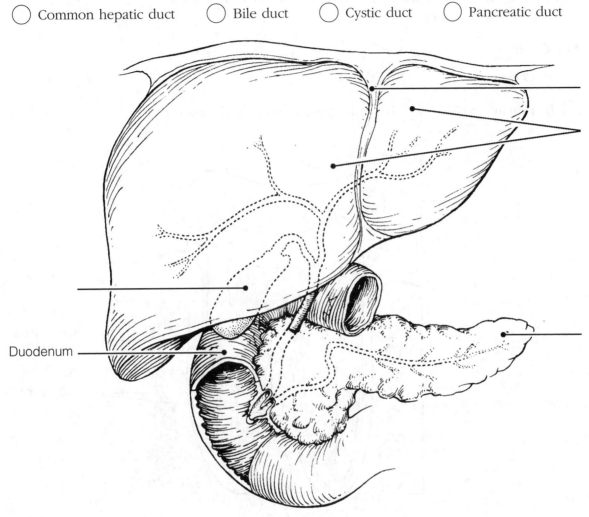

Duodenum

Figure 14–7

12. Complete the following statements referring to human dentition by inserting your answers in the answer blanks.

_____ 1.

_____ 2.

_____ 3.

_____ 4.

_____ 5.

The first set of teeth, called the __(1)__ teeth, begin to appear around the age of __(2)__ and usually have begun to be replaced by the age of __(3)__. The __(4)__ teeth are more numerous; that is, there are __(5)__ teeth in the second set as opposed to a total of __(6)__ teeth in the first set. If an adult has a full set of teeth, you can expect to find two __(7)__, one __(8)__, two __(9)__, and three __(10)__ in one side of each jaw. The most posterior molars in each jaw are commonly called __(11)__ teeth.

_____ 6.   _____ 7.   _____ 8.

_____ 9.   _____ 10.   _____ 11.

13. First, use the key choices to label the tooth diagrammed in Figure 14–8. Second, select different colors to represent the key choices and use them to color in the coding circles and corresponding structures in the figure. Third, add labels to the figure to identify the crown, gingiva, and root of the tooth. Last, choose terms from the key choices to match the descriptions below the figure.

Key Choices

◯ A. Cement   ◯ C. Enamel   ◯ E. Pulp

◯ B. Dentin   ◯ D. Periodontal membrane (ligament)

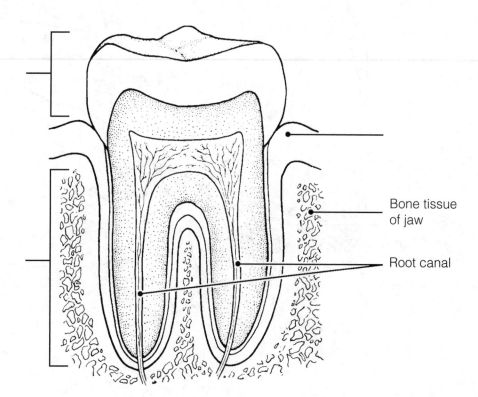

Bone tissue of jaw

Root canal

Figure 14–8

_____ 1. Material covering the tooth root

_____ 2. Forms the bulk of tooth structure

_____ 3. A collection of blood vessels, lymphatics, and nerve fibers

_____ 4. Cells that produce this substance degenerate after tooth eruption

PHYSIOLOGY OF THE DIGESTIVE SYSTEM

14. Match the descriptions in Column B with the appropriate terms referring
to digestive processes in Column A.

Column A	Column B
_____ 1. Ingestion	A. Transport of nutrients from lumen to blood
_____ 2. Propulsion	B. Enzymatic breakdown
_____ 3. Mechanical digestion	C. Elimination of feces
_____ 4. Chemical digestion	D. Eating
_____ 5. Absorption	E. Chewing
_____ 6. Defecation	F. Churning
	G. Includes swallowing
	H. Segmentation and peristalsis

15. Identify the pathological conditions described below by using terms from the
key choices. Insert the correct term or letter in the answer blanks.

Key Choices

A. Appendicitis   C. Diarrhea   E. Heartburn   G. Peritonitis

B. Constipation   D. Gallstones   F. Jaundice   H. Ulcer

_____ 1. Inflammation of the abdominal serosa

_____ 2. Condition resulting from the reflux of acidic gastric juice into
the esophagus

_____ 3. Usually indicates liver problems or blockage of the biliary ducts

_____ 4. An erosion of the stomach or duodenal mucosa

_____ 5. Passage of watery stools

_____ 6. Causes severe epigastric pain; associated with prolonged storage
of bile in the gallbladder

_____ 7. Inability to pass feces; often a result of poor bowel habits

16. This section relates to food breakdown in the digestive tract. Using the key choices, select the appropriate terms to complete the following statements. Insert the correct letter or term in the answer blanks.

Key Choices

A. Bicarbonate-rich fluid	F. HCl	K. Mucus
B. Bile	G. Hormonal stimulus	L. Pepsin
C. Brush border enzymes	H. Lipases	M. Psychological stimulus
D. Chewing	I. Mechanical stimulus	N. Rennin
E. Churning	J. Mouth	O. Salivary amylase

_____ 1. Starch digestion begins in the mouth when __(1)__ is ducted in by the salivary glands.

_____ 2. Gastrin, which prods the stomach glands to produce more enzymes and HCl, represents a __(2)__ .

_____ 3. The fact that the mere thought of a relished food can make your mouth water is an example of __(3)__ .

_____ 4. Many people chew gum to increase saliva formation when their mouths are dry. This type of stimulus is a __(4)__ .

_____ 5. Protein foods are largely acted on in the stomach by __(5)__ .

_____ 6. For the stomach protein-digesting enzymes to become active, __(6)__ is needed.

_____ 7. Considering living cells of the stomach (and everywhere) are largely protein, it is amazing that they are not digested by the activity of stomach enzymes. The most important means of stomach protection is the __(7)__ it produces.

_____ 8. A milk protein–digesting enzyme found in children but uncommon in adults is __(8)__ .

_____ 9. The third layer of smooth muscle found in the stomach wall allows mixing and mechanical breakdown by __(9)__ .

_____ 10. Important intestinal enzymes are the __(10)__ .

_____ 11. The small intestine is protected from the corrosive action of hydrochloric acid in chyme by __(11)__ , which is ducted in by the pancreas.

_____ 12. The pancreas produces protein-digesting enzymes, amylase, and nucleases. It is the only important source of __(12)__ .

_____ 13. A nonenzyme substance that causes fat to be dispersed into smaller globules is __(13)__ .

17. Hormonal stimuli are important in digestive activities that occur in the stomach and small intestine. Using the key choices, identify the hormones that function as described in the following statements. Insert the correct term or letter response in the answer blanks.

Key Choices

A. Cholecystokinin   B. Gastrin   C. Secretin

_____ 1. These two hormones stimulate the pancreas to release its secretions.

_____ 2. This hormone stimulates increased production of gastric juice.

_____ 3. This hormone causes the gallbladder to release stored bile.

_____ 4. This hormone causes the liver to increase its output of bile.

18. Various types of foods are ingested in the diet and broken down to their building blocks. Use the key choices to complete the following statements according to these understandings. Insert the correct term or letter in the answer blanks. In some cases, more than one choice applies.

Key Choices

A. Amino acids   D. Galactose   G. Maltose

B. Fatty acids   E. Glucose   H. Starch

C. Fructose   F. Lactose   I. Sucrose

_____ 1. The building blocks of carbohydrates are monosaccharides,
_____   or simple sugars. The three common simple sugars in our diet
_____   are ____, ____, and ____.

_____ 2. Disaccharides include ____, ____, and ____.

_____

_____

_____ 3. Protein foods must be digested to ____ before they can be absorbed.

_____ 4. Fats are broken down to two types of building blocks, ____ and
                        glycerol.

_____ 5. Of the simple sugars, ____ is most important; it is the sugar
                        referred to as "blood sugar."

19. Dietary substances capable of being absorbed are listed next. If the substance is *most often* absorbed from the digestive tract by active transport processes, put an *A* in the blank. If it is usually absorbed passively (by diffusion or osmosis), put a *P* in the blank. In addition, circle the substance that is *most likely* to be absorbed into a lacteal rather than into the capillary bed of the villus.

_____ 1. Water   _____ 3. Simple sugars   _____ 5. Electrolytes

_____ 2. Amino acids   _____ 4. Fatty acids

20. Complete the following statements that describe mechanisms of food mixing and movement. Insert your responses in the answer blanks.

_____ 1.

_____ 2.

_____ 3.

_____ 4.

_____ 5.

_____ 6.

_____ 7.

_____ 8.

_____ 9.

_____ 10.

_____ 11.

_____ 12.

_____ 13.

_____ 14.

_____ 15.

_____ 16.

_____ 17.

Swallowing, or __(1)__, occurs in two major phases—the __(2)__ and __(3)__. During the voluntary phase, the __(4)__ is used to push the food into the throat, and the __(5)__ rises to close off the nasal passageways. As food is moved involuntarily through the pharynx, the __(6)__ rises to ensure that its passageway is covered by the __(7)__ so that ingested substances do not enter respiratory passages. It is possible to swallow water while standing on your head because the water is carried along the esophagus involuntarily by the process of __(8)__. The pressure exerted by food on the __(9)__ valve causes it to open so that food can enter the stomach.

The two major types of movements that occur in the small intestine are __(10)__ and __(11)__. One of these movements, the __(12)__, acts to continually mix the food with digestive juices, and (strangely) also plays a major role in propelling foods along the tract. Still another type of movement seen only in the large intestine, __(13)__ occurs infrequently and acts to move feces over relatively long distances toward the anus. Presence of feces in the __(14)__ excites stretch receptors so that the __(15)__ reflex is initiated. Irritation of the gastrointestinal tract by drugs or bacteria might stimulate the __(16)__ center in the medulla, causing __(17)__, which is essentially a reverse peristalsis.

NUTRITION AND METABOLISM
Nutrients Used by Body Cells

21. Using the key choices, identify the foodstuffs used by cells in the cellular functions described below. Insert the correct term or key letter in the answer blanks.

Key Choices

A. Amino acids   B. Carbohydrates   C. Fats

_____ 1. The most used substance for producing the energy-rich ATP

_____ 2. Important in building myelin sheaths and cell membranes

_____ 3. Tend to be conserved by cells

_____ 4. The second most important food source for making cellular energy

_____ 5. Form insulating deposits around body organs and beneath the skin

_____ 6. Used to make the bulk of cell structure and functional substances such as enzymes

22. Identify the nutrients described by using the key choices. Insert the correct letter(s) in the answer blanks.

Key Choices

A. Bread/pasta   D. Fruits   G. Starch

B. Cheese/cream   E. Meat/fish   H. Vegetables

C. Cellulose   F. Minerals   I. Vitamins

_____ 1. Examples of *carbohydrate-rich foods* in the diet.

_____ 2. Fatty foods ingested in the normal diet include _____.

_____ 3. The only important *digestible* polysaccharide.

_____ 4. An *indigestible* polysaccharide that aids elimination because it adds bulk to the diet is _____.

_____ 5. *Protein-rich foods* include _____ and _____.

_____ 6. Most examples of these nutrients, which are found largely in vegetables and fruits, are used as coenzymes.

_____ 7. Include copper, iron, and sodium.

Metabolic Processes

23. Figure 14–9 depicts the three stages of cellular respiration. Label the figure by placing the following terms on the appropriate answer blanks. Color the diagram as suits your fancy, and then answer the questions below the figure. (Terms may be used more than once.)

ATP	Glucose	Mitochondrion
Carbon dioxide	Glycolysis	Pyruvic acid
Chemical energy	Electron transport chain	Water
Cytosol	Krebs cycle	

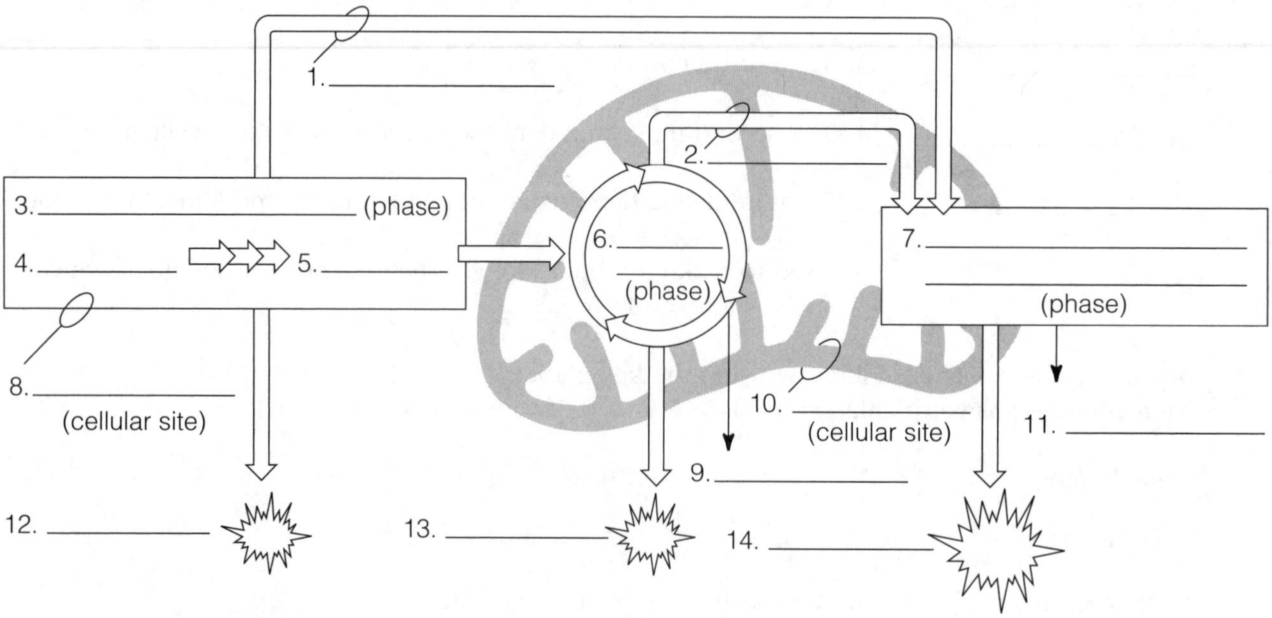

Figure 14–9

1. Which of the oxidative phases does not require oxygen?

_____

2. Which phases do require oxygen? _____

_____

3. In what form is chemical energy transferred from the first two phases to the third phase?

_____

4. Which of the phases produces the largest amount of ATP?

_____

5. Which phase combines energetic H atoms with molecular oxygen?

_____

24. This section considers the process of cellular metabolism. Insert the correct word(s) from the key choices in the answer blanks.

Key Choices

A. ATP	G. Basal metabolic rate (BMR)	M. Ketosis
B. Acetic acid	H. Carbon dioxide	N. Monosaccharides
C. Acetoacetic acid	I. Essential	O. Oxygen
D. Acetone	J. Fatty acids	P. Total metabolic rate (TMR)
E. Amino acids	K. Glucose	Q. Urea
F. Ammonia	L. Glycogen	R. Water

_____ 1.

_____ 2.

_____ 3.

_____ 4.

_____ 5.

_____ 6.

_____ 7.

_____ 8.

_____ 9.

_____ 10.

_____ 11.

_____ 12.

The key "fuel" used by body cells is __(1)__. The cells break this fuel molecule apart piece by piece. The hydrogen removed is combined with __(2)__ to form __(3)__, while its carbon leaves the body in the form of __(4)__ gas. The importance of this process is that it provides __(5)__, a form of energy that the cells can use to power all their activities. For carbohydrates to be oxidized, or burned for energy, they must first be broken down to __(6)__. When carbohydrates are unavailable to prime the metabolic pump, intermediate products of fat metabolism such as __(7)__ and __(8)__ accumulate in the blood, causing __(9)__ and low blood pH. Amino acids are actively accumulated by cells because protein cannot be made unless all amino acid types are present. The amino acids that *must* be taken in the diet are called __(10)__ amino acids. When amino acids are oxidized to form cellular energy, their amino groups are removed and liberated as __(11)__. In the liver, this is combined with carbon dioxide to form __(12)__, which is removed from the body by the kidneys.

25. Circle the term that does not belong in each of the following groupings.

1. BMR   TMR   Rest   Postabsorptive state

2. Thyroxine   Iodine   ↓Metabolic rate   ↑Metabolic rate

3. Obese person   ↓Metabolic rate   Women   Child

4. 4 kcal/gram   Fats   Carbohydrates   Proteins

5. Radiation   Vasoconstriction   Evaporation   Vasodilation

26. The liver has many functions in addition to its digestive function. Complete the following statements that elaborate on the liver's function by inserting the correct terms in the answer blanks.

_____ 1.

_____ 2.

_____ 3.

_____ 4.

_____ 5.

_____ 6.

_____ 7.

_____ 8.

_____ 9.

_____ 10.

_____ 11.

_____ 12.

_____ 13.

_____ 14.

_____ 15.

_____ 16.

_____ 17.

_____ 18.

_____ 19.

_____ 20.

The liver is the most important metabolic organ in the body. In its metabolic role, the liver uses amino acids from the nutrient-rich hepatic portal blood to make many blood proteins such as __(1)__, which helps to hold water in the bloodstream, and __(2)__, which prevent blood loss when blood vessels are damaged. The liver also makes a steroid substance that is released to the blood. This steroid, __(3)__, has been implicated in high blood pressure and heart disease. Additionally, the liver acts to maintain homeostatic blood glucose levels. It removes glucose from the blood when blood glucose levels are high, a condition called __(4)__, and stores it as __(5)__. Then, when blood glucose levels are low, a condition called __(6)__, liver cells break down the stored carbohydrate and release glucose to the blood once again. This latter process is termed __(7)__. When the liver makes glucose from noncarbohydrate substances such as fats or proteins, the process is termed __(8)__. In addition to its processing of amino acids and sugars, the liver plays an important role in the processing of fats. Other functions of the liver include the __(9)__ of drugs and alcohol. Its __(10)__ cells protect the body by ingesting bacteria and other debris.

The liver forms small complexes called __(11)__, which are needed to transport fatty acids, fats, and cholesterol in the blood because lipids are __(12)__ in a watery medium. The function of __(13)__ is transport of cholesterol to peripheral tissues, where cells use it to construct their plasma __(14)__ or to synthesize __(15)__. The function of high-density lipoproteins (HDLs) is transport of cholesterol to the __(16)__, where it is degraded and secreted as __(17)__, which are eventually excreted. High levels of cholesterol in the plasma are of concern because of the risk of __(18)__.

Two other important functions of the liver are the storage of vitamins (such as vitamin __(19)__ needed for vision) and of the metal __(20)__ (as ferritin).

27. Using the key choices, select the terms identified in the following descriptions. Insert the appropriate term(s) or letter(s) in each answer blank.

Key Choices

A. Blood

B. Constriction of skin blood vessels

C. Frostbite

D. Heat

E. Hyperthermia

F. Hypothalamus

G. Hypothermia

H. Perspiration

I. Radiation

J. Pyrogens

K. Shivering

_____ 1. By-product of cell metabolism

_____ 2. Means of conserving/increasing body heat

_____ 3. Means by which heat is distributed to all body tissues

_____ 4. Site of the body's thermostat

_____ 5. Chemicals released by injured tissue cells and bacteria, causing resetting of the thermostat

_____ 6. Death of cells deprived of oxygen and nutrients, resulting from withdrawal of blood from the skin circulation when the external temperature is low

_____ 7. Means of liberating excess body heat

_____ 8. Extremely low body temperature

_____ 9. Fever

DEVELOPMENTAL ASPECTS OF THE DIGESTIVE SYSTEM

28. Using the key choices, select the terms identified in the following descriptions. Insert the correct term(s) or letter(s) in each answer blank.

Key Choices

A. Accessory organs   F. Gallbladder problems   K. Rooting

B. Alimentary canal   G. Gastritis   L. Sucking

C. Appendicitis   H. PKU (phenylketonuria)   M. Stomach

D. Cleft palate/lip   I. Periodontal disease   N. Tracheoesophageal fistula

E. Cystic fibrosis   J. Peristalsis   O. Ulcers

_____ 1. Internal tubelike cavity of the embryo

_____ 2. Glands formed by branching from the digestive mucosa

_____ 3. Most common congenital defect; aspiration of feeding common

_____ 4. Congenital condition characterized by a connection between digestive and respiratory passageways

_____ 5. Congenital condition in which large amounts of mucus are produced, clogging respiratory passageways and pancreatic ducts

_____ 6. Metabolic disorder characterized by an inability to properly use the amino acid phenylalanine

_____ 7. Reflex aiding the newborn baby to find the nipple

_____ 8. Vomiting is common in infants because this structure is small

_____ 9. Most common adolescent digestive system problem

_____ 10. Inflammations of the gastrointestinal tract

_____ 11. Condition of loose teeth and inflamed gums; generally seen in elderly people

INCREDIBLE JOURNEY

A Visualization Exercise for the Digestive System

. . . the passage beneath you opens, and you fall into a huge chamber with mountainous folds.

29. Where necessary, complete statements by inserting the missing word(s) in the answer blanks.

_____ 1.

_____ 2.

_____ 3.

_____ 4.

_____ 5.

_____ 6.

_____ 7.

_____ 8.

_____ 9.

_____ 10.

_____ 11.

_____ 12.

_____ 13.

_____ 14.

_____ 15.

In this journey you are to travel through the digestive tract as far as the appendix and then await further instructions. You are miniaturized as usual and provided with a wet suit to protect you from being digested during your travels. You have a very easy entry into your host's open mouth. You look around and notice the glistening pink lining, or __(1)__, and the perfectly cared-for teeth. Within a few seconds, the lips part and you find yourself surrounded by bread. You quickly retreat to the safety of the __(2)__ between the teeth and the cheek to prevent getting chewed. From there you watch with fascination as a number of openings squirt fluid into the chamber, and the __(3)__ heaves and rolls, mixing the bread with the fluid.

As the bread begins to disappear, you decide that the fluid contains the enzyme __(4)__. You then walk toward the back of the oral cavity. Suddenly, you find yourself being carried along by a squeezing motion of the walls around you. The name given to this propelling motion is __(5)__. As you are carried helplessly downward, you see two openings—the __(6)__ and the __(7)__—below you. Just as you are about to straddle the solid area between them to stop your descent, the structure to your left moves quickly upward, and a trapdoor-like organ, the __(8)__, flaps over its opening. Down you go in the dark, seeing nothing. Then the passage beneath you opens, and you fall into a huge chamber with mountainous folds. Obviously, you have reached the __(9)__. The folds are very slippery, and you conclude that it must be the __(10)__ coat that you read about earlier. As you survey your surroundings, juices begin to gurgle into the chamber from pits in the "floor," and your face begins to sting and smart. You cannot seem to escape this caustic fluid and conclude that it must be very dangerous to your skin since it contains __(11)__ and __(12)__. You reach down and scoop up some of the slippery substance from the folds and smear it on your face, confident that if it can protect this organ it can protect you as well! Relieved, you begin to slide toward the organ's far exit and squeeze through the tight __(13)__ valve into the next organ. In the dim light, you see lumps of cellulose lying at your feet and large fat globules dancing lightly about. A few seconds later, your observations are interrupted by a wave of fluid pouring into the chamber from an opening high in the wall above you. The large fat globules begin to fall apart, and you decide that this enzyme flood has to contain __(14)__, and the opening must be the duct from the __(15)__. As you move quickly away to escape the deluge, you lose your footing and find

_____16.   yourself on a roller-coaster ride—twisting, coiling, turning, and diving through the lumen of this active organ. As you
_____17.   move, you are stroked by velvety, finger-like projections of the wall, the __(16)__. Abruptly your ride comes to a halt as you are catapulted through the __(17)__ valve and fall into the appendix. Headquarters informs you that you are at the end of your journey. Your exit now depends on your own ingenuity.

AT THE CLINIC

30. Mary Maroon comes to the clinic to get information on a vegetarian diet. What problems may arise when people make uninformed decisions on what to eat for a vegetarian diet? What combinations of vegetable foods will provide Mary with all the essential amino acids?

31. Mr. Ashe, a man in his mid-60s, comes to the clinic complaining of heartburn. Questioning by the clinic staff reveals that the severity of his attacks increases when he lies down after eating a heavy meal. The man is about 50 pounds overweight. What is your diagnosis? Without treatment, what conditions might develop?

32. There has been a record heat wave lately, and many elderly people are coming to the clinic complaining that they "feel poorly." In most cases, their skin is cool and clammy, and their blood pressure is low. What is their problem? What can be done to alleviate it?

33. During the same period, Bert Winchester, a construction worker, is rushed in unconscious. His skin is hot and dry, and his coworkers say that he just suddenly keeled over on the job. What is Bert's condition and how should it be handled?

34. Mrs. Ironfield is brought to an emergency room complaining of severe pain in her left iliac region. She claims previous episodes and says that the condition is worse when she is constipated and is relieved by defecation. A large tender mass is palpated in the left iliac fossa, and a barium study reveals a large number of diverticula in her descending and sigmoid colon. What are diverticula, and what is believed to promote their formation? Does this woman have diverticulitis or diverticulosis? Explain.

35. A woman in her 50s complains of bloating, cramping, and diarrhea when she drinks milk. What is the cause of her complaint and what is a solution?

36. Clients are instructed not to eat before having blood tests run. How would a lab technician know if someone "cheated" and ate a fatty meal a few hours before having his blood drawn?

37. Zena, a teenager, has gone to the sports clinic for the past 2 years to have her fat content checked. This year, her percentage of body fat is up, and tissue protein has not increased. Questioning reveals that Zena has been on crash diets four times since the last checkup, only to regain the weight (and more) each time. She also admits sheepishly that she "detests" exercise. How does cyclic dieting, accompanied by lack of exercise, cause an increase in fat and a decrease in protein?

38. Mrs. Rodriguez has a bleeding ulcer and has lost her appetite. She appears pale and lethargic when she comes in for a physical. She proves to be anemic, and her RBCs are large and pale. What mineral supplements should be ordered?

39. Mr. Roddick, a 21-year-old man with severe appendicitis, did not seek treatment in time and died a week after his abdominal pain and fever began. Explain why appendicitis can quickly lead to death.

40. In the mid-1960s, a calorie-free substitute (olestra) that is neither digested nor absorbed hit the market shelves in the United States. At that time there was concern that vitamin deficiencies might result from its use. What type of vitamins caused this concern and why?

✓ THE FINALE: MULTIPLE CHOICE

41. Select the best answer or answers from the choices given.

1. Which of the following terms are synonyms?

   A. Gastrointestinal tract

   B. Digestive system

   C. Digestive tract

   D. Alimentary canal

2. A digestive organ that is *not* part of the alimentary canal is the:

   A. stomach   D. large intestine

   B. liver   E. pharynx

   C. small intestine

3. The GI tube layer responsible for the actions of segmentation and peristalsis is:

   A. serosa   C. muscularis externa

   B. mucosa   D. submucosa

4. Which alimentary canal tunic has the greatest abundance of lymph nodules?

   A. Mucosa   C. Serosa

   B. Muscularis   D. Submucosa

5. Proteins secreted in saliva include:

   A. mucin   C. lysozyme

   B. amylase   D. IgA

6. The closure of which valve is assisted by the diaphragm?

   A. Ileocecal

   B. Pyloric

   C. Gastroesophageal

   D. Upper esophageal

7. Smooth muscle is found in the:

   A. tongue

   B. pharynx

   C. esophagus

   D. external anal sphincter

8. Which of these organs lies in the right hypochondriac region of the abdomen?

   A. Stomach   C. Cecum

   B. Spleen   D. Liver

9. Which phases of gastric secretion depend (at least in part) on the vagus nerve?

   A. Cephalic

   B. Gastric

   C. Intestinal (stimulatory)

   D. Intestinal (inhibitory)

10. Which of the following are tied to sodium transport?

    A. Glucose

    B. Fructose

    C. Galactose

    D. Amino acids

11. Excess iron is stored primarily in the:

    A. liver

    B. bone marrow

    C. duodenal epithelium

    D. blood

12. A 3-year-old girl was rewarded with a hug because she was now completely toilet trained. Which muscle had she learned to control?

    A. Levator ani

    B. Internal anal sphincter

    C. Internal and external obliques

    D. External anal sphincter

13. Which cell type fits this description? It occurs in the stomach mucosa, contains abundant mitochondria and many microvilli, and pumps hydrogen ions.

    A. Absorptive cell   C. Goblet cell

    B. Parietal cell   D. Mucous neck cell

14. Which of the following are "essential" nutrients?

    A. Glucose   C. Cholesterol

    B. Linoleic acid   D. Leucine

15. Deficiency of which of these vitamins results in anemia?

    A. Thiamin   C. Biotin

    B. Riboflavin   D. Folic acid

16. Vitamins that act as coenzymes in the Krebs cycle include:

    A. riboflavin   C. biotin

    B. niacin   D. pantothenic acid

17. Substrate-level phosphorylation occurs during:

    A. glycolysis   C. Krebs cycle

    B. beta-oxidation   D. electron transport

18. Chemicals that can be used for gluconeo-genesis include:

    A. amino acids

    B. glycerol

    C. fatty acids

    D. alpha-ketoglutaric acid

19. The chemiosmotic process involves:

    A. buildup of hydrogen ion concentration

    B. electron transport

    C. oxidation and reduction

    D. ATP synthase

20. Only the liver functions to:

    A. store iron

    B. form urea

    C. produce plasma proteins

    D. form ketone bodies

21. Which events occur during the absorptive state?

    A. Use of amino acids as a major source of energy

    B. Lipogenesis

    C. Beta-oxidation

    D. Increased uptake of glucose by skeletal muscles

22. Hormones that act to decrease blood glucose level include:

    A. insulin   C. epinephrine

    B. glucagon   D. growth hormone

23. During the postabsorptive state:

    A. glycogenesis occurs in the liver

    B. fatty acids are used for fuel

    C. amino acids are converted to glucose

    D. lipolysis occurs in adipose tissue

24. Which transport particles carry cholesterol destined for excretion from the body?

    A. HDL   C. LDL

    B. Chylomicron   D. VLDL (very low-density lipoprotein)

25. Glucose (or its metabolites) can be converted to:

    A. glycogen

    B. triglycerides

    C. nonessential amino acids

    D. starch

26. Basal metabolic rate:

    A. is the lowest metabolic rate of the body

    B. is the metabolic rate during sleep

    C. is measured as kcal per square meter of skin per hour

    D. increases with age

27. Which of the following types of heat transfer involves heat loss in the form of infrared waves?

    A. Conduction   C. Evaporation

    B. Convection   D. Radiation

28. PKU is the result of inability to metabolize:

    A. tyrosine   C. ketone bodies

    B. melanin   D. phenylalanine

15 THE URINARY SYSTEM

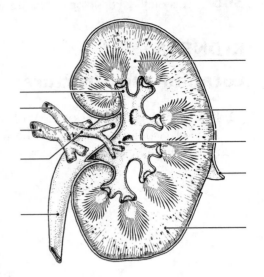

Metabolism of nutrients by body cells produces various wastes such as carbon dioxide and nitrogenous wastes (creatinine, urea, and ammonia), as well as imbalances of water and essential ions. The metabolic wastes and excesses must be eliminated from the body. Essential substances are retained to ensure proper body functioning.

Although several organ systems are involved in excretory processes, the urinary system bears the primary responsibility for removing nitrogenous wastes from the blood. In addition to this purely excretory function, the kidneys maintain the electrolyte, acid-base, and fluid balances of the blood. Thus, kidneys are major homeostatic organs of the body. Malfunction of the kidneys leads to a failure of homeostasis, resulting (unless corrected) in death.

Activities in this chapter are concerned with identification of urinary system structures, urine composition, and physiological processes involved in urine formation. It also focuses on the composition of the body's fluid compartments and the water, electrolyte, and acid-base balance of these compartments.

1. Complete the following statements by inserting your answers in the answer blanks.

_____ 1.

_____ 2.

_____ 3.

_____ 4.

_____ 5.

_____ 6.

_____ 7.

_____ 8.

The kidney is referred to as an excretory organ because it excretes __(1)__ wastes. It is also a major homeostatic organ because it maintains the electrolyte, __(2)__, and __(3)__ balance of the blood. Urine is continuously formed by the __(4)__ and is routed down the __(5)__ by the mechanism of __(6)__ to a storage organ called the __(7)__. Eventually the urine is conducted to the body exterior by the __(8)__. In males, this tube-like structure is about __(9)__ inches long; in females, it is approximately __(10)__ inches long.

_____ 9.

_____ 10.

KIDNEYS

Location and Structure

2. Figure 15–1 is an anterior view of the entire urinary system. Identify and select different colors for the following organs. Use them to color the coding circles and the corresponding organs on the figure.

◯ Kidney   ◯ Bladder   ◯ Ureters   ◯ Urethra

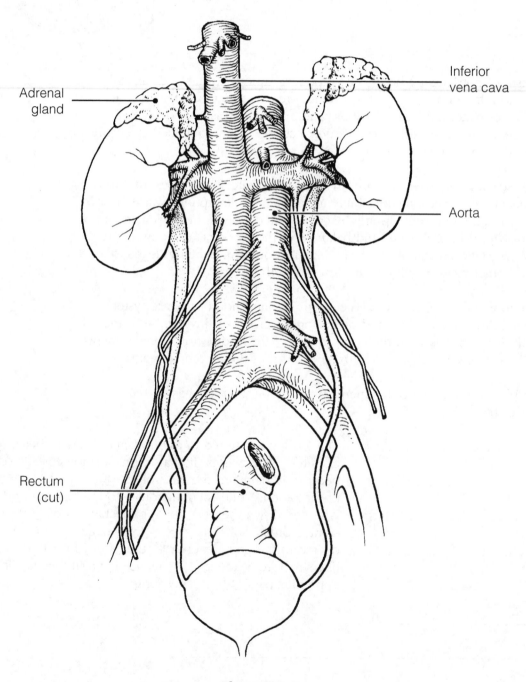

Adrenal gland

Inferior vena cava

Aorta

Rectum (cut)

Figure 15–1

3. Figure 15–2 is a longitudinal section of a kidney. First, using the correct anatomical terminology, label the following regions/structures indicated by leader lines on the figure.

- Fibrous membrane immediately surrounding the kidney

- Basin-like area of the kidney that is continuous with the ureter

- Cuplike extension of the pelvis that drains the apex of a pyramid

- Area of cortex-like tissue running through the medulla

Then, excluding the color red, select different colors to identify the following areas and structures. Then color in the coding circles and the corresponding area/structures on the figure; label these regions using the correct anatomical terms.

◯ Area of the kidney that contains the greatest proportion of nephron structures

◯ Striped-appearing structures formed primarily of collecting ducts

Finally, beginning with the renal artery, *draw in* the vascular supply to the cortex on the figure. Include and label the interlobar artery, arcuate artery, and cortical radiate artery. Color the vessels bright red.

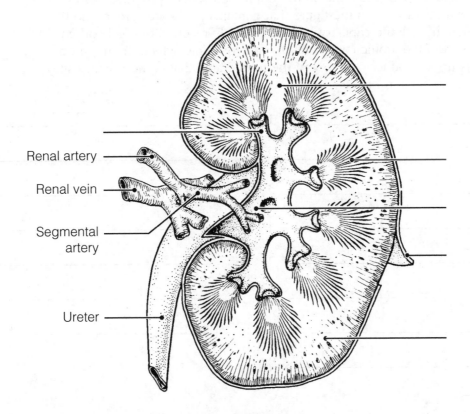

Renal artery

Renal vein

Segmental artery

Ureter

Figure 15–2

4. Circle the term that does not belong in each of the following groupings.

1. Intraperitoneal   Kidney   Retroperitoneal   Superior lumbar region

2. Drains kidney   Ureter   Urethra   Renal pelvis

3. Peritubular capillaries   Reabsorption   Glomerulus   Low-pressure vessels

4. Juxtaglomerular apparatus   Distal tubule   Glomerulus   Afferent arteriole

5. Glomerulus   Peritubular capillaries   Blood vessels   Collecting duct

6. Cortical nephrons   Juxtamedullary nephrons   Cortex/medulla junction

   Long nephron loops

7. Nephron   Proximal convoluted tubule   Distal convoluted tubule   Collecting duct

8. Medullary pyramids   Glomeruli   Renal pyramids   Collecting ducts

9. Glomerular capsule   Podocytes   Nephron loop   Glomerulus

Nephrons, Urine Formation, and Control of Blood Composition

5. Figure 15–3 is a diagram of the nephron and associated blood supply. First, match each of the numbered structures on the figure to one of the terms below the figure. Place the terms in the numbered spaces provided below. Then color the structure on the figure that contains podocytes green; the filtering apparatus red; the capillary bed that directly receives the reabsorbed substances from the tubule cells blue; the structure into which the nephron empties its urine product yellow; and the tubule area that is the primary site of tubular reabsorption orange.

_____ 1.   _____ 9.

_____ 2.   _____ 10.

_____ 3.   _____ 11.

_____ 4.   _____ 12.

_____ 5.   _____ 13.

_____ 6.   _____ 14.

_____ 7.   _____ 15.

_____ 8.

Figure 15-3

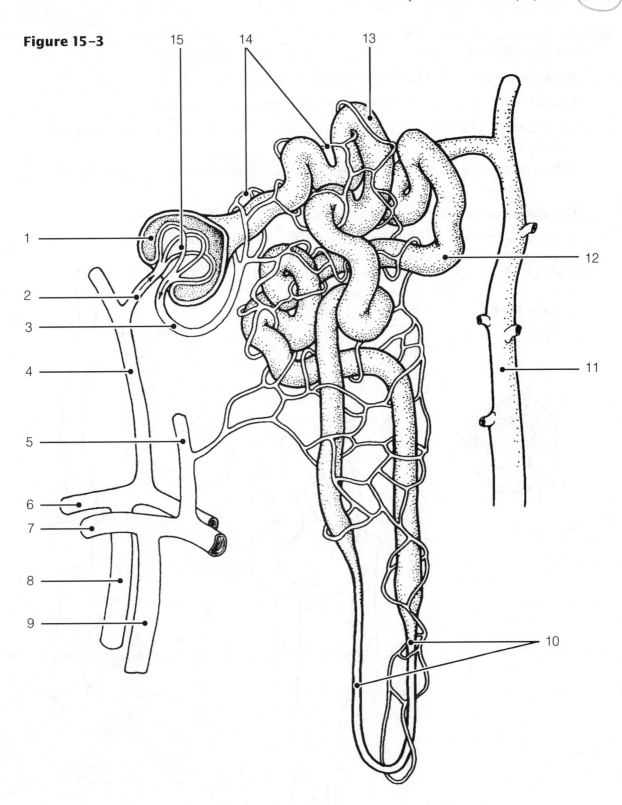

Afferent arteriole

Arcuate artery

Arcuate vein

Glomerular capsule

Collecting duct

Distal convoluted tubule

Efferent arteriole

Glomerulus

Interlobar artery

Interlobar vein

Cortical radiate artery

Cortical radiate vein

Nephron loop

Proximal convoluted tubule

Peritubular capillaries

6. Figure 15–4 is a diagram of a nephron. Add colored arrows on the figure as instructed to show the location and direction of the following processes.

1. **Black arrows** at the site of filtrate formation

2. **Red arrows** at the major site of amino acid and glucose reabsorption

3. **Green arrows** at the sites most responsive to action of ADH (show direction of water movement)

4. **Yellow arrows** at the sites most responsive to the action of aldosterone (show direction of Na^+ movement)

5. **Blue arrows** at the major site of tubular secretion

Then, label the proximal convoluted tubule (PCT), distal convoluted tubule (DCT), nephron loop, glomerular capsule, and glomerulus on the figure. Also label the collecting duct (not part of the nephron).

Figure 15–4

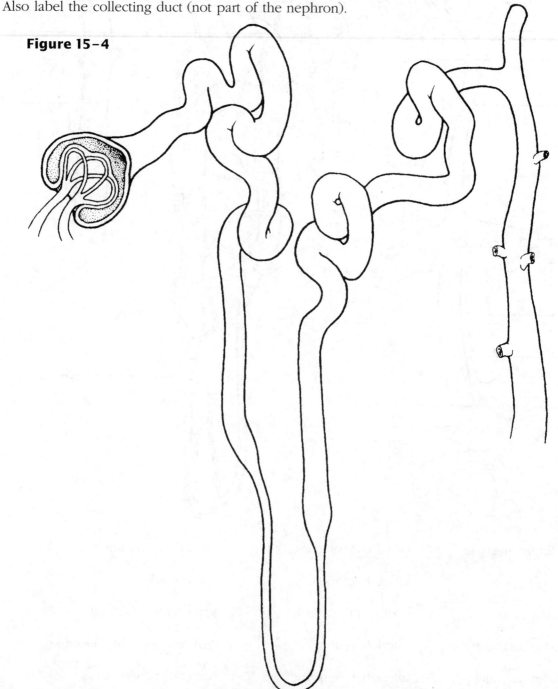

7. Complete the following statements by inserting your answers in the answer blanks.

_____ 1.

_____ 2.

_____ 3.

_____ 4.

_____ 5.

_____ 6.

_____ 7.

_____ 8.

_____ 9.

_____10.

_____11.

_____12.

_____13.

_____14.

_____15.

_____16.

The glomerulus is a unique high-pressure capillary bed because the __(1)__ arteriole feeding it is larger in diameter than the __(2)__ arteriole draining the bed. Glomerular filtrate is very similar to __(3)__, but it has fewer proteins. Mechanisms of tubular reabsorption include __(4)__ and __(5)__. As an aid for the reabsorption process, the cells of the proximal convoluted tubule have dense __(6)__ on their luminal surface, which increase the surface area dramatically. Other than reabsorption, an important tubule function is __(7)__, which is important for ridding the body of substances not already in the filtrate. Blood composition depends on __(8)__, __(9)__, and __(10)__. In a day's time, 180 L of blood plasma are filtered into the kidney tubules, but only about __(11)__ L of urine are actually produced. __(12)__ is responsible for the normal yellow color of urine. The three major nitrogenous wastes found in the blood, which must be disposed of, are __(13)__, __(14)__, and __(15)__. The kidneys are the final "judges" of how much water is to be lost from the body. When water loss via vaporization from the __(16)__ or __(17)__ from the skin is excessive, urine output __(18)__. If the kidneys become nonfunctional, __(19)__ is used to cleanse the blood of impurities.

_____17.

_____18.

_____19.

8. Decide whether the following conditions would cause urine to become more acidic or more basic. If more acidic, insert an *A* in the blank; if more basic, insert a *B* in the blank.

_____ 1. Protein-rich diet

_____ 2. Bacterial infection

_____ 3. Starvation

_____ 4. Diabetes mellitus

_____ 5. Vegetarian diet

9. Decide whether the following conditions would result in an increase or decrease in urine specific gravity. Insert *I* in the answer blank to indicate an increase and *D* to indicate a decrease.

_____ 1. Drinking excessive fluids

_____ 2. Chronic renal failure

_____ 3. Pyelonephritis

_____ 4. Using diuretics

_____ 5. Limited fluid intake

_____ 6. Fever

10. Assuming *normal* conditions, note whether each of the following substances would be (*G*) in greater concentration in the urine than in the glomerular filtrate, (*L*) in lesser concentration in the urine than in the glomerular filtrate, or (*A*) absent in both urine and glomerular filtrate. Place the correct letter in the answer blanks.

_____ 1. Water   _____ 5. Glucose   _____ 9. Potassium ions

_____ 2. Urea   _____ 6. Albumin   _____ 10. Red blood cells

_____ 3. Uric acid   _____ 7. Creatinine   _____ 11. Sodium ions

_____ 4. Pus (white   _____ 8. Hydrogen ions   _____ 12. Amino acids
      blood cells)

11. Several specific terms are used to indicate the presence of abnormal urine constituents. Identify each of the following abnormalities by inserting the term that names the condition in the spaces provided. Then for each condition, provide one possible cause of the condition in the remaining spaces.

1. Presence of red blood cells: _____. Cause: _____

2. Presence of ketones: _____. Cause: _____

3. Presence of albumin: _____. Cause: _____

4. Presence of pus: _____. Cause: _____

5. Presence of bile: _____. Cause: _____

6. Presence of "sand": _____. Cause: _____

7. Presence of glucose: _____. Cause: _____

12. Glucose and albumin are both normally absent from urine, but the reason for their exclusion differs. Respond to the following questions in the spaces provided.

1. Explain the reason for the absence of glucose in urine. _____

_____

2. Explain the reason for the absence of albumin in urine. _____

_____

13. By what three methods is H^+ concentration in body fluids regulated?

1. _____

2. _____

3. _____

Which of these methods is the fastest? _____

Which acts slowly but is most important for acid-base balance? _____

Which method removes CO_2 from the body? _____

14. Circle the term that does not belong in each of the following groupings
(ECF = extracellular fluid compartment).

1. Female adult   Male adult   About 50% water   Less muscle

2. Obese adult   Lean adult   Less body water   More adipose tissue

3. ECF   Interstitial fluid   Intracellular fluid   Plasma

4. Electric charge   Nonelectrolyte   Ions   Conducts a current

5. ↑Water output   ↓Na^+ concentration   ↑ADH   ↓ADH

6. Aldosterone   ↑Na^+ reabsorption   ↑K^+ reabsorption   ↑Blood pressure (BP)

URETERS, URINARY BLADDER, AND URETHRA

15. Circle the term that does not belong in each of the following groupings.

1. Bladder   Kidney   Transitional epithelium   Detrusor muscle

2. Trigone   Ureter openings   Urethral opening   Bladder   Forms urine

3. Surrounded by prostate gland   Contains internal and external sphincters

   Continuous with renal pelvis   Urethra

4. Prostatic   Male   Female   Membranous   Spongy

16. Using the key choices, identify the structures that best fit the following descriptions. Insert the correct term(s) or corresponding letter(s) in the answer blanks.

Key Choices

A. Bladder   B. Urethra   C. Ureter

_____ 1. Drains the bladder

_____ 2. Storage area for urine

_____ 3. Contains the trigone

_____ 4. In males has prostatic, membranous, and spongy parts

_____ 5. Conducts urine by peristalsis

_____ 6. Substantially longer in males than in females

_____ 7. A common site of "trapped" renal calculi

_____ 8. Contains transitional epithelium

_____ 9. Also transports sperm in males

17. Complete the following statements by inserting your answers in the answer blanks.

_____ 1.

_____ 2.

_____ 3.

_____ 4.

_____ 5.

_____ 6.

_____ 7.

_____ 8.

_____ 9.

_____ 10.

_____ 11.

_____ 12.   _____ 13.

Another term that means voiding or emptying of the bladder is __(1)__. Voiding has both voluntary and involuntary aspects. As urine accumulates in the bladder, __(2)__ are activated. This results in a reflex that causes the muscular wall of the bladder to __(3)__, and urine is forced past the __(4)__ sphincter. The more distal __(5)__ sphincter is controlled __(6)__; thus an individual can temporarily postpone emptying the bladder until it has accumulated __(7)__ mL of urine. __(8)__ is a condition in which voiding cannot be voluntarily controlled. It is normal in __(9)__ because nervous control of the voluntary sphincter has not been achieved. Other conditions that might result in an inability to control the sphincter include __(10)__ and __(11)__. __(12)__ is essentially the opposite of incontinence and often is a problem in elderly men because of __(13)__ enlargement.

18. Match the terms in Column B with the appropriate descriptions provided in Column A. Insert the correct term or letter in the answer blanks.

Column A

_____ 1. Inflammatory condition common in women with poor toileting habits

_____ 2. Backup of urine into the kidney; often a result of a blockage in the urinary tract

_____ 3. Toxic condition caused by renal failure

_____ 4. Inflammation of a kidney

_____ 5. A condition in which excessive amounts of urine are produced because of a deficiency of antidiuretic hormone (ADH)

_____ 6. Dropping of the kidney to a more inferior position in the abdomen; may result from a rapid weight loss that decreases the fatty cushion surrounding the kidney

Column B

A. Cystitis

B. Diabetes insipidus

C. Hydronephrosis

D. Ptosis

E. Pyelonephritis

F. Uremia

FLUID, ELECTROLYTE, AND ACID-BASE BALANCE

19. Determine if the following descriptions refer to electrolytes (*E*) or to nonelectrolytes (*N*).

_____ 1. Lipids, monosaccharides, and neutral fats

_____ 2. Have greater osmotic power at equal concentrations

_____ 3. The most numerous solutes in the body's fluid compartments

_____ 4. Salts, acids, and bases

_____ 5. Most of the *mass* of dissolved solutes in the body's fluid compartments

_____ 6. Each molecule dissociates into two or more ions

20. Circle the term that does not belong in each of the following groupings.

1. Hypothalamus   ADH   Aldosterone   Osmoreceptors

2. Glomerulus   Secretion   Filtration $\uparrow$ BP

3. Aldosterone $\uparrow$ Na$^+$ reabsorption $\uparrow$ K$^+$ reabsorption $\uparrow$ BP

4. ADH $\downarrow$ BP $\uparrow$ Blood volume $\uparrow$ Water reabsorption

5. $\downarrow$ Aldosterone   Edema $\downarrow$ Blood volume $\downarrow$ K$^+$ retention

6. $\downarrow$ Urine pH $\uparrow$ H$^+$ in urine $\uparrow$ HCO$_3^-$ in urine $\uparrow$ Ketones

7. $\uparrow$ ADH   Dilute urine $\uparrow$ Water absorption by collecting ducts   Dehydration

8. Renin   Angiotensin $\downarrow$ BP   JG (juxtaglomerular) apparatus

21. Figure 15–5 illustrates the three major fluid compartments of the body. Arrows indicate direction of fluid flow. First, select three different colors and color the coding circles and the fluid compartments on the figure. Then, referring to Figure 15–5, respond to the statements that follow. If a statement is true, write *T* in the answer blank. If a statement is false, change the underlined word(s) and write the correct word(s) in the answer blank.

◯ Interstitial fluid   ◯ Intracellular fluid   ◯ Plasma

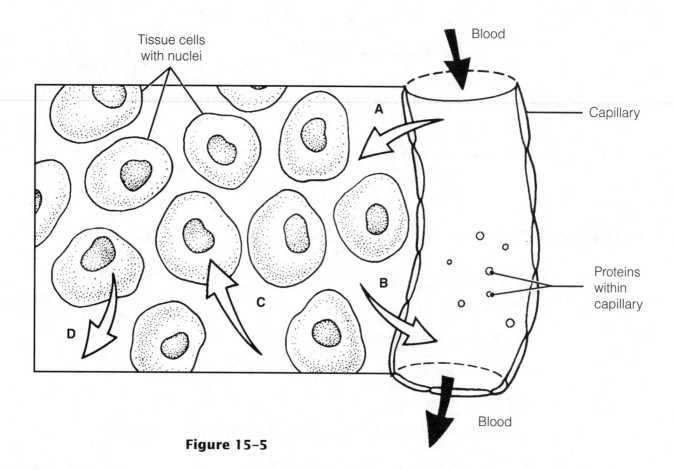

Figure 15–5

_____ 1. Exchanges between plasma and interstitial fluid compartments take place across the <u>capillary</u> membranes.

_____ 2. The fluid flow indicated by arrow A is driven by <u>active transport</u>.

_____ 3. If the osmolarity of the ECF is increased, the fluid flow indicated by arrow <u>C</u> will occur.

_____ 4. The excess of fluid flow at arrow A over that at arrow B normally enters the <u>tissue cells</u>.

_____ 5. Exchanges between the interstitial and intracellular fluid compartments occur across <u>capillary</u> membranes.

_____ 6. <u>Interstitial fluid</u> serves as the link between the body's external and internal environments.

22. Name three sources of body water and specify which source accounts for the bulk of body water.

_____

_____

23. Name four routes by which water is lost from the body and specify which route accounts for the greatest water loss.

_____

_____

_____

24. Match the pH values in Column B with the conditions described in Column A.

Column A	Column B
_____ 1. Normal pH of arterial blood	A. pH < 7.00
_____ 2. Physiological alkalosis (arterial blood)	B. pH = 7.00
_____ 3. Physiological acidosis (arterial blood)	C. pH < 7.35
_____ 4. Chemical neutrality; neither acidic nor basic	D. pH = 7.35
_____ 5. Chemical acidity	E. pH = 7.40
	F. pH > 7.45

25. Use the terms in Column B to complete the statements in Column A.

Column A	Column B
_____ 1. Acids are proton _____.	A. Acceptors
_____ 2. A strong acid dissociates _____.	B. Donors
_____ 3. A weak acid dissociates_____.	C. Completely
_____ 4. Strong bases bind _____ quickly	D. Hydrogen ions
	E. Incompletely

26. The activity of the bicarbonate buffer system of the blood is shown by the equation:

$$CO_2 + H_2O \rightleftharpoons H_2CO_3 \rightleftharpoons H^+ + HCO_2^-$$

1. Which chemical formulas refer to ions? _____

2. Which formula refers to a weak acid? _____ Which is a weak base? _____

3. If more CO_2 enters the blood, the reaction shifts up to the (right/left) _____.

DEVELOPMENTAL ASPECTS OF THE URINARY SYSTEM

27. Complete the following statements by inserting your responses in the answer blanks.

_____ 1.

_____ 2.

_____ 3.

_____ 4.

_____ 5.

_____ 6.

_____ 7.

_____ 8.

_____ 9.

_____ 10.

_____ 11.

_____ 12.

Three separate sets of renal tubules develop in the embryo; however, embryonic nitrogenous wastes are actually disposed of by the __(1)__. A congenital condition typified by blister-like sacs in the kidneys is __(2)__. __(3)__ is a congenital condition seen in __(4)__, when the urethral opening is located ventrally on the penis. A newborn baby voids frequently, which reflects its small __(5)__. Daytime control of the voluntary urethral sphincter is usually achieved by approximately __(6)__ months. Urinary tract infections are fairly common and not usually severe with proper medical treatment. A particularly problematic condition, called __(7)__, may result later in life as a sequel to childhood streptococcal infection. In this disease, the renal filters become clogged with __(8)__ complexes, urine output decreases, and __(9)__ and __(10)__ begin to appear in the urine. In old age, progressive __(11)__ of the renal blood vessels results in the death of __(12)__ cells. The loss of bladder tone leads to __(13)__ and __(14)__ and is particularly troublesome to elderly people.

_____ 13.

_____ 14.

INCREDIBLE JOURNEY

A Visualization Exercise for the Urinary System

You see the kidney looming brownish red through the artery wall.

28. Where necessary, complete statements by inserting the missing word(s) in the answer blanks.

_____ 1.

_____ 2.

For your journey through the urinary system, you must be made small enough to filter through the filtration membrane from the bloodstream into a renal __(1)__. You will be injected into the subclavian vein and must pass through the heart before entering the arterial circulation. As you travel through the systemic circulation, you have at least 2 minutes to relax before reaching the __(2)__ artery, feeding a kidney. You see the kidney looming brownish red through the artery wall. Once you have entered the kidney, the blood vessel conduits become increasingly smaller until you finally reach

_____ 3.

_____ 4.

_____ 5.

_____ 6.

_____ 7.

_____ 8.

_____ 9.

_____ 10.

_____ 11.

_____ 12.

_____ 13.

_____ 14.

_____ 15.

_____ 16.

_____ 17.

_____ 18.

_____ 19.

_____ 20.

_____ 21.

_____ 22.

_____ 23.

_____ 24.

the __(3)__ arteriole, feeding into the filtering device, or __(4)__. Once in the filter, you maneuver yourself so that you are directly in front of a pore. Within a fraction of a second, you are swept across the filtration membrane into the __(5)__ part of the nephron. Drifting along, you lower the specimen cup to gather your first filtrate sample for testing. You study the readout from the sample and note that it is very similar in composition to __(6)__ with one exception: There are essentially no __(7)__. Your next sample doesn't have to be collected until you reach the "hairpin," or, using the proper terminology, the __(8)__ part of the tubule. As you continue your journey, you notice that the tubule cells have dense finger-like projections extending from their surfaces into the lumen of the tubule. These are __(9)__, which increase the surface area of tubules because this portion of the tubule is very active in the process of __(10)__. Soon you collect your second sample, and then later, in the distal convoluted tubule, your third sample. When you read the computer's summary of the third sample, you make the following notes in your register.

- Virtually no nutrients such as __(11)__ and __(12)__ are left in the filtrate.

- The pH is acidic, 6.0. This is quite a change from the pH of __(13)__ recorded for the newly formed filtrate.

- There is a much higher concentration of __(14)__ wastes here.

- There are many fewer __(15)__ ions but more of the __(16)__ ions noted.

- Color of the filtrate is yellow, indicating a high relative concentration of the pigment __(17)__.

Gradually you become aware that you are moving along much more quickly. You see that the water level has dropped dramatically and that the stream is turbulent and rushing. As you notice this, you realize that the hormone __(18)__ must have been released recently to cause this water drop. You take an abrupt right turn and then drop straight downward. You realize that you must be in a __(19)__. Within a few seconds, you are in what appears to be a large tranquil sea with a tide flowing toward a darkened area at the far shore. You drift toward the darkened area, confident that you are in the kidney __(20)__. As you reach and enter the dark tubelike structure seen from the opposite shore, your progress becomes rhythmic—something like being squeezed through a sausage skin. Then you realize that your progress is being regulated by the process of __(21)__. Suddenly, you free-fall and land in the previously stored __(22)__ in the bladder, where the air is very close. Soon the walls of the bladder begin to gyrate, and you realize you are witnessing a __(23)__ reflex. In a moment, you are propelled out of the bladder and through the __(24)__ to exit from your host.

AT THE CLINIC

29. A man was admitted to the hospital after being trampled by his horse. He received crushing blows to his lower back, on both sides. He is in considerable pain, and his chart shows a urine output of 70 mL in the last 24 hours. What is this specific symptom called? What will be required if the renal effects of his trauma persist?

30. Four-year-old Eddie is a chronic bed wetter. He wets the bed nearly every night. What might explain his problem?

31. If a tumor of the glucocorticoid-secreting cells of the adrenal cortex crowds out the cells that produce aldosterone, what is the likely effect on urine composition and volume?

32. Jimmy has been stressed out lately as he has been juggling two jobs while taking classes at a local college. He appears at the clinic complaining of a pounding headache. Tests show that he has high blood pressure, and his cortiscosteroid levels are elevated. What is the relationship between his stress and his signs and symptoms?

33. Mr. O'Toole is very drunk when he is brought to the emergency room after falling down City Hall steps during a political rally. He is complaining about his "cotton mouth." Knowing that alcohol inhibits ADH's action, you explain to him why his mouth is so dry. What do you tell him?

34. Mrs. Rodriques is breathing rapidly and is slurring her speech when her husband calls the clinic in a panic. Shortly after, she becomes comatose. Tests show that her blood glucose and ketone levels are high, and her husband said that she was urinating every few minutes before she became lethargic. What is Mrs. Rodriques's problem? Would you expect her blood pH to be acidic or alkaline? What is the significance of her rapid breathing? Are her kidneys reabsorbing or secreting bicarbonate ions during this crisis?

35. Many employers now require that prospective employees' urine be tested before they will consider hiring them. What aspect of kidney function is involved here and what is being investigated?

36. Conn's syndrome results from adrenocortical tumors that secrete aldosterone in an unregulated way. What would you "guesstimate" would be the major symptom of this syndrome?

✅ THE FINALE: MULTIPLE CHOICE

37. Select the best answer or answers from the choices given.

1. A radiologist is examining an X-ray of the lumbar region of a patient. Which of the following is (are) indicative of normal positioning of the right kidney?

   A. Slightly lower than the left kidney

   B. More medial than the left kidney

   C. Closer to the inferior vena cava than the left kidney

   D. Anterior to the 12th rib

2. Which of the following encloses both kidney and adrenal gland?

   A. Renal fascia

   B. Perirenal fat capsule

   C. Fibrous capsule

   D. Visceral peritoneum

3. Microscopic examination of a section of the kidney shows a thick-walled vessel with renal corpuscles scattered in the tissue on one side of the vessel but not on the other side. What vessel is this?

   A. Interlobar artery

   B. Cortical radiate artery

   C. Cortical radiate vein

   D. Arcuate artery

4. Structures that are at least partly composed of simple squamous epithelium include:

   A. collecting ducts

   B. glomerulus

   C. glomerular capsule

   D. nephron loop

5. Which structures are freely permeable to water?

   A. Distal convoluted tubule

   B. Thick segment of ascending limb of the nephron loop

   C. Descending limb of the nephron loop

   D. Proximal convoluted tubule

6. A major function of the collecting ducts is:

   A. secretion

   B. filtration

   C. concentrating urine

   D. lubrication with mucus

7. What is the glomerulus?

   A. The same as the renal corpuscle

   B. The same as the renal tubule

   C. The same as the nephron

   D. Capillaries

8. Urine passes through the ureters by which mechanism?

   A. Ciliary action

   B. Peristalsis

   C. Gravity alone

   D. Suction

9. Sodium deficiency hampers reabsorption of:

   A. glucose

   B. albumin

   C. creatinine

   D. water

10. The main function of transitional epithelium in the ureter is:

    A. protection against kidney stones

    B. secretion of mucus

    C. reabsorption

    D. stretching

11. Jim was standing at a urinal in a crowded public restroom and a long line was forming behind him. He became anxious (sympathetic response) and found he could not micturate no matter how hard he tried. Use logic to deduce Jim's problem.

    A. His internal urethral sphincter was constricted and would not relax.

    B. His external urethral sphincter was constricted and would not relax.

    C. His detrusor muscle was contracting too hard.

    D. He almost certainly had a burst bladder.

12. Which of the following are normal values?

    A. Urine output of 1.5 L/day

    B. Specific gravity of 1.5

    C. pH of 6

    D. GFR of 125 mL/hour

13. The ureter:

    A. is continuous with the renal pelvis

    B. is lined by the renal capsule

    C. exhibits peristalsis

    D. is much longer in the male than in the female

14. The urinary bladder:

    A. is lined with transitional epithelium

    B. has a thick, muscular wall

    C. receives the ureteral orifices at its superior aspect

    D. is innervated by the renal plexus

15. Which of the following are controlled voluntarily?

    A. Detrusor muscle

    B. Internal urethral sphincter

    C. External urethral sphincter

    D. Levator ani muscle

16. In movement between IF and ICF:

    A. water flow is bidirectional

    B. nutrient flow is unidirectional

    C. ion flow is selectively permitted

    D. ion fluxes are not permitted

17. Which is a normal value for percentage of body weight that is water for a middle-aged man?

    A. 73%   C. 45%

    B. 50%   D. 60%

18. The smallest fluid compartment is the:

    A. ICF   C. plasma

    B. ECF   D. IF

19. Which of the following are electrolytes?

    A. Glucose

    B. Lactic acid

    C. Urea

    D. Bicarbonate

20. Chloride ion reabsorption:

    A. exactly parallels sodium ion reabsorption

    B. fluctuates according to blood pH

    C. increases during acidosis

    D. is controlled directly by aldosterone

21. Respiratory acidosis occurs in:

    A. asthma

    B. emphysema

    C. barbiturate overdose

    D. cystic fibrosis

22. Hyperkalemia:

    A. triggers secretion of aldosterone

    B. may result from severe alcoholism

    C. disturbs acid-base balance

    D. results from widespread tissue injury

23. Renal tubular secretion of potassium is:

    A. obligatory

    B. increased by aldosterone

    C. balanced by tubular reabsorption

    D. increased in alkalosis

24. Which buffer system(s) is (are) not important urine buffers?

    A. Phosphate   C. Protein

    B. Ammonium   D. Bicarbonate

16 THE REPRODUCTIVE SYSTEM

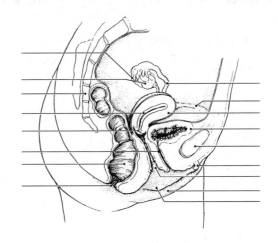

The biological function of the reproductive system is to produce offspring. The essential organs are those producing the germ cells (testes in males and ovaries in females). The male manufactures sperm and delivers them to the female's reproductive tract. The female, in turn, produces eggs. If the time is suitable, the egg and sperm fuse, producing a fertilized egg, which is the first cell of the new individual. Once fertilization has occurred, the female uterus protects and nurtures the developing embryo.

In this chapter, student activities concern the structures of the male and female reproductive systems, germ cell formation, the menstrual cycle, and embryonic development.

ANATOMY OF THE MALE REPRODUCTIVE SYSTEM

1. Using the following terms, trace the pathway of sperm from the testis to the urethra: rete testis, epididymis, seminiferous tubule, ductus deferens. List the terms in the proper order in the spaces provided.

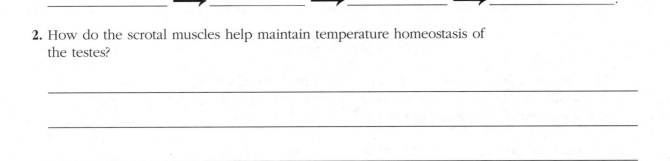

2. How do the scrotal muscles help maintain temperature homeostasis of the testes?

_____

_____

_____

3. Using the key choices, select the terms identified in the following descriptions. Insert the appropriate term(s) or corresponding letter(s) in the answer blanks.

Key Choices

A. Bulbo-urethral glands   E. Penis   I. Scrotum

B. Epididymis   F. Prepuce   J. Spermatic cord

C. Ductus deferens   G. Prostate   K. Testes

D. Glans penis   H. Seminal vesicles   L. Urethra

_____ 1. Organ that delivers semen to the female reproductive tract

_____ 2. Site of testosterone production

_____ 3. Passageway from the epididymis to the ejaculatory duct

_____ 4. Conveys both sperm and urine down the length of the penis

_____ 5. Organs that contribute to the formation of semen

_____ 6. External skin sac that houses the testes

_____ 7. Tubular storage site for sperm; hugs the lateral aspect of the testes

_____ 8. Cuff of skin encircling the glans penis

_____ 9. Surrounds the urethra at the base of the bladder; produces a milky fluid

_____ 10. Produces more than half of the seminal fluid

_____ 11. Produces a lubricating mucus that cleanses the urethra

_____ 12. Connective tissue sheath enclosing the ductus deferens, blood vessels, and nerves.

4. Figure 16–1 is a sagittal view of the male reproductive structures. First, identify the following organs on the figure by placing each term at the end of the appropriate leader line.

Bulbo-urethral gland   Erectile tissue   Urethra

Ductus deferens   Glans penis   Scrotum

Ejaculatory duct   Prepuce   Seminal vesicle

Epididymis   Prostate   Testis

Next, select different colors for the structures that correspond to the following descriptions, and color in the coding circles and the corresponding structures on the figure.

◯ Spongy tissue that is engorged with blood during erection

◯ Portion of the duct system that also serves the urinary system

◯ Structure that provides the ideal temperature conditions for sperm formation

◯ Structure removed in circumcision

◯ Gland whose secretion contains sugar to nourish sperm

◯ Structure cut or cauterized during a vasectomy

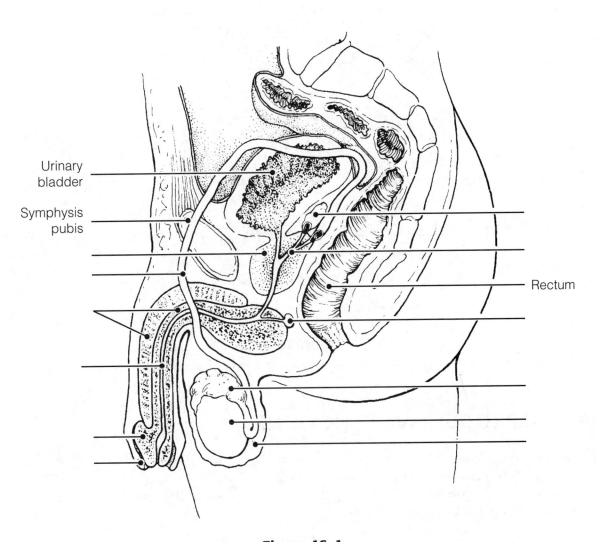

Figure 16–1

5. Figure 16–2 is a longitudinal section of a testis. First, select different colors for the structures that correspond to the following descriptions. Then color the coding circles and color and label the corresponding structures on the figure. *Complete the labeling* of the figure by adding the following terms: lobule, rete testis, and septum.

◯ Site(s) of spermatogenesis

◯ Tubular structure in which sperm mature and become motile

◯ Fibrous coat protecting the testis

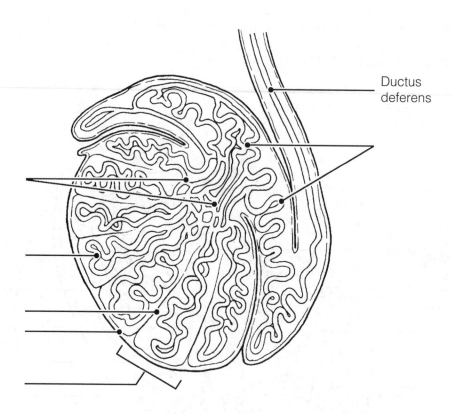

Ductus deferens

Figure 16–2

MALE REPRODUCTIVE FUNCTIONS

6. This section considers the process of sperm production in the testis. Figure 16–3 is a cross-sectional view of a seminiferous tubule in which spermatogenesis is occurring. First, using the key choices, select the terms identified in the following descriptions.

Key Choices

A.   Follicle-stimulating hormone (FSH)   E. ◯   Sperm

B. ◯ Primary spermatocyte   F. ◯   Spermatid

C. ◯ Secondary spermatocyte   G.   Testosterone

D. ◯ Spermatogonium

_____ 1. Primitive stem cell

_____ 2. Contain 23 chromosomes (3 answers)

_____ and _____

_____ 3. Product of meiosis I

_____ 4. Product of meiosis II

_____ 5. Functional motile gamete

_____ 6. Two hormones necessary for sperm production

_____

Then label the cells with leader lines. Select different colors for the cell types
with color-coding circles listed in the key choices and color in the coding circles and corresponding structures on the figure. In addition, label and color
the cells that produce testosterone.

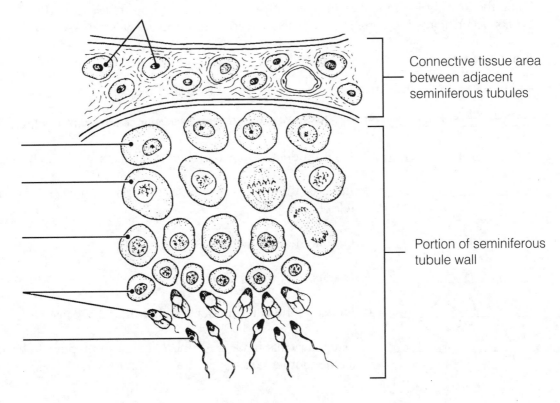

Connective tissue area
between adjacent
seminiferous tubules

Portion of seminiferous
tubule wall

Figure 16–3

7. Figure 16–4 illustrates a single sperm. On the figure, bracket and label the head and the midpiece and circle and label the tail. Select different colors for the structures that correspond to the following descriptions. Color the coding circles and corresponding structures on the figure. Then label the structures, using correct terminology.

◯ The DNA-containing area

◯ The enzyme-containing sac that aids sperm penetration of the egg

◯ Metabolically active organelles that provide ATP to energize sperm movement

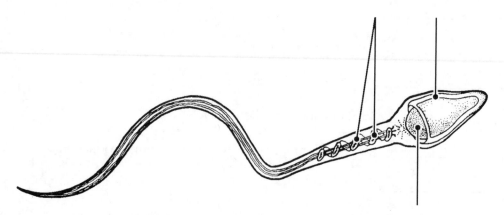

Figure 16–4

8. The following statements refer to events that occur during cellular division. Using the key choices, indicate in which type of cellular division the described events occur. Place the correct term or letter response in the answer blanks.

Key Choices

A. Mitosis   B. Meiosis   C. Both mitosis and meiosis

_____ 1. Final product is two daughter cells, each with 46 chromosomes

_____ 2. Final product is four daughter cells, each with 23 chromosomes

_____ 3. Involves the phases prophase, metaphase, anaphase, and telophase

_____ 4. Occurs in all body tissues

_____ 5. Occurs only in the gonads

_____ 6. Increases the cell number for growth and repair

_____ 7. Daughter cells have the same number and types of chromosomes as the mother cell

_____ 8. Daughter cells are different from the mother cell in their chromosomal makeup

_____ 9. Chromosomes are replicated before the division process begins

_____ 10. Provides cells for the reproduction of offspring

_____ 11. Consists of two consecutive divisions of the nucleus; chromosomes are not replicated before the second division

9. Name four of the male secondary sex characteristics. Insert your answers on the lines provided.

_____

_____

_____

_____

ANATOMY OF THE FEMALE REPRODUCTIVE SYSTEM

10. Identify the female structures described by inserting your responses in the answer blanks.

_____ 1. Chamber that houses the developing fetus

_____ 2. Canal that receives the penis during sexual intercourse

_____ 3. Usual site of fertilization

_____ 4. Erects during sexual stimulation

_____ 5. Duct through which the ovum travels to reach the uterus

_____ 6. Membrane that partially closes the vaginal canal

_____ 7. Primary female reproductive organ

_____ 8. Move to create fluid currents to draw the ovulated egg into the uterine (fallopian) tube

11. Figure 16–5 is a sagittal view of the female reproductive organs. First, label all structures on the figure provided with leader lines. Then select different colors for the following structures, and use them to color the coding circles and corresponding structures on the figure.

○ Lining of the uterus, endometrium

○ Muscular layer of the uterus, myometrium

○ Pathway along which an egg travels from the time of its release to its implantation

○ Ligament helping to anchor the uterus

○ Structure producing female hormones and gametes

○ Homologue of the male scrotum

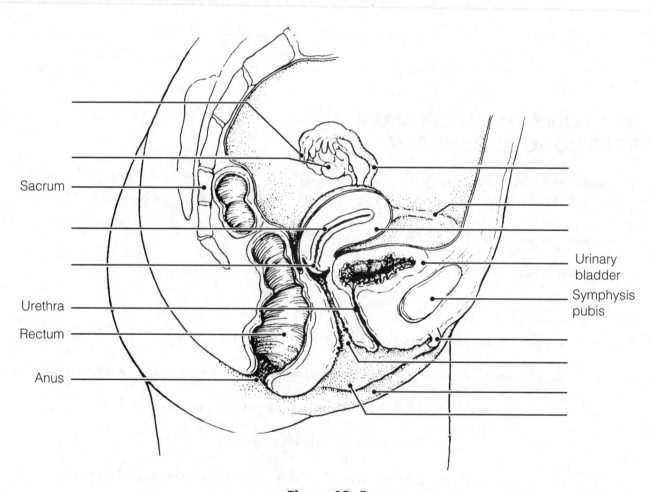

Sacrum

Urethra

Rectum

Anus

Urinary bladder

Symphysis pubis

Figure 16–5

12. Figure 16–6 is a ventral view of the female external genitalia. Label the clitoris, labia minora, urethral orifice, hymen, mons pubis, and vaginal orifice on the figure. These structures are indicated with leader lines. Then color the homologue of the male penis blue, color the membrane that partially obstructs the vagina yellow, and color the distal end of the birth canal red.

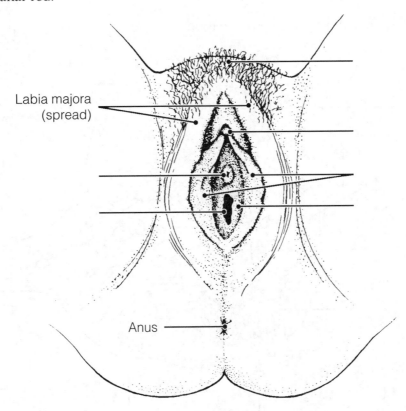

Labia majora (spread)

Anus

Figure 16–6

FEMALE REPRODUCTIVE FUNCTIONS AND CYCLES

13. Using the key choices, identify the cell type you would expect to find in the following structures. Insert the correct term or letter response in the answer blanks.

Key Choices

A. Oogonium   C. Secondary oocyte

B. Primary oocyte   D. Ovum

_____ 1. Forming part of the primary follicle in the ovary

_____ 2. In the uterine tube before fertilization

_____ 3. In the mature, or Graafian, follicle of the ovary

_____ 4. In the uterine tube shortly after sperm penetration

14. Figure 16–7 is a sectional view of the ovary. First, identify all structures indicated with leader lines on the figure. Second, select different colors for the following structures, and use them to color the coding circles and corresponding structures on the figure.

◯ Cells that produce estrogen

◯ Glandular structure that produces progesterone

◯ All oocytes

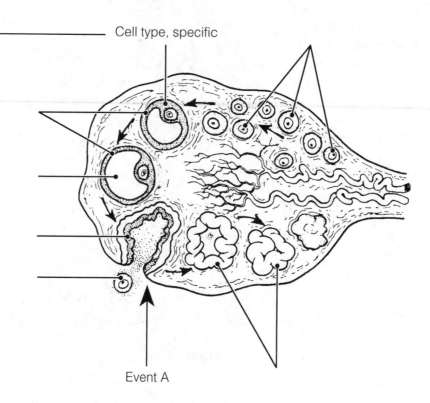

Cell type, specific

Event A

Figure 16–7

Third, in the space provided, name the event depicted as "Event A" on the figure. _____

Fourth, answer the following questions by inserting your answers in the spaces provided.

1. Are there any oogonia in a mature female's ovary? _____

2. Into what area is the ovulated cell released? _____

3. When is a mature ovum (egg) produced in humans? _____

4. What structure in the ovary becomes a corpus luteum? _____

5. What are the four final cell types produced by oogenesis in the female? (Name the cell

type and number of each.) _____

6. How does this compare with the final product of spermatogenesis in males? _____

_____

7. What happens to the tiny cells nearly devoid of cytoplasm ultimately produced during

oogenesis? _____

8. Why? _____

9. What name is given to the period of a woman's life when her ovaries begin to become

nonfunctional? _____

15. What is the significance of the fact that the uterine tubes are not structurally
continuous with the ovaries? Address this question from both reproductive
and health aspects.

_____

_____

_____

16. The following statements deal with anterior pituitary and ovarian hormonal
interrelationships. Name the hormone(s) described in each statement.
Place your answers in the answer blanks.

_____ 1. Promotes growth of ovarian follicles and production of estrogen

_____ 2. Triggers ovulation

_____ 3. Inhibit follicle-stimulating hormone (FSH) release by the anterior
pituitary

_____ 4. Stimulates luteinizing hormone (LH) release by the anterior
pituitary

_____ 5. Converts the ruptured follicle into a corpus luteum and causes it
to produce progesterone and estrogen

_____ 6. Maintains the hormonal production of the corpus luteum

17. Name four of the secondary sex characteristics of females. Place your answers
in the spaces provided.

_____

_____

_____

18. Use the key choices to identify the ovarian hormone(s) responsible for the following events. Insert the correct term(s) or letter(s) in the answer blanks.

Key Choices

A. Estrogens   B. Progesterone

_____ 1. Lack of this (these) causes the blood vessels to kink and the endometrium to slough off (menses)

_____ 2. Causes the endometrial glands to begin the secretion of nutrients

_____ 3. The endometrium is repaired and grows thick and velvety

_____ 4. Maintains the myometrium in an inactive state if implantation of an embryo has occurred

_____ 5. Glands are formed in the endometrium

_____ 6. Responsible for the secondary sex characteristics of females

19. The following exercise refers to Figure 16–8 A–D.

On Figure 16–8A, the blood levels of two gonadotropic hormones (FSH and LH) of the anterior pituitary are indicated. Identify each hormone by appropriately labeling the blood level lines on the figure. Then select different colors for each of the blood level lines and color them in on the figure.

On Figure 16–8B, identify the blood level lines for the ovarian hormones, estrogens and progesterone. Then select different colors for each blood level line, and color them in on the figure.

On Figure 16–8C, select different colors for the following structures and use them to color in the coding circles and corresponding structures in the figure.

◯ Primary follicle   ◯ Secondary (growing) follicle

◯ Vesicular follicle   ◯ Corpus luteum

◯ Ovulating follicle   ◯ Atretic (deteriorating) corpus luteum

On Figure 16–8D, identify the endometrial changes occurring during the menstrual cycle by color-coding and coloring the areas depicting the three phases of that cycle.

◯ Secretory phase   ◯ Menses   ◯ Proliferative phase

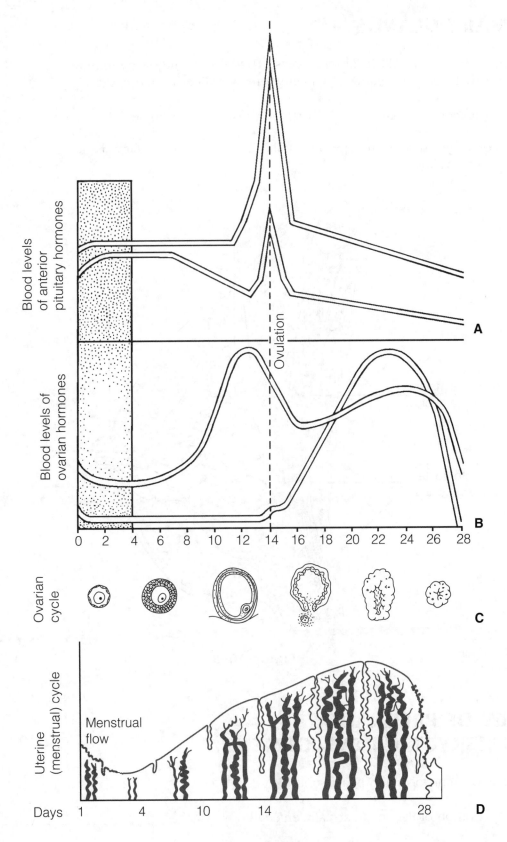

Figure 16-8

MAMMARY GLANDS

20. Figure 16–9 is a sagittal section of a breast. First, use the following terms to correctly label all structures provided with leader lines on the figure.

Alveolar glands   Areola   Lactiferous duct   Nipple

Then color the structures that produce milk blue and color the fatty tissue of the breast yellow.

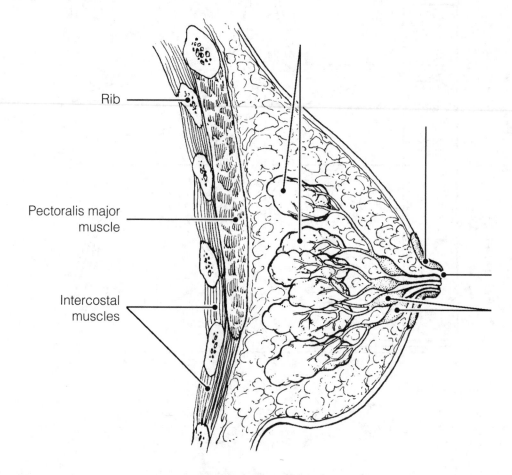

Rib

Pectoralis major
muscle

Intercostal
muscles

Figure 16–9

SURVEY OF PREGNANCY AND EMBRYONIC DEVELOPMENT

21. Relative to events of sperm penetration:

1. What portion of the sperm actually enters the oocyte? _____

2. What is the functional importance of the acrosomal reaction? _____

_____

_____

22. Figure 16–10 depicts early embryonic events. In questions #1–5, identify the events, cell types, or processes referring to the figure. Then respond to question #6. Place your answers in the spaces provided.

1. Event A _____

2. Cell resulting from event A _____

3. Process B _____

4. Embryonic structure B_1 _____

5. Completed process C _____

6. Assume that a sperm has entered a polar body instead of a secondary oocyte and their nuclei fuse. Why would it be unlikely for that "fertilized cell" to develop into an embryo?

_____

Figure 16–10

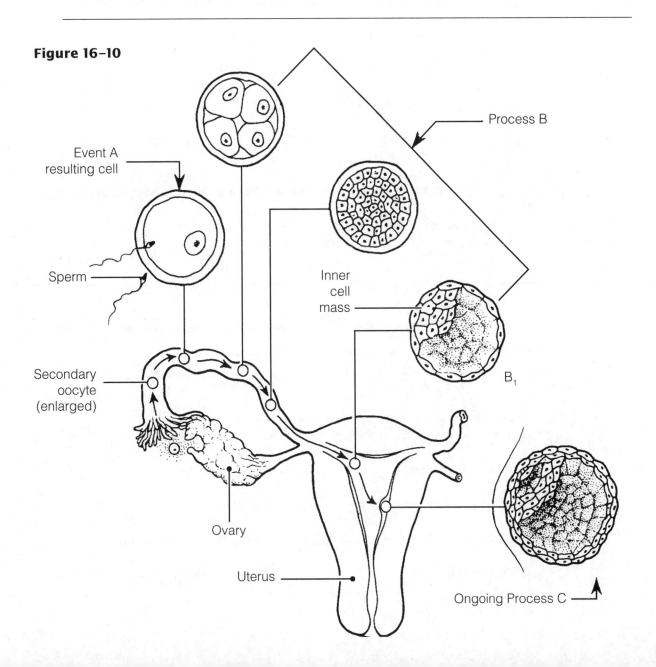

23. Using the key choices, select the terms that are identified in the following descriptions. Insert the correct term or letter response in the answer blanks.

Key Choices

A. Amnion   D. Fertilization   G. Umbilical cord

B. Chorionic villi   E. Fetus   H. Zygote

C. Endometrium   F. Placenta

_____ 1. The fertilized egg

_____ 2. Secretes estrogen and progesterone to maintain the pregnancy

_____ 3. Cooperate to form the placenta

_____ 4. Fluid-filled sac surrounding the developing embryo/fetus

_____ 5. Attaches the embryo to the placenta

_____ 6. Finger-like projections of the blastocyst

_____ 7. The embryo after 8 weeks

_____ 8. The organ that delivers nutrients to and disposes of wastes for the fetus

_____ 9. Event leading to combination of ovum and sperm "genes"

24. Explain why the corpus luteum does not stop producing its hormones (estrogens and progesterone) when fertilization has occurred.

_____

_____

25. The first "tissues" of the embryo's body are the primary germ layers:

A. Ectoderm   B. Mesoderm   C. Endoderm

Indicate which germ layer gives rise to each of the following structures by placing the corresponding letter in the answer blank.

_____ 1. Heart and blood vessels   _____ 5. Skin epidermis

_____ 2. Digestive system mucosa   _____ 6. Bones

_____ 3. Brain and spinal cord   _____ 7. Respiratory system mucosa

_____ 4. Skeletal muscles   _____ 8. Liver and pancreas

26. What two hormones are essential to initiate labor in humans?

_____

27. 1. What hormone is responsible for milk production? _____

2. For milk ejection? _____

28. A pregnant woman undergoes numerous changes during her pregnancy—anatomical, metabolic, and physiological. Several such possibilities are listed below. Check (✓) all that are commonly experienced during pregnancy.

_____ 1. Diaphragm descent is impaired

_____ 2. Breasts decline in size

_____ 3. Pelvic ligaments are relaxed by relaxin

_____ 4. Vital capacity decreases

_____ 5. Lordosis

_____ 6. Blood pressure and pulse rates decline

_____ 7. Metabolic rate declines

_____ 8. Increased mobility of GI tract

_____ 9. Blood volume and cardiac output increase

_____ 10. Nausea, heartburn, constipation

_____ 11. Dyspnea may occur

_____ 12. Urgency and stress incontinence

29. What are Braxton Hicks contractions, and why do they occur?

_____

_____

_____

_____

30. Name the three phases of parturition, and briefly describe each phase.

1. _____

_____

2. _____

_____

3. _____

_____

31. The very simple flowchart in Figure 16–11 illustrates the sequence of events that occur during labor. Complete the flowchart by filling in the missing terms in the boxes. Use color as desired.

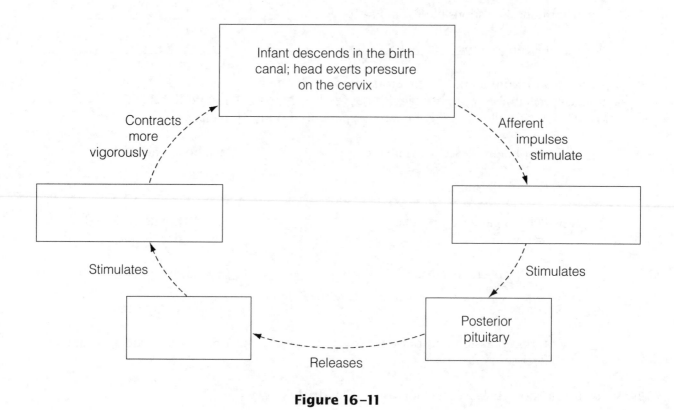

Figure 16–11

32. How long will the cycle illustrated in Figure 16–11 continue to occur?

_____

33. Labor is an example of a positive feedback mechanism. What does that mean?

_____

_____

_____

DEVELOPMENTAL ASPECTS OF THE REPRODUCTIVE SYSTEM

34. Complete the following statements by inserting your responses in the answer blanks.

_____ 1.

_____ 2.

_____ 3.

_____ 4.

_____ 5.

_____ 6.

_____ 7.

_____ 8.

_____ 9.

_____ 10.

_____ 11.

_____ 12.

_____ 13.

_____ 14.

_____ 15.

_____ 16.

_____ 17.

_____ 18.

_____ 19.

_____ 20.

_____ 21.

_____ 22.

A male embryo has __(1)__ sex chromosomes, whereas a female has __(2)__. During early development, the reproductive structures of both sexes are identical, but by the 8th week, __(3)__ begin to form if testosterone is present. In the absence of testosterone, __(4)__ form. The testes of a male fetus descend to the scrotum shortly before birth. If this does not occur, the resulting condition is called __(5)__.

The most common problem affecting the reproductive organs of women are infections, particularly __(6)__, __(7)__, and __(8)__. When the entire pelvis is inflamed, the condition is called __(9)__. Most male problems involve inflammations, resulting from __(10)__. A leading cause of cancer death in adult women is cancer of the __(11)__; the second most common female reproductive system cancer is cancer of the __(12)__. Thus a yearly __(13)__ is a very important preventive measure for early detection of this latter cancer type. The cessation of ovulation in an aging woman is called __(14)__. Intense vasodilation of blood vessels in the skin lead to uncomfortable __(15)__. Additionally, bone mass __(16)__ and blood levels of cholesterol __(17)__ when levels of the hormone __(18)__ wane. In contrast, healthy men are able to father children well into their 8th decade of life. Postmenopausal women are particularly susceptible to __(19)__ inflammations. The single most common problem of elderly men involves the enlargement of the __(20)__, which interferes with the functioning of both the __(21)__ and __(22)__ systems.

INCREDIBLE JOURNEY

A Visualization Exercise for the Reproductive System

. . . you hear a piercing sound coming from the almond-shaped organ as its wall ruptures.

35. Where necessary, complete statements by inserting the missing word(s) in the answer blanks.

_____ 1.

_____ 2.

_____ 3.

_____ 4.

_____ 5.

_____ 6.

_____ 7.

_____ 8.

This is your final journey. You are introduced to a hostess this time, who has agreed to have her cycles speeded up by megahormone therapy so that all of your observations can be completed in less than a day. Your instructions are to observe and document as many events of the two female cycles as possible.

You are miniaturized to enter your hostess through a tiny incision in her abdominal wall (this procedure is called a laparotomy, or, more commonly, "belly button surgery") and end up in her peritoneal cavity. You land on a large and pear-shaped organ in the abdominal cavity midline, the __(1)__. You survey the surroundings and begin to make organ identifications and notes of your observations. Laterally and way above you on each side is an almond-shaped __(2)__, which is suspended by a ligament and almost touched by "feather-duster-like" projections of a tube snaking across the abdominal cavity toward the almond-shaped organs. The projections appear to be almost still, which is puzzling because you thought that they were the __(3)__, or finger-like projections of the uterine tubes, which are supposed to be in motion. You walk toward the end of one of the uterine tubes to take a better look. As you study the ends of the uterine tube more closely, you discover that the feather-like projections are now moving more rapidly, as if they are trying to coax something into the uterine tube. Then you spot a reddened area on the almond-shaped organ, which seems to be enlarging even as you watch. As you continue to observe the area, you waft gently up and down in the peritoneal fluid. Suddenly you feel a gentle but insistent sucking current, drawing you slowly toward the uterine tube. You look upward and see that the reddened area now looks like an angry boil, and the uterine tube projections are gyrating and waving frantically. You realize that you are about to witness __(4)__. You try to get still closer to the opening of the uterine tube when you hear a piercing sound coming from the almond-shaped organ as its wall ruptures. Then you see a ball-like structure, with a "halo" of tiny cells enclosing it, being drawn into the uterine tube. You have just seen the __(5)__, surrounded by its capsule of __(6)__ cells, entering the uterine tube. You hurry into the uterine tube behind it and, holding onto one of the tiny cells, follow it to the uterus. The cell mass that you have attached to has no way of propelling itself, yet you are being squeezed along toward the uterus by a process called __(7)__. You also notice that there are __(8)__, or tiny hair-like projections of the tubule cells, that are all waving in the same direction as you are moving.

9. _____

10. _____

11. _____

12. _____

13. _____

14. _____

15. _____

16. _____

17. _____

18. _____

19. _____

20. _____

Nothing seems to change as you are carried along until finally you are startled by a deafening noise. Suddenly there are thousands of tadpole-like __(9)__ swarming all around you and the sphere of cells. Their heads seem to explode as their __(10)__ break and liberate digestive enzymes. The cell mass now has hundreds of openings in it, and some of the small cells are beginning to fall away. As you peer through the rather transparent cell "halo," you see that one of the tadpole-like structures has penetrated the large central cell. Chromosomes then appear, and that cell begins to divide. You have just witnessed the second __(11)__ division. The products of this division are one large cell, the __(12)__, and one very tiny cell, a __(13)__, which is now being ejected. This cell will soon be __(14)__ because it has essentially no cytoplasm or food reserves. As you continue to watch, the sperm nucleus and that of the large central cell fuse, an event called __(15)__. You note that the new cell just formed by this fusion is called a __(16)__, the first cell of the embryonic body.

As you continue to move along the uterine tube, the central cell divides so fast that no cell growth occurs between the divisions. Thus the number of cells forming the embryonic body increases, but the cells become smaller and smaller. This embryonic division process is called __(17)__.

Finally, the uterine chamber looms before you. As you drift into its cavity, you scrutinize its lining, the __(18)__. You notice that it is thick and velvety in appearance and that the fluids you are drifting in are slightly sweet. The embryo makes its first contact with the lining, detaches, and then makes a second contact at a slightly more inferior location. This time it sticks, and as you watch, the lining of the organ begins to erode away. The embryo is obviously beginning to burrow into the rich cushiony lining, and you realize that __(19)__ is occurring.

You now leave the embryo and propel yourself well away from it. As you float in the cavity fluids, you watch the embryo disappear from sight beneath the lining. Then you continue to travel downward through your hostess's reproductive tract, exiting her body at the external opening of the __(20)__.

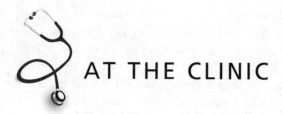

AT THE CLINIC

36. A 28-year-old primigravida (in first pregnancy) has been in the first stage of labor for several hours. Her uterine contractions are weak, and her labor is not progressing normally. Since the woman insists upon a vaginal delivery, the physician orders that Pitocin (a synthetic oxytocin) be infused. What will be the effect of Pitocin? What is the normal mechanism by which oxytocin acts to promote birth?

37. A 38-year-old male is upset about his low sperm count and visits a "practitioner" who commonly advertises his miracle cures for sterility. In fact, the practitioner is a quack who treats conditions of low sperm count with megadoses of testosterone. Although his patients experience a huge surge in libido, their sperm count is even lower after hormone treatment. Explain why.

38. Mr. and Mrs. John Cary, a young couple who had been trying unsuccessfully to have a family for years, underwent a series of tests with a fertility clinic to try to determine the problem. Mr. Cary was found to have a normal sperm count, sperm morphology, and motility.

Mrs. Cary's history sheet revealed that she had two episodes of pelvic inflammatory disease (PID) during her early 20s, and the time span between successive menses ranged from 21 to 30 days. She claimed that her family was "badgering" her about not giving them grandchildren and that she was frequently despondent. A battery of hormonal tests was ordered, and Mrs. Cary was asked to perform cervical mucus testing and daily basal temperature recordings. Additionally, gas was blown through her uterine tubes to determine their patency. Her tubes proved to be closed, and she was determined to be anovulatory. What do you suggest might have caused the closing of her tubes? Which of the tests done or ordered would have revealed her anovulatory condition?

39. A man swam in a cold lake for an hour and then noticed that his scrotum was shrunken and wrinkled. His first thought was that he had lost his testicles. What had really happened?

40. Mary is a heavy smoker and has ignored a friend's advice to stop smoking during her pregnancy. On the basis of what you know about the effect of smoking on physiology, describe how Mary's smoking might affect her fetus.

41. Mrs. Ginko's Pap smear shows some abnormal cells. What possibility should be investigated?

42. Mrs. Weibel has just given birth to an infant with a congenital deformity of the stomach. She is convinced that a viral infection she suffered during the third trimester of her pregnancy is responsible. Do you think she is right? Why or why not?

43. Julio is infected with gonorrhea and chlamydia. What clinical name is given to this general class of infections, and why is it crucial to inform his partners of his infection?

44. By what procedure was Julius Caesar *supposedly* born?

45. Jane started taking estradiol and progesterone immediately after the start of her menstrual period. What effect on ovulation should she expect?

46. Mary and Jim, fraternal twins, were enjoying a lunch together when she revealed that she had just had a tubal ligation procedure. "Oh, my gosh," he said, "we *do* think alike. I just had a vasectomy." How are these procedures alike structurally and functionally?

✓ THE FINALE: MULTIPLE CHOICE

47. Select the best answer or answers from the choices given.

1. Which of the following structures have a region called the ampulla?

   A. Ductus deferens

   B. Uterine tube

   C. Ejaculatory duct

   D. Lactiferous duct

2. Seminal vesicle secretions have:

   A. a low pH

   B. fructose

   C. a high pH

   D. sperm-activating enzymes

3. If the uterine tube is a trumpet ("salpinx"), what part of it represents the wide, open end of the trumpet?

   A. Isthmus   C. Infundibulum

   B. Ampulla   D. Flagellum

4. The myometrium is the muscular layer of the uterus, and the endometrium is the _____ layer.

   A. serosa   C. submucosa

   B. adventitia   D. mucosa

5. All of the following are true of the gonadotropins *except* that they are:

   A. secreted by the pituitary gland

   B. LH and FSH

   C. hormones with important functions in both males and females

   D. the sex hormones secreted by the gonads

6. The approximate area between the anus and clitoris in the female is the:

   A. peritoneum   C. vulva

   B. perineum   D. labia

7. A test to detect cancerous changes in cells of the uterus and cervix is:

   A. pyelogram   C. D&C

   B. Pap smear   D. laparoscopy

8. In humans, separation of the cells at the two-cell stage following fertilization may lead to the production of twins, which in this case, would be:

   A. of different sexes   C. fraternal

   B. identical   D. dizygotic

9. Human ova and sperm are similar in that:

    A. about the same number of each is produced per month

    B. they have the same degree of motility

    C. they are about the same size

    D. they have the same number of chromosomes

10. Which of the following attach to the ovary?

    A. Fimbriae

    B. Mesosalpinx

    C. Suspensory ligaments

    D. Broad ligament

11. As a result of crossover:

    A. maternal genes can end up on a paternal chromosome

    B. synapsis occurs

    C. a tetrad is formed

    D. no two spermatids have exactly the same genetic makeup

12. The first mitotic division in the zygote occurs as soon as:

    A. male and female pronuclei fuse

    B. male and female chromosomes are replicated

    C. meiosis II in the oocyte nucleus is completed

    D. the second polar body is ejected

13. The acrosomal reaction:

    A. allows degradation of the corona radiata

    B. involves release of hyaluronidase

    C. occurs in the male urogenital tract

    D. involves only one sperm, which penetrates the oocyte membrane

14. Which contain cells that ultimately become part of the embryo?

    A. Blastocyst   C. Cytotrophoblast

    B. Trophoblast   D. Inner cell mass

15. The blastocyst:

    A. is the earliest stage at which differentiation is clearly evident

    B. is the stage at which implantation occurs

    C. has a three-layered inner cell mass

    D. can detect "readiness" of uterine endometrium

16. Human chorionic gonadotropin is secreted by the:

    A. trophoblast

    B. 5-month placenta

    C. chorion

    D. corpus luteum

17. The first major event in organogenesis is:

    A. gastrulation

    B. appearance of the notochord

    C. neurulation

    D. development of blood vessels in the umbilical cord

18. Which of the following appears first in the development of the nervous system?

    A. Neural crest cells

    B. Neural folds

    C. Neural plate

    D. Neural tube

19. Which of these digestive structures develops from ectoderm?

    A. Midgut

    B. Liver

    C. Lining of the mouth and anus

    D. Lining of esophagus and pharynx

20. Mesodermal derivatives include:

    A. somites

    B. mesenchyme

    C. most of the intestinal wall

    D. sweat glands

21. On day 17 of a woman's monthly cycle:

    A. FSH levels are rising

    B. progesterone is being secreted

    C. the ovary is in the ovulatory phase

    D. the uterus is in the proliferative phase

22. A sudden decline in estrogen and progesterone levels:

    A. causes spasms of the spiral arteries

    B. triggers ovulation

    C. ends inhibition of FSH release

    D. causes fluid retention

23. A sexually transmitted infection (STI) that is more easily detected in males than females, is treatable with penicillin, and can cause lesions in the nervous and cardiovascular systems is:

    A. gonorrhea   C. syphilis

    B. chlamydia   D. herpes

24. Which of the following are hormones associated with lactation?

    A. Placental lactogen

    B. Colostrum

    C. Prolactin

    D. Oxytocin

25. The outer layer of the blastocyst, which attaches to the uterine wall, is the:

    A. yolk sac   C. amnion

    B. inner cell mass   D. trophoblast

26. The notochord:

    A. develops from the primitive streak

    B. develops from mesoderm beneath the primitive streak

    C. becomes the vertebral column

    D. persists as the nucleus pulposis in the intervertebral discs

27. Amniotic fluid:

    A. prevents fusion of embryonic parts

    B. contains cells and chemicals derived from the embryo

    C. is derived from embryonic endoderm

    D. helps maintain a constant temperature for the developing fetus

28. Which of the following is a shunt to bypass the fetal liver?

    A. Ductus arteriosus

    B. Ductus venosus

    C. Ligamentum teres

    D. Umbilical vein

29. The usual and most desirable presentation for birth is:

    A. vertex   C. nonvertex

    B. breech   D. head first

Answers

Chapter 1   The Human Body: An Orientation

An Overview of Anatomy and Physiology

1. 1. D or physiology.   2. A or anatomy.   3. B or homeostasis.   4. C or metabolism.

2. Physiological study: C, D, E, F, G, H, J, K.   Anatomical study: A, B, I, K, L, M.

Levels of Structural Organization

3. Cells, tissues, organs, organ systems.

4. 1. Electron.   2. Epithelium.   3. Heart.   4. Digestive system.

5. 1. K or urinary.   2. C or endocrine.   3. J or skeletal.   4. A or cardiovascular.   5. D or integumentary.
6. E or lymphatic/immune.   7. B or digestive.   8. I or respiratory.   9. A or cardiovascular.   10. F or muscular.
11. K or urinary.   12. H or reproductive.   13. C or endocrine.   14. D or integumentary.

6. 1. A or cardiovascular.   2. C or endocrine.   3. K or urinary.   4. H or reproductive.   5. B or digestive.
6. J or skeletal.   7. G or nervous.

7.

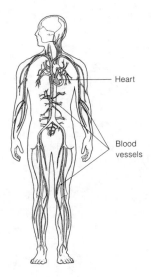

Figure 1–1: Cardiovascular system

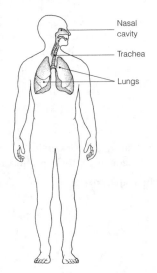

Figure 1–2: Respiratory system

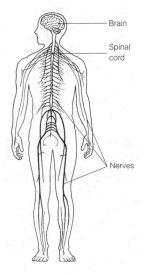

Figure 1–3: Nervous system

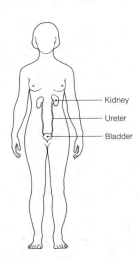

Figure 1–4: Urinary system

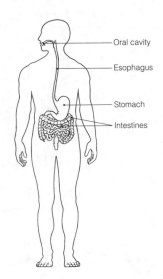

Figure 1–5: Digestive system

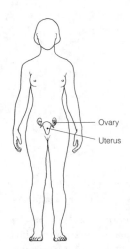

Figure 1–6: Reproductive system

Maintaining Life

8. 1. D or maintenance of boundaries.   2. H or reproduction.   3. C or growth.   4. A or digestion.   5. B or excretion.
6. G or responsiveness.   7. F or movement.   8. E or metabolism.   9. D or maintenance of boundaries.

9. 1. C or nutrients.   2. B or atmospheric pressure.   3. E or water.   4. D or oxygen.   5. E or water.
6. A or appropriate body temperature.

Homeostasis

10. 1. Receptor.   2. Control center.   3. Afferent.   4. Control center.   5. Effector.   6. Efferent.   7. Negative.
8. Positive.   9. Negative.

The Language of Anatomy

11. 1. Ventral.   2. Dorsal.   3. Dorsal.

12. 1. Distal.   2. Antecubital.   3. Brachial.   4. Left upper quadrant.   5. Ventral cavity.

13. **Figure 1–7:**

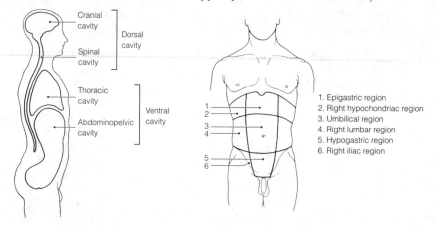

1. Epigastric region
2. Right hypochondriac region
3. Umbilical region
4. Right lumbar region
5. Hypogastric region
6. Right iliac region

14. 1. C or axillary.   2. G or femoral.   3. H or gluteal.   4. F or cervical.   5. P or umbilical.   6. M or pubic.
7. B or antecubital.   8. K or occipital.   9. I or inguinal.   10. J or lumbar.   11. E or buccal.

15. **Figure 1–8:** Section A: Midsagittal.   Section B: Transverse.

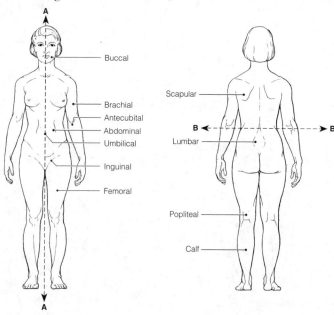

16. 1. G or ventral, D or pelvic.   2. G or ventral, F or thoracic.   3. C or dorsal, B or cranial.   4. G or ventral,
D or pelvic.   5. G or ventral, A or abdominal.

17. 1. A or anterior.   2. G or posterior.   3. J or superior.   4. J or superior.   5. E or lateral.   6. A or anterior.
7. F or medial.   8. H or proximal.   9. B or distal.   10. G or posterior.   11. J or superior.   12. I or sagittal.
13. C or frontal.   14. C or frontal.   15. K or transverse.

18. 1.–5. A or abdominopelvic.   6. C or spinal.   7. A or abdominopelvic.   8. and 9. D or thoracic.   10. B or
cranial.   11. and 12. A or abdominopelvic.

19. A. 2.   B. 3.   C. 1.   D. 4.

At the Clinic

20. Skeletal, muscular, cardiovascular, integumentary, nervous.

21. The need for nutrients and water.

22. The anterior and lateral aspects of the abdomen have no bony (skeletal) protection.

23. John has a hernia in the area where his thigh and trunk meet, pain from his infected kidney radiating to his lower back, and bruises in his genital area.

24. Negative feedback causes the initial stimulus, TSH in this case, to decline.

25. The high blood pressure increases the workload on the heart. Circulation of blood decreases, and the heart itself begins to receive an inadequate blood supply. As the heart weakens further, the backup in the veins worsens, and the blood pressure rises even higher. Without intervention, circulation becomes so sluggish that organ failure sets in. A heart-strengthening medication will increase the force of the heartbeat so that more blood is pumped out with each beat. More blood can then flow into the heart, reducing backflow and blood pressure. The heart can then pump more blood, further reducing the backup and increasing circulation. The blood supply to the heart musculature improves, and the heart becomes stronger.

26. CT and DSA utilize X-rays. MRI employs radio waves and magnetic fields. PET uses radioisotopes. CT, MRI, and PET scans can display body regions in sections.

27. Right side, below the rib cage.

28. He will apply the splint to his right wrist.

The Finale: Multiple Choice

29. 1. B.   2. C.   3. A, B, C, D.   4. B.   5. B.   6. A, B, D.   7. A, B, D.   8. C.   9. C, E.   10. D.
11. A, C, E.   12. B, C, D, E.   13. A.   14. B.   15. B.

Chapter 2   Basic Chemistry

Concepts of Matter and Energy

1. 1. B, D.   2. A, B, C, D.   3. A, B.

2. 1. C or mechanical.   2. B, D or electrical, radiant.   3. C or mechanical.   4. A or chemical.   5. D or radiant.

Composition of Matter

3.

Particle	Location	Electrical charge	Mass
Proton	Nucleus	+1	1 amu
Neutron	Nucleus	0	1 amu
Electron	Orbitals	−1	0 amu

4. 1. O.   2. C.   3. K.   4. I.   5. H.   6. N.   7. Ca.   8. Na.   9. P.   10. Mg.   11. Cl.   12. Fe.

5. 1. E or ion.   2. F or matter.   3. C or element.   4. B or electrons.   5. B or electrons.   6. D or energy.   7. A or atom.   8. G or molecule.   9. I or protons.   10. J or valence.   11. and 12. H and I or neutrons and protons.

6. 1. *T.*   2. Protons.   3. More.   4. *T.*   5. Radioactive.   6. *T.*   7. Chlorine.   8. Iodine.   9. *T.*

Molecules, Chemical Bonds, and Chemical Reactions

7. 1. C or synthesis.   2. B or exchange.   3. A or decomposition.

8. **Figure 2–1:** The nucleus is the innermost circle containing 6P and 6N; the electrons are indicated by the small circles in the orbits.   1. Atomic number is 6.   2. Atomic mass is 12 amu.   3. Carbon.   4. Isotope.   5. Chemically active.   6. Four electrons.   7. Covalent because it would be very difficult to gain or lose four electrons.

9. H_2O_2 is one molecule of hydrogen peroxide (a compound). $2OH^-$ represents two hydroxide ions.

10. Figure 2–2: A represents an ionic bond; **B** shows a covalent bond.

11. Figure 2–3

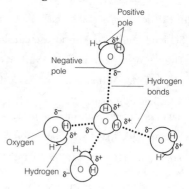

12. Circle B, C, E.

13. 1. H_2CO_3.   2. H^+ and HCO_3^-.   3. The ions should be circled.   4. An additional arrow going to the left should be added between H_2CO_3 and H^+.

Biochemistry: The Composition of Living Matter

14. 1.–3. A or acid(s), B or base(s), and D or salt(s).   4. B or base(s).   5. A or acid(s).   6. D or salt(s).
7. D or salt(s).   8. A or acid(s).   9. C or buffer.

15. 1. Heat capacity.   2. Water.   3. 70% (60–80%).   4. Hydrogen.   5. and 6. Hydrolysis and dehydration.
7. Polarity.   8. Lubricants.

16. *X* carbon dioxide, oxygen, KCl, and H_2O.

17. Weak acid: B, C, E. Strong acid: A, D, E, F, G.

18. 1. G or monosaccharides.   2. D or fatty acids and E or glycerol.   3. A or amino acids.   4. F or nucleotides.
5. H or proteins.   6. G or monosaccharides (B or carbohydrates).   7. C or lipids.   8. G or monosaccharides
(B or carbohydrates).   9. C or lipids.   10. and 11. F or nucleotides and A or amino acids.   12. F or nucleotides.
13. C or lipids.   14. H or proteins.   15. B or carbohydrates.   16. C or lipids.   17. H or proteins.
18. C or lipids.   19. H or proteins.

19. 1. B or collagen, H or keratin.   2. D or enzyme, F or hemoglobin, some of G or hormones.   3. D or enzyme.
4. L or starch.   5. E or glycogen.   6. C or DNA.   7. A or cholesterol (some G or hormones are steroids).
8. I or lactose, J or maltose.

20. Figure 2–4: A. Monosaccharide.   B. Globular protein.   C. Nucleotide.   D. Fat.   E. Polysaccharide.

21. 1. Glucose.   2. Ribose.   3. Glycogen.   4. Glycerol.   5. Glucose.

22. 1. *T*.   2. Neutral fats.   3. *T*.   4. Polar.   5. *T*.   6. ATP.   7. *T*.   8. O.

23. Unnamed nitrogen bases: thymine (T) and guanine (G). 1. Hydrogen bonds.   2. Double helix.   3. 12.
4. Complementary.

Figure 2–5:

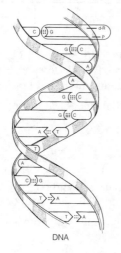

DNA

Note that the stippled parts of the backbones represent phosphate units (P) while the unaltered (white) parts of the backbones that are attached to the bases are deoxyribose sugar (d-R) units.

24. The polymer is to the left of the arrow; the monomers (5) are to the right.   1. C or glucose.
2. C or enter between the monomers, etc.   3. B or hydrolysis.   4. A or R group.

Incredible Journey

25. 1. Negatively.   2. Positive.   3. Hydrogen bonds.   4. Red blood cells.   5. Protein.   6. Amino acids.
7. Peptide.   8. H^+ and OH^-.   9. Hydrolysis.   10. Enzyme.   11. Glucose.   12. Glycogen.
13. Dehydration synthesis.   14. H_2O.   15. Increase.

At the Clinic

26. Acidosis means blood pH is below the normal range. The patient should be treated with something to raise
the pH.

27. Each of the 20 amino acids has a different chemical group called the R group. The R group on each amino acid determines how it will fit in the folded, three-dimensional, tertiary structure of the protein and the bonds it may form. If the
wrong amino acid is inserted, its R group might not fit into the tertiary structure properly, or required bonds might not
be made; hence the entire structure might be altered. Because function depends on structure, this means the protein
will not function properly.

28. Heat increases the kinetic energy of molecules. Vital biological molecules, like proteins and nucleic acids, are
denatured (rendered nonfunctional) by excessive heat because intramolecular bonds essential to their functional
structure are broken. Because all enzymes are proteins, their destruction is lethal.

29. An MRI because it allows visualization of soft structures enclosed by bone (e.g., the skull).

30. Stomach discomfort is frequently caused by excess stomach acidity ("acid indigestion"). An antacid contains a weak
base that will neutralize the excess acid (H^+).

31. Breaking ATP down to ADP and P_i releases the energy stored in the bonds. Only part of that potential energy is
actually used by the cell. The rest is lost as heat. Nonetheless, the total amount of energy released (plus activation
energy) must be absorbed to remake the bonds of ATP.

The Finale: Multiple Choice

32. 1. A, C, D.   2. E.   3. A, B, C, D, E.   4. D.   5. A.   6. C, E.   7. C, D.   8. C.   9. A.   10. A, D.
11. B, C.   12. B, C.   13. D.   14. B.   15. B, D.   16. A, B, D.

Chapter 3   Cells and Tissues

Cells

1. 1.–4. (in any order): Carbon, oxygen, nitrogen, hydrogen.   5. Water.   6. Calcium.   7. Iron.   8.–12. (five of
the following, in any order): Metabolism, reproduction, irritability, mobility, ability to grow, ability to digest foods,
ability to excrete waste.   13.–15. (three of the following, in any order): Cubelike, tilelike, disk shaped, round
spheres, branching, cylindrical.   16. Interstitial (tissue) fluid.   17. Squamous epithelial.

2. Figure 3–1:

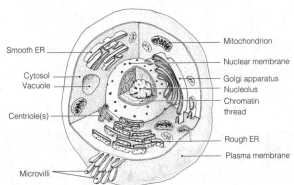

3. Figure 3–2: 1. Glycocalyx.   2. C.   3. Hydrophobic.   4. Enzymes, receptors, recognition sites, etc.

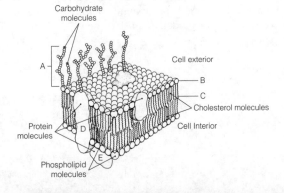

4. **Figure 3–3:** 1. Microvilli are found on cells involved in secretion and/or absorption. 2. Tight junction. 3. Desmosome. 4. Desmosome. 5. Gap junctions allow cells to communicate by allowing ions and other chemicals to pass from cell to cell via protein channels. 6. Gap junctions and desmosomes.

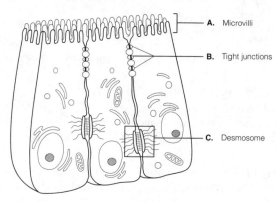

5. 1. Centrioles. 2. Cilia. 3. Smooth ER. 4. Vitamin A storage. 5. Mitochondria. 6. Ribosomes. 7. Lysosomes.

6. 1. Microtubules. 2. Intermediate filaments. 3. Microtubules. 4. Microfilaments. 5. Intermediate filaments. 6. Microtubules.

7. 1. B. 2. F. 3. D. 4. E. 5. C, H. 6. G. 7. A.

8. 1. A. 2. B. 3. C. 4. A.

9. **Figure 3–5:** 1. A; Crenated. 2. B; The same solute concentration inside and outside the cell. 3. C; They are bursting (lysis); water is moving by osmosis from its site of higher concentration (cell exterior) into the cell where it is in lower concentration, causing the cells to swell.

10. **Figure 3–6:** Arrow for Na^+ should be red and shown leaving the cell; those for glucose, Cl^-, O_2, fat, and steroids (except cholesterol, which enters by receptor-mediated endocytosis) should be blue and entering the cell. CO_2 (blue arrow) should be leaving the cell and moving into the extracellular fluid. Amino acids and K^+ (red arrows) should be entering the cell. Water (H_2O) moves passively (blue arrows) through the membrane (in or out) depending on local osmotic conditions.
    1. Fat, steroid, O_2, CO_2. 2. Glucose. 3. H_2O, (probably) Cl^-. 4. Na^+, K^+, amino acid.

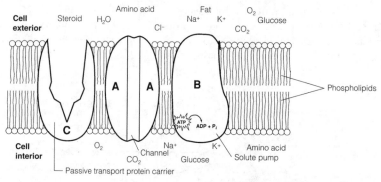

11. 1. D or exocytosis, G or phagocytosis, H or pinocytosis, I or receptor-mediated endocytosis. 2. B or diffusion, simple; C or diffusion, osmosis; E or facilitated diffusion. 3. F or filtration. 4. B or diffusion, simple; C or diffusion, osmosis; E or facilitated diffusion. 5. A or active transport. 6. B or Diffusion, simple. 7. A or active transport. 8. D or exocytosis, G or phagocytosis, H or pinocytosis, I or receptor-mediated endocytosis. 9. G or phagocytosis. 10. D or exocytosis. 11. E or facilitated diffusion.

12. 1. P or proteins. 2. K or helix. 3. O or phosphate. 4. T or sugar. 5. C or bases. 6. B or amino acids. 7. E or complementary. 8. F or cytosine. 9. V or thymine. 10. S or ribosome. 11. Q or replication. 12. M or nucleotides. 13. U or template, or model. 14. L or new. 15. N or old. 16. H or genes. 17. I or growth. 18. R or repair.

13. **Figure 3–7:** A. Prophase.   B. Anaphase.   C. Telophase.   D. Metaphase.

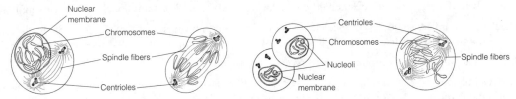

14. 1. C or prophase.   2. A or anaphase.   3. D or telophase.   4. D or telophase.   5. B or metaphase.
    6. C or prophase.   7. C or prophase.   8. E or none of these.   9. C or prophase.   10. C or prophase.
    11. D or telophase.   12. A or anaphase, B or metaphase.   13. E or none of these.

15. 1. Nucleus.   2. Cytoplasm.   3. Coiled.   4. Centromeres.   5. Binucleate cell.   6. Spindle.   7. Interphase.

16. **Figure 3–8:** 1. Transcription.   2. Translation.   3. Anticodon; triplet.

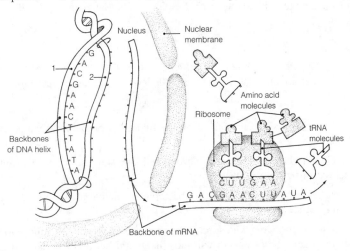

Body Tissues

17. **Figure 3–9:**

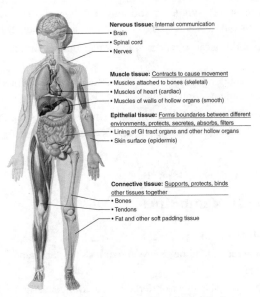

18. **Figure 3–10:** A. Simple squamous epithelium.   B. Simple cuboidal epithelium.   C. Cardiac muscle.
    D. Dense fibrous connective tissue.   E. Bone.   F. Skeletal muscle.   G. Nervous tissue.
    H. Hyaline cartilage.   I. Smooth muscle tissue.   J. Adipose (fat) tissue.   K. Stratified squamous epithelium.
    L. Areolar connective tissue. The noncellular portions of D, E, H, J, and L are matrix.

19. The neuron has long cytoplasmic extensions that promote its ability to transmit impulses over long distances
    within the body.

20. 1. B or epithelium.   2. C or muscle.   3. D or nervous.   4. A or connective.   5. B or epithelium.
    6. D or nervous.   7. C or muscle.   8. B or epithelium.   9. A or connective.   10. A or connective.
    11. C or muscle.   12. A or connective.   13. D or nervous.

21. 1. E or stratified squamous.   2. B or simple columnar.   3. E or stratified squamous.   4. A or pseudostratified columnar (ciliated).   5. A or pseudostratified columnar (ciliated).   6. F or transitional.   7. D or simple squamous.

22. 1. Skeletal.   2. Cardiac, smooth.   3. Skeletal, cardiac.   4. Smooth (most cardiac).   5.–7. Skeletal. 8. and 9. Smooth.   10. Cardiac.   11. Skeletal.   12. Cardiac.   13. Skeletal.   14. Smooth, cardiac.   15. Cardiac.

23. 1. Cell.   2. Elastic fibers.   3. Bones.   4. Nervous.   5. Blood.

24. 1. C or dense fibrous.   2. A or adipose.   3. C or dense fibrous.   4. D or osseous tissue.   5. B or areolar. 6. F or hyaline cartilage.   7. A or adipose.   8. F or hyaline cartilage.   9. D or osseous tissue.   10. E or reticular.

25. 1. Inflammation.   2. Clotting proteins.   3. Granulation.   4. Regeneration.   5. *T.*   6. Collagen.   7. *T.*

Developmental Aspects of Cells and Tissues

26. 1. Tissues.   2. Growth.   3. Nervous.   4. Muscle.   5. Connective (scar).   6. Chemical.   7. Physical. 8. Genes (DNA).   9.–11. Connective tissue changes; Decreased endocrine system activity; Dehydration of body tissues. 12. Division.   13. and 14. Benign, malignant.   15. Benign.   16. Malignant.   17. Biopsy.   18. Surgical removal. 19. Hyperplasia.   20. Atrophy.

Incredible Journey

27. 1. Cytoplasm (cytosol).   2. Nucleus.   3. Mitochondrion.   4. ATP.   5. Ribosomes.   6. Rough endoplasmic reticulum.   7. Pores.   8. Chromatin.   9. DNA.   10. Nucleoli.   11. Golgi apparatus.   12. Lysosome.

At the Clinic

28. The oxidases of ruptured peroxisomes were converting the hydrogen peroxide to water and (free) oxygen gas (which causes the bubbling).

29. Generally speaking, stratified epithelia consisting of several cell layers are more effective where abrasion is a problem than are simple epithelia (consisting of one cell layer).

30. Streptomycin inhibits bacterial protein synthesis. If the bacteria are unable to synthesize new proteins (many of which would be essential enzymes), they will die.

31. Considering connective tissue is the most widespread tissue in the body and is found either as part of or is associated with every body organ, the physician will most likely tell her that she can expect the effects of lupus to be very diffuse and widespread.

32. Granulation tissue secretes substances that kill bacteria.

33. Mitochondria are the site of most ATP synthesis, and muscle cells use tremendous amounts of ATP during contraction. After ingesting bacteria or other debris, phagocytes must digest them, explaining the abundant lysosomes.

34. Recovery will be long and painful because tendons, like other dense connective tissue structures, are poorly vascularized.

35. Edema will occur because the filtration pressure exerted by the blood forces blood proteins into the interstitial space, and water follows down its concentration gradient.

36. Phagocytes engulf and remove debris from body tissues. A smoker's lung would be expected to have carbon particles.

The Finale: Multiple Choice

37. 1. B.   2. A.   3. B.   4. C.   5. A, B, C, D.   6. C.   7. E.   8. A.   9. D.   10. C.   11. D.   12. C.   13. E. 14. C.   15. B.   16. C.   17. A, C, D.   18. A, B, C, D, E.   19. B.

Chapter 4   Skin and Body Membranes

Classification of Body Membranes

1. The mucous, serous, and cutaneous membranes are all composite membranes composed of an epithelial layer underlaid by a connective tissue layer.

- A mucous membrane is an epithelial sheet underlaid by a connective tissue layer called the lamina propria. Mucosae line the respiratory, digestive, urinary, and reproductive tracts; functions include protection, lubrication, secretion, and absorption.

- Serous membranes consist of a layer of simple squamous epithelium resting on a scant layer of fine connective tissue. Serosae line internal ventral body cavities and cover their organs; their function is to produce a lubricating fluid that reduces friction.

- The cutaneous membrane, or skin, is composed of the epithelial epidermis and the connective tissue dermis. It covers the body exterior and protects deeper body tissues from external insults.

The synovial membranes, which line joint cavities of synovial joints, are composed entirely of connective tissue. They function to produce lubrication to decrease friction within the joint cavity.

2. Figure 4–1: In each case, the visceral layer of the serosa covers the external surface of the organ, and the parietal layer lines the body cavity walls.

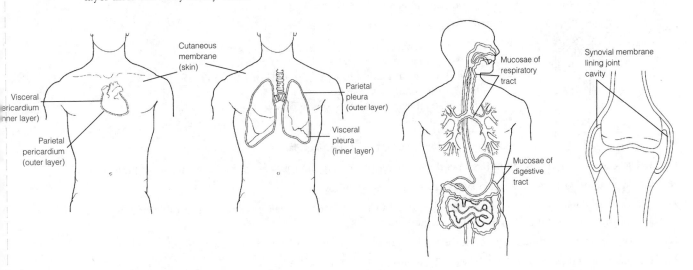

Integumentary System (Skin)

3.   1. B.   2. M.   3. C.   4. C, M.   5. C.   6. C.

4.   Sunburn inhibits the immune response by depressing macrophage activity.

5.   When body temperature begins to rise to undesirable levels, the sweat glands are activated by nerve fibers of the (sympathetic) nervous system. As sweat is evaporated from the skin surface, it carries body heat with it.

6.   1. Nervous.   2. Temperature (heat and cold).   3. Pain.   4. Light pressure.   5. Deep pressure.   6. Cholesterol.
7. UV light.   8. Calcium.

7.   Figure 4–2:

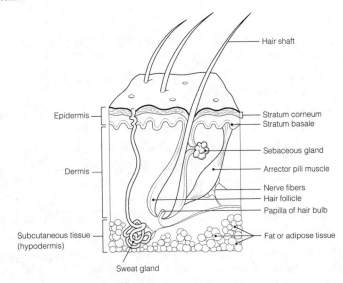

8.   1. As the basal cells continue to divide, the more superficial cells are pushed farther and farther from the nutrient supply diffusing from the dermis.   2. Waterproofing substances (keratin and others) made by the keratinocytes effectively limit nutrient entry into the cells.

9.   1. D or stratum lucidum.   2. B or stratum corneum, D or stratum lucidum.   3. F or papillary layer.
4. I or dermis as a whole.   5. A or stratum basale.   6. B or stratum corneum.   7. I or dermis as a whole.
8. A or stratum basale.   9. H or epidermis as a whole.   10. B or stratum corneum.   11. I or dermis as a whole.

10.   1. Keratin.   2. Wart.   3. Stratum basale.   4. Arrector pili.

11.   1. C or melanin.   2. A or carotene.   3. C or melanin.   4. B or hemoglobin.   5. C or melanin.   6. A or carotene.
7. B or hemoglobin.

12.   1. Heat.   2. Subcutaneous.   3. Vitamin D.   4. Elasticity.   5. Oxygen (blood flow).   6. Cyanosis.

13.   1. Sweat.   2. Keratin.   3. *T.*   4. Shaft.   5. Dermis.

14. **Figure 4–3:**

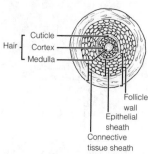

Hair — Cuticle, Cortex, Medulla
Follicle wall
Epithelial sheath
Connective tissue sheath

15. 1. Poor nutrition.   2. Keratin.
3. Stratum corneum.
4. Eccrine glands.   5. Vellus hair.

16. Alopecia.

17. 1. E or sebaceous glands.   2. A or arrector pili.
3. G or sweat gland (eccrine).   4. D or hair follicle(s).
5. F or sweat gland (apocrine).   6. C or hair.
7. B or cutaneous receptors.
8. E or sebaceous glands, and F or sweat gland (apocrine).
9. G or sweat gland (eccrine).   10. E or sebaceous gland.

18. 1. Arrector pili.   2. Absorption.   3. Epithelial sheath.   4. Eccrine glands.   5. Wrinkles.

19. 1. The cuticle.   2. The stratum basale is thicker here, preventing the rosy cast of blood from flushing through.

20. Water/protein/electrolyte loss, circulatory collapse, renal shutdown.

21. 1. C or third-degree burn.   2. B or second-degree burn.   3. A or first-degree burn.   4. B or second-degree burn.
5. C or third-degree burn.   6. C or third-degree burn.

22. It allows estimation of the extent of burns so that fluid volume replacement can be correctly calculated.

23. 1. Squamous cell carcinoma.   2. Basal cell carcinoma.   3. Malignant melanoma.

24. Pigmented areas that are <u>A</u>symmetrical, have irregular <u>B</u>orders, exhibit several <u>C</u>olors, and have a <u>D</u>iameter greater than 6 mm are likely to be cancerous.

Developmental Aspects of the Skin and Body Membranes

25. 1. C or dermatitis.   2. D or delayed-action gene.   3. F or milia.   4. B or cold intolerance.   5. A or acne.
6. G or vernix caseosa.   7. E or lanugo.

Incredible Journey

26. 1. Collagen.   2. Elastin (or elastic).   3. Dermis.   4. Phagocyte (macrophage).   5. Hair follicle connective tissue.
6. Epidermis.   7. Stratum basale.   8. Melanin.   9. Keratin.   10. Squamous (stratum corneum) cells.

At the Clinic

27. Chemotherapy drugs used to treat cancer kill the most rapidly dividing cells in the body, including many matrix cells in the hair follicles; thus, the hair falls out.

28. The baby has seborrhea, or cradle cap, a condition of overactive sebaceous glands. It is not serious; the oily deposit is easily removed with attentive washing and soon stops forming.

29. Bedridden patients are turned at regular intervals so that no region of their body is pressed against the bed long enough to deprive the blood supply to that skin; thus, bedsores are avoided.

30. The baby was cyanotic from lack of oxygen when born, a problem solved by breathing. Vernix caseosa, a cheesy substance made by the sebaceous glands covered her skin. This substance helps to protect the fetus's skin in utero.

31. Norwegians in the United States. They are originally from a region of the world where the sun is always far away from them and have very fair skin; hence they have little protective melanin.

32. Besides storing fat as a source of nutrition, the hypodermis anchors the skin to underlying structures (such as muscles) and acts as an insulator against heat loss.

33. The body of a nail is its visible, attached part (not its white free edge). The root is the proximal part that is embedded in skin. The bed is the part of the epidermis upon which the nail lies. The matrix is the proximal part of the nail bed, and it is responsible for nail growth. The cuticle is the skin fold around the perimeter of the nail body. Because the matrix is gone, the nail will not grow back.

34. The peritoneum will be inflamed and infected. Because the peritoneum encloses so many richly vascularized organs, a spreading peritoneal infection can be life threatening.

35. He probably told her that regeneration would occur, and grafts would not be needed if infection was avoided.

36. Replacing lost fluid and electrolytes and prevention of infection.

37. Fat is a good insulator, so its lack or decrease results in a greater sensitivity to cold.

The Finale: Multiple Choice

38. 1. B, D.   2. D.   3. B.   4. B.   5. D.   6. C.   7. B.   8. C, D.   9. C.   10. C.   11. A, B, D.   12. D.   13. A, B.
14. B.

Chapter 5   The Skeletal System

Bones—An Overview

1. 1. P.   2. P.   3. D.   4. D.   5. P.   6. D.   7. P.   8. P.   9. P.

2. 1. S.   2. F.   3. L.   4. L.   5. F.   6. L.   7. L.   8. F.   9. I.

3. 1. C or epiphysis.   2. A or diaphysis.   3. C or epiphysis, D or red marrow.   4. A or diaphysis.   5. E or yellow marrow cavity.   6. B or epiphyseal plate.

4. 1. G or parathyroid hormone.   2. F or osteocytes.   3. A or atrophy.   4. H or stress/tension.   5. D or osteoblasts.   6. B or calcitonin.   7. E or osteoclasts.   8. C or gravity.

5. Figure 5–1:

1. B or concentric lamellae.
2. C or lacunae.
3. A or central (Haversian) canal.
4. E or bone matrix.
5. D or canaliculi.

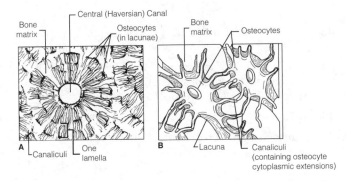

6. 1. Yellow marrow.   2. Osteoblasts.   3. Marrow cavity.   4. Periosteum.

7. Figure 5–2: The epiphyseal plate is the white band shown in the center region of the head; the articular cartilage is the white band on the external surface of the head. Red marrow is found within the spongy bone cavities; yellow marrow is found within the cavity of the diaphysis.

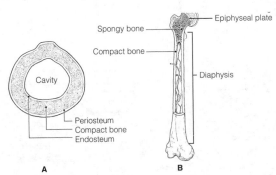

8. 1. 4.   2. 3.   3. 2.   4. 1.   5. 5.   6. 6.   NOTE: Events 2 and 3 may occur simultaneously.

Axial Skeleton

9. 1. B or frontal.   2. N or zygomatic.   3. E or mandible.   4. G or nasals.   5. I or palatines.   6. J or parietals.
7. H or occipital.   8. K or sphenoid.   9. D or lacrimals.   10. F or maxillae.   11. A or ethmoid.
12. L or temporals.   13. K or sphenoid.   14. A or ethmoid.   15. E or mandible.   16. L or temporals.
17.–20. A or ethmoid, B or frontal, F or maxillae, and K or sphenoid.   21. H or occipital.   22. H or occipital.
23. L or temporals.   24. M or vomer.   25. A or ethmoid.   26. L or temporals.

10. 1. Membranous.   2. *T.*   3. Osteoblasts.   4. Secondary.   5. Hyaline cartilage.   6. Endosteal.   7. *T.*

11. Figure 5–3:

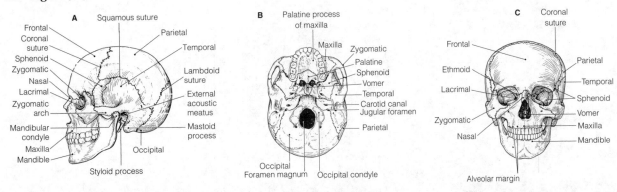

12. Figure 5–4: 1. Mucosa-lined, air-filled cavities in bone.   2. They lighten the skull and serve as resonance chambers for speech.   3. Their mucosa is continuous with that of the nasal passages into which they drain.

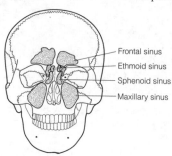

Frontal sinus
Ethmoid sinus
Sphenoid sinus
Maxillary sinus

13. 1. F or vertebral arch.   2. A or body.   3. C or spinous process, E or transverse process.   4. A or body, E or transverse process.   5. B or intervertebral foramina.

14. 1. A or atlas, B or axis, C or cervical vertebra—typical.   2. B or axis.   3. G or thoracic vertebra.   4. F or sacrum.   5. E or lumbar vertebra.   6. D or coccyx.   7. A or atlas.   8. A or atlas, B or axis, and C or cervical vertebra—typical.   9. G or thoracic vertebra.

15. 1. Kyphosis.   2. Scoliosis.   3. Fibrocartilage.   4. Springiness or flexibility.

16. Figure 5–5: A. Cervical; atlas.   B. Cervical.   C. Thoracic.   D. Lumbar.

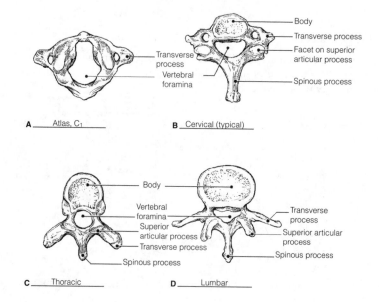

Transverse process
Vertebral foramina
A Atlas, C₁

Body
Transverse process
Facet on superior articular process
Spinous process
B Cervical (typical)

Body
Vertebral foramina
Superior articular process
Transverse process
Spinous process
C Thoracic

Transverse process
Superior articular process
Spinous process
D Lumbar

17. Figure 5–6: 1. Cervical, C_1–C_7.   2. Thoracic, T_1–T_{12}.   3. Lumbar, L_1–L_5.   4. Sacrum, fused.   5. Coccyx, fused.   6. Atlas, C_1.   7. Axis, C_2.

18. 1. Lungs.   2. Heart.   3. True.   4. False.   5. Floating.   6. Thoracic vertebrae.   7. Sternum.   8. An inverted cone.

19. Figure 5–7: Ribs #1–#7 on each side are true ribs; ribs #8–#12 on each side are false ribs.

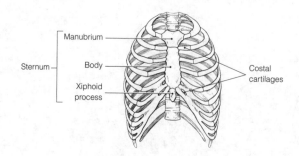

Manubrium
Sternum
Body
Xiphoid process
Costal cartilages

Appendicular Skeleton

20. Figure 5–8: Scapula.

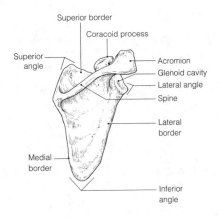

Superior border
Coracoid process
Superior angle
Acromion
Glenoid cavity
Lateral angle
Spine
Lateral border
Medial border
Inferior angle

21. Figure 5–9: A. Humerus.   B. Ulna.   C. Radius

22. Figure 5–10:

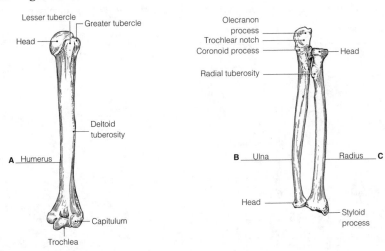

Lesser tubercle
Greater tubercle
Head
Deltoid tuberosity
A Humerus
Capitulum
Trochlea

Olecranon process
Trochlear notch
Coronoid process
Radial tuberosity
Head
B Ulna
Radius C
Head
Styloid process

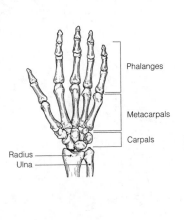

Phalanges
Metacarpals
Carpals
Radius
Ulna

23. Pectoral: A, C, D. Pelvic: B, E, F.

24. 1. G or deltoid tuberosity.   2. I or humerus.   3. and 4. D or clavicle, P or scapula.   5. and 6. O or radius, T or ulna.
7. A or acromion.   8. P or scapula.   9. D or clavicle.   10. H or glenoid cavity.   11. E or coracoid process.
12. D or clavicle.   13. S or trochlea.   14. T or ulna.   15. B or capitulum.   16. F or coronoid fossa.
17. T or ulna.   18. and 19. P or scapula, Q or sternum.   20. C or carpals.   21. M or phalanges.   22. J or metacarpals.

25. 1. Female inlet is larger and more circular.   2. Female sacrum is less curved; pubic arch is rounder.
3. Female ischial spines are shorter; pelvis is shallower/lighter.

Figure 5–11:

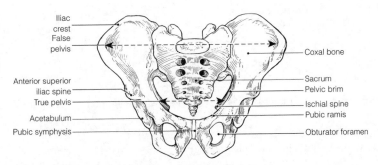

Iliac crest
False pelvis
Anterior superior iliac spine
True pelvis
Acetabulum
Pubic symphysis
Coxal bone
Sacrum
Pelvic brim
Ischial spine
Pubic ramis
Obturator foramen

26. 1. Ulna.   2. Pelvis.   3. Scapula.   4. Mandible.   5. Carpals.

27. 1. I or ilium, K or ischium, S or pubis.   2. J or ischial tuberosity.   3. R or pubic symphysis.   4. H or iliac crest.
5. A or acetabulum.   6. T or sacroiliac joint.   7. C or femur.   8. D or fibula.   9. W or tibia.   10. C or femur,
Q or patella, W or tibia.   11. X or tibial tuberosity.   12. Q or patella.   13. W or tibia.   14. N or medial
malleolus.   15. L or lateral malleolus.   16. B or calcaneus.   17. V or tarsals.   18. O or metatarsals.   19. P or
obturator foramen.   20. G or greater and lesser trochanters, E or gluteal tuberosity.   21. U or talus.

28. 1. Pelvic. 2. Phalanges. 3. *T.* 4. Acetabulum. 5. Sciatic. 6. *T.* 7. Coxal bones (hip bones). 8. *T.* 9. Femur. 10. *T.* 11. Kyphosis.

29. Figure 5–12:

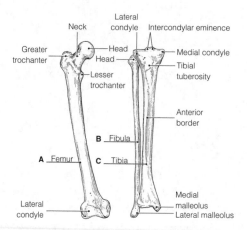

30. Figure 5–13: Bones of the skull, vertebral column, and bony thorax are parts of the axial skeleton. All others belong to the appendicular skeleton.

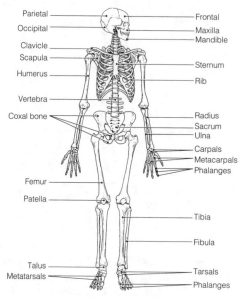

Bone Fractures

31. 1. G or simple fracture. 2. A or closed reduction. 3. E or greenstick fracture. 4. B or compression fracture. 5. C or compound fracture. 6. F or open reduction. 7. H or spiral fracture.

Figure 5–14:

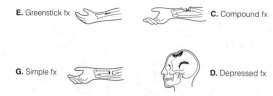

32. 1. *T.* 2. *T.* 3. Phagocytes (macrophages). 4. *T.* 5. Periosteum. 6. *T.* 7. Spongy.

Joints

33. 1. Synovial fluid. 2. Articular cartilage. 3. Ligaments.

Figure 5–15:

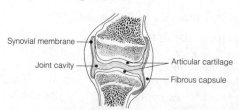

34. 1. A or cartilaginous.   2. C or synovial.   3. B or fibrous and 2 or suture.   4. B or fibrous and 2 or suture.
5. C or synovial.   6. C or synovial.   7. C or synovial.   8. A or cartilaginous and 3 or symphysis.   9. C or synovial.   10. B or fibrous and 2 or suture.   11. C or synovial.   12. A or cartilaginous and 1 or epiphyseal disk.
13. C or synovial.   14. C or synovial.   15. C or synovial.

35. Synovial joints, which are diarthroses or freely movable joints. The axial skeleton supports and protects internal organs; thus, strength is more important than mobility for joints of the axial skeleton.

36. 1. *T.*   2. Osteoarthritis.   3. Acute.   4. Vascularized.   5. *T.*   6. Gouty arthritis or gout.   7. Rickets.   8. *T.*

Developmental Aspects of the Skeleton

37. 1. D or nervous.   2. F or urinary.   3. A or endocrine.   4. C or muscular.   5. A or endocrine.   6. B or integumentary.

38. 1. Fontanels.   2. Compressed.   3. Growth.   4. Sutures.   5. Thoracic.   6. Sacral.   7. Primary.   8. Cervical.   9. Lumbar.

Incredible Journey

39. 1. Femur.   2. Spongy.   3. Stress (or tension).   4. Red blood cells (RBCs).   5. Red marrow.   6. Nerve.
7. Central or Haversian.   8. Compact.   9. Canaliculi.   10. Lacunae (osteocytes).   11. Matrix.   12. Osteoclast.

At the Clinic

40. Seven bones contribute to the orbit: frontal, sphenoid, zygomatic, maxilla, palatine, lacrimal, and ethmoid bones.

41. Mrs. Bruso has severe osteoporosis in which her bones have become increasingly fragile. The postmenopausal deficit of estrogen has placed her bones at risk. Weight-bearing exercise and supplemental calcium will probably be prescribed.

42. The cribriform plates of the ethmoid bone, which surround the olfactory nerves. These plates are quite fragile and are often crushed by a blow to the front of the skull. This severs the olfactory nerve fibers, which cannot grow back.

43. Rheumatoid arthritis, fairly common in middle-aged women, causes this type of deformity.

44. James has all the classic signs and symptoms of osteoarthritis.

45. Janet will be watched for signs of scoliosis because of injury to thoracic vertebrae (and probably associated muscles) on *one* side of the body.

46. The serving arm is subjected to much greater physical (mechanical) stress because of the additional requirement to serve the ball. Consequently, the bones grow thicker to respond to the greater stress.

47. The sternum is compressed during CPR.

48. Osteoporosis is the deterioration and breakdown of bone matrix. Osteoclasts are the cells that cause this breakdown.

The Finale: Multiple Choice

49. 1. A, B, C.   2. D.   3. D.   4. B.   5. A, C.   6. D.   7. A, C, D.   8. B.   9. B, D, E.   10. C.   11. B, D, E.
12. A, C, D.   13. B, D.   14. B, D.   15. A, B.   16. A, B, C, D.   17. D.   18. B.   19. B.   20. D.   21. C, D.
22. B, C.

Chapter 6   The Muscular System

Overview of Muscle Tissues

1. 1. A or cardiac, B or smooth.   2. A or cardiac, C or skeletal.   3. B or smooth.   4. C or skeletal.   5. A or cardiac.
6. A or cardiac.   7. C or skeletal.   8. C or skeletal.   9. C or skeletal.

2. A. Smooth muscle.   B. Cardiac muscle.

3. 1. Bones.   2. Promotes labor during birth.   3. Contractility.   4. Stretchability.   5. Promotes growth.

Microscopic Anatomy of Skeletal Muscle

4. 1. G or perimysium.   2. B or epimysium.   3. I or sarcomere.   4. D or fiber.   5. A or endomysium.
6. H or sarcolemma.   7. F or myofibril.   8. E or myofilament.   9. K or tendon.   10. C or fascicle.

Figure 6–2: The endomysium is the connective tissue that surrounds each muscle cell (fiber).

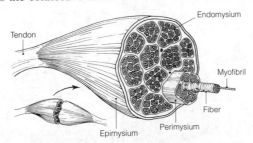

5. Figure 6–3: In the student art of a contracted sarcomere, the myosin filaments should nearly touch the Z discs and the opposing actin filaments should nearly touch each other. The area of the myosin filaments should be labeled *dark band,* and the reduced area containing actin filaments labeled *light band.* Only the light band shortens during contraction.

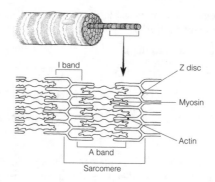

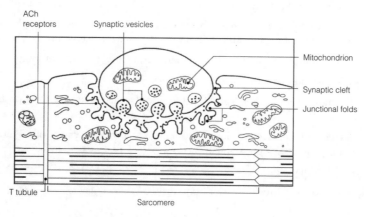

Skeletal Muscle Activity

6. 1. Motor unit.   2. Axon terminals.   3. Synaptic cleft.   4. Acetylcholine.   5. Nerve impulse (or action potential). 6. Depolarization.

7. Figure 6–4:

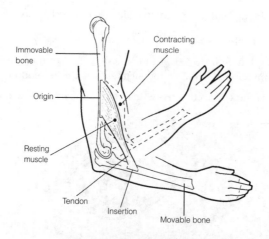

8. 1. 1.   2. 4.   3. 7.   4. 2.   5. 5.   6. 3.   7. 6.

9. 1. F.   2. E.   3. C.   4. B.   5. H.   6. G.   7. I.

10. 1. G or fused tetanus.   2. B or isotonic contraction.   3. I or many motor units.   4. H or few motor units. 5. A or fatigue.   6. E or isometric contraction.

11. 1. B.   2. C.   3. A.   4. A, B.   5.–7. C.   8. B.   9. A.

12. Your rate of respiration (breathing) is much faster, and you breathe more deeply.

13. Check 1, 3, 4, and 7.

Muscle Movements, Types, and Names

14. Figure 6–5:

15. 1. Plantar flexion.   2. Dorsiflexion.   3. Circumduction.   4. Adduct.   5. Flexion.   6. Extension.   7. Extension.
8. Flexed.   9. Flexion.   10. Rotation.   11. Circumduction.   12. Rotation.   13. Pronation.   14. Abduction.

16. 1. C or prime mover.   2. B or fixator.   3. D or synergist.   4. D or synergist.   5. A or antagonist.   6. B or fixator.

17. 1. E, G.   2. A, G.   3. D, E.   4. E, F.   5. A, E.   6. B.   7. E, F.   8. E, F.

Gross Anatomy of the Skeletal Muscles

18. **Figure 6–6:** 1. I.   2. A.   3. D.   4. B.   5. E.   6. C.   7. G.   8. F.

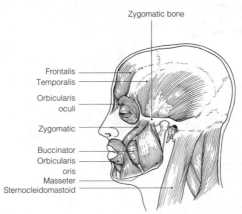

19. 1. D.   2. B.   3. G.   4. E.   5. F.   6. C.

20. **Figure 6–7:** 1. I.   2. H.   3. A.   4. D.   5. J.   6. F.   7. K.   8. C.   9. B.

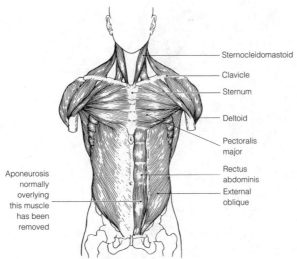

21. **Figure 6–8:** 1. G.   2. E.   3. A.   4. B.   5. E.   6. F.

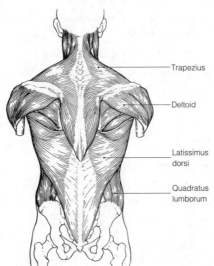

22. Figure 6–9: 1. H.　2. E.　3. D.　4. O.　5. A.　6. I.　7. G.　8. F.　9. C.　10. K.　11. N.

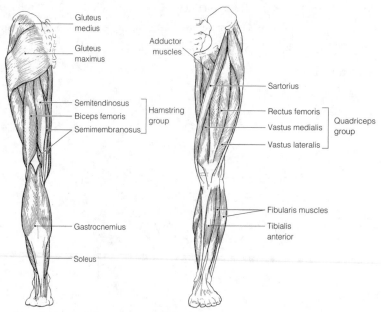

23. The calf muscles must work against gravity whereas the ventral leg muscles do not.

24. Figure 6–10: 1. E.　2. D.　3. F.　4. A.　5. G.　6. B.

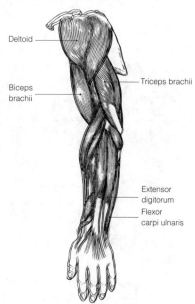

25. 1.–3. Deltoid, Gluteus maximus, Gluteus medius.　4. Quadriceps.　5. Calcaneal (Achilles).　6. Proximal.
　　7. Forearm.　8. Anterior.　9. Posteriorly.　10. Knee.　11. Flex.

26. 1. Biceps femoris.　2. Antagonists.　3. Frontalis.　4. Vastus medialis.

27. The iliopsoas and rectus femoris flex the hip. The quadriceps extends the knee. The tibialis anterior is the main dorsiflexor of the foot.

28. 1. 4.　2. 5.　3. 17.　4. 16.　5. 7.　6. 6.　7. 19.　8. 14.　9. 18.　10. 12.　11. 11.　12. 10.　13. 21.
　　14. 1.　15. 2.　16. 3.　17. 15.　18. 20.　19. 13.　20. 9.　21. 8.

29. 1. 2.　2. 1.　3. 5.　4. 9.　5. 7.　6. 4.　7. 12.　8. 3.　9. 8.　10. 10.　11. 11.　12. 6.

Developmental Aspects of the Muscular System

30. 1. Quickening.　2. Muscular dystrophy.　3.–4. Proximal-distal and cephalocaudal.　5. Gross.　6. Fine.
　　7. Exercised.　8. Atrophy.　9. Myasthenia gravis.　10. Weight.　11. Size and mass.　12. Connective (scar).

Incredible Journey

31. 1. Endomysium.　2. Motor unit.　3. Neuromuscular.　4. Acetylcholine.　5. Sodium.　6. Action potential.
　　7. Calcium.　8. Actin.　9. Myosin.　10. Calcium.

At the Clinic

32. When we are in the fully bent-over position, the erector spinae are relaxed. When we reverse this hip flexion, they are totally inactive, leaving the gluteus maximus and hamstrings to initiate the action. Thus, sudden or improper lifting techniques are likely to injure both back ligaments and the erector spinae, causing them to go into painful spasms.

33. The hamstrings can be strained (pulled) when the hip is flexed and the knee is vigorously extended at the same time.

34. The rectus abdominis is a narrow, medially placed muscle that does not extend completely across the iliac regions. No, if the incision was made as described, the rectus abdominis was not cut.

35. The latissimus dorsi and the trapezius, which together cover most of the superficial surface of the back, are receiving most of the massage therapist's attention.

36. The chances are good that the boy has Duchenne's muscular dystrophy. This condition is fatal when it impairs the respiratory muscles.

37. By reducing the size of the abdomen, the abdominal contents are forced into a smaller space which would increase the intra-abdominal pressure. The rise in intra-abdominal pressure would, in turn, force the vertebrae to move farther apart, reducing vertebral compression and pressure on the nerve fibers that transmit pain.

38. The pesticide is a chemical that inhibits the enzyme that destroys acetylcholine. Acetylcholine remains in the synapse and stimulates muscle activity.

39. The pulled muscles are the adductor muscles.

40. Some muscles attach to fascia (connective tissue) or skin as well.

The Finale: Multiple Choice

41. 1. C.   2. B.   3. B, C.   4. C.   5. A, B, C, D.   6. C.   7. C.   8. C.   9. A, B, C, D.   10. A, C, D.
11. A, C, D.   12. C.   13. A, B, D.   14. C.   15. D.   16. D.   17. D.   18. A.   19. A, D.

Chapter 7   The Nervous System

1. 1. It monitors all information about changes occurring both inside and outside the body.   2. It processes and interprets the information received and integrates it in order to make decisions.   3. It commands responses by activating muscles, glands, and other parts of the nervous system.

Organization of the Nervous System

2. 1. B or CNS.   2. D or somatic nervous system.   3. C or PNS.   4. A or autonomic nervous system.
5. B or CNS.   6. C or PNS.

Nervous Tissue—Structure and Function

3. 1. B or neuroglia.   2.–4. A or neurons.   5. B or neuroglia.

4. 1. B or axon terminal.   2. C or dendrite.   3. D or myelin sheath.   4. E or cell body.   5. A or axon.   6. F or Nissl bodies.

5. 1. A or bare nerve endings, D or muscle spindle.   2. A or bare nerve endings, E or Pacinian corpuscle.
3. E or Pacinian corpuscle (perhaps also B and D).   4. B or Golgi tendon organ, D or muscle spindle.
5. C or Meissner's (tactile) corpuscle.

6. 1. C or cutaneous sense organs.   2. L or Schwann cells.   3. M or synapse.   4. O or tract.   5. B or association neuron.   6. I or nodes of Ranvier.   7. E or ganglion.   8. D or efferent neuron.   9. K or proprioceptors.
10. N or stimuli.   11. A or afferent neuron.   12. G or neurotransmitters.

7. Figure 7–1:

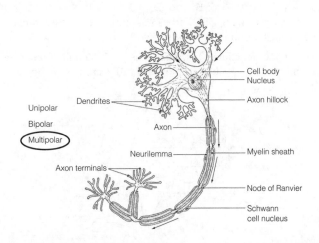

8. 1. Stimulus.   2. Receptor.   3. Afferent neuron.   4. Efferent neuron.   5. Effector organ.

9. Figure 7–2:

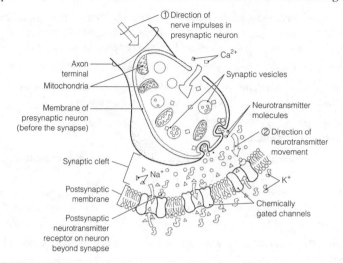

10. 1. E or refractory period.   2. B or depolarization.   3. C or polarized.   4. F or repolarization.   5. A or action potential.   6. D or potassium ions.   7. I or sodium-potassium pump.   8. G or resting period.

11. 1. A or somatic reflex(es).   2. B or autonomic reflex(es).   3. A or somatic reflex(es).   4. B or autonomic reflex(es).   5. A or somatic reflex(es).   6. B or autonomic reflex(es).   7. B or autonomic reflex(es).

12. 1. Pinprick pain.   2. Skeletal muscle.   3. Two (third with muscle).

Figure 7–3:

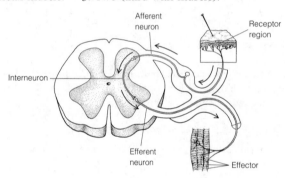

13. 1. Neurons.   2. K^+ enters the cell.   3. Unmyelinated.   4. Voluntary act.   5. Microglia.   6. Stretch.   7. High Na^+.

Central Nervous System

14. 1. Cerebral hemispheres.   2. Brain stem.   3. Cerebellum.   4. Ventricles.   5. Cerebrospinal fluid.

15. Circle: Cerebral hemispheres, cerebellum, diencephalon.

16. 1. Gyrus.   2. Surface area.   3. Neuron cell bodies and unmyelinated fibers.   4. Myelinated fibers.   5. Basal nuclei.

17. Figure 7–4: 1. D.   2. L.   3. F.   4. C.   5. K.   6. B.   7. E.   8. A.   9. I.   10. H.   11. J.   12. G.   Areas B and C should be striped.

18. Figure 7–5: 1. J.   2. L.   3. O.   4. M.   5. B.   6. D.   7. A.   8. K.   9. G.   10. I.   11. E.   12. N.   13. F.   14. H.   15. C.   Structures #4, #6, #10, and #14 should be blue. Structure #2, the cavity enclosed by #15, #2, and #8, and the entire gray area around the brain should be colored yellow.

19. 1. Hypothalamus.   2. Pons.   3. Cerebellum.   4. Thalamus.   5. Medulla oblongata.   6. Corpus callosum.   7. Cerebral aqueduct.   8. Thalamus.   9. Choroid plexus.   10. Cerebral peduncle.   11. Hypothalamus.

20. 1. G.   2. G.   3. W.   4. W.   5. W.   6. G.   7. W.

21. Figure 7–6:

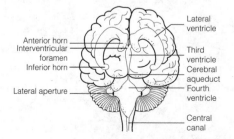

22. 1. Postcentral.   2. Temporal.   3. Frontal.   4. Broca's.   5. Left.   6. *T.*   7. Precentral.   8. Premotor.
9. Fingers.   10. General interpretation area.   11. Occipital.   12. *T.*   13. *T.*   14. Alert.

23. 1. Dura mater.   2. Pia mater.   3. Arachnoid villi.   4. Arachnoid mater.   5. Dura mater.

24. Figure 7–7:

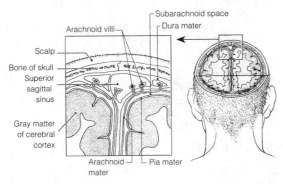

25. 1. Choroid plexuses.   2. Ventricles.   3. Cerebral aqueduct.   4. Central canal.   5. Subarachnoid space.
6. Fourth ventricle.   7. Hydrocephalus.

26. 1. E or concussion.   2. F or contusion.   3. D or coma.   4. G or intracranial hemorrhage.   5. B or cerebral
edema.   6. C or CVA.   7. A or Alzheimer's disease.   8. H or multiple sclerosis.   9. I or TIA.

27. 1. Foramen magnum.   2. Lumbar.   3. Lumbar tap, or puncture.   4. 31.   5. 8.   6. 12.   7. 5.   8. 5.
9. Cauda equina.

28. 1. D or association neurons.   2. B or efferent.   3. A or afferent.   4. B or efferent.   5. A or afferent.
6. C or both afferent and efferent.   7. C or both afferent and efferent.   8. A or afferent.   9. B or efferent.

29. Figure 7–8:

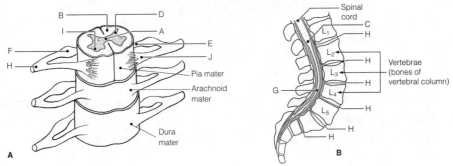

30. Figure 7–9:

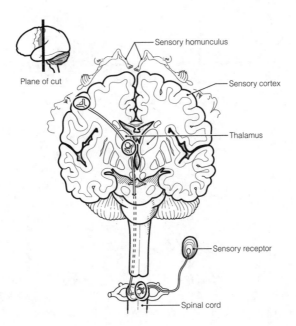

Peripheral Nervous System

31. Figure 7–10:

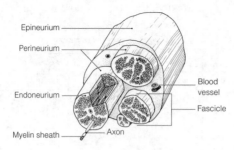

Epineurium

Perineurium

Endoneurium

Blood vessel

Fascicle

Myelin sheath — Axon

32. 1. Nerve (or fascicle).   2. Mixed.   3. Afferent.

33. Figure 7–11:

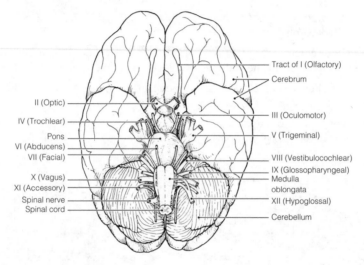

Tract of I (Olfactory)

Cerebrum

II (Optic)

IV (Trochlear)

Pons

VI (Abducens)

VII (Facial)

X (Vagus)

XI (Accessory)

Spinal nerve

Spinal cord

III (Oculomotor)

V (Trigeminal)

VIII (Vestibulocochlear)

IX (Glossopharyngeal)

Medulla oblongata

XII (Hypoglossal)

Cerebellum

34. 1. XI-Accessory.   2. I-Olfactory.   3. III-Oculomotor.   4. X-Vagus.   5. VII-Facial.   6. V-Trigeminal.
    7. VIII-Vestibulocochlear.   8. VII-Facial.   9. III, IV, VI.   10. V-Trigeminal.   11. II-Optic.   12. I, II, VIII.

35. 1. Plexuses.   2. Limbs and anterolateral body trunk.   3. Thorax.   4. Posterior body trunk.

36. Figure 7–12:

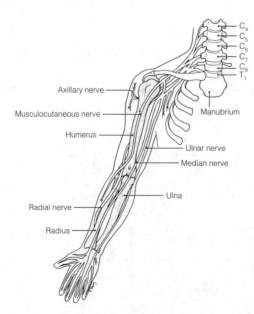

C₄
C₅
C₆
C₇
C₈
T₁

Axillary nerve

Musculocutaneous nerve

Humerus

Manubrium

Ulnar nerve

Median nerve

Ulna

Radial nerve

Radius

37. 1. Cervical plexus.   2. Lumbar plexus.   3. Femoral nerve.   4. Phrenic nerve.   5. Sciatic nerve.
    6. Fibular and tibial nerves.

38. Figure 7–13:

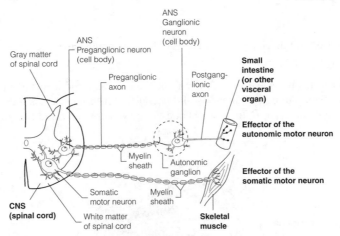

39. Check sympathetic for 1, 4, 6, 8, and 10. Check parasympathetic for 2, 3, 5, 7, 9, and 11.

1. Increased respiratory rate.   2. Increased heart rate and blood pressure.   3. Increased availability of blood glucose.   4. Pupils dilate; increased blood flow to heart, brain, and skeletal muscles.

Developmental Aspects of the Nervous System

40. 1. Hypothalamus.   2. Oxygen.   3. Cephalocaudal.   4. Gross.   5. Blood pressure.   6. Decreased oxygen (blood) to brain.   7. Senility.   8. Stroke (CVA).

Incredible Journey

41. 1. Cerebellum.   2. Medulla.   3. Hypothalamus.   4. Memories.   5. Temporal.   6. Broca's area.   7. Reasoning.   8. Frontal.   9. Vagus (X).   10. Dura mater.   11. Subarachnoid space.   12. Fourth.

At the Clinic

42. Parasympathetic.

43. Considering the nerve cells are amitotic, the tumor is most likely a glioma, developing from one of the types of neuroglia.

44. During sympathetic activation, large amounts of epinephrine from the adrenal medulla pour into the blood. It will take time for the hormone to be broken down throughout the body.

45. The stroke has destroyed the trunk, hip, and lower limb region of the primary motor cortex that corresponds to those paralyzed areas on the *left* side of the body. (Remember, the motor pathways are crossed.)

46. 1. Cerebellum.   2. Basal nuclei.   3. Meningitis.   4. III (oculomotor).   5. Somatosensory cortex.   6. Broca's area.   7. Electroencephalogram.

47. Marie has ataxia, indicating problems of the cerebellum.

48. Paresthesia because the fiber tracts in the dorsal white matter are sensory tracts.

49. Sympathetic; the "fight or flight" response was activated.

50. The self-propagating change in membrane potential that travels along the membrane from the point of stimulation.

51. Peripheral nerves of the somatic nervous system carry both motor and sensory fibers, so you might expect the cut to affect both types of function.

52. Schwann cells which myelinate the peripheral nerve fibers.

53. The somatic division is involved in stretching, sit-ups, walking, and brushing her teeth. The autonomic division causes stomach gurgling.

The Finale: Multiple Choice

54. 1. A, B.   2. C.   3. C.   4. A.   5. C.   6. B.   7. C.   8. A, C, D.   9. A, B, C.   10. C.   11. A, C.
12. A, C, D.   13. B, D.   14. B.   15. A.   16. D.   17. B.   18. A.   19. A.   20. C.   21. C.   22. B.   23. A.
24. B.   25. A.   26. A.   27. A.   28. B.   29. C.

Chapter 8 Special Senses

The Eye and Vision

1. 1. Extrinsic, or external eye.   2. Eyelids.   3. Tarsal glands.   4. Conjunctivitis.

2. 1. 2.   2. 4.   3. 3.   4. 1.

3. Figure 8–1: 1. Superior rectus turns eye superiorly and medially.   2. Inferior rectus turns eye inferiorly and medially.   3. Superior oblique turns eye inferiorly and laterally.   4. Lateral rectus turns eye laterally.   5. Medial rectus turns eye medially.   6. Inferior oblique turns eye superiorly and laterally.

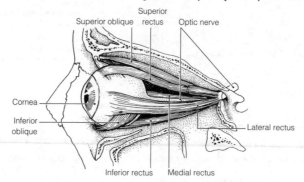

4. 1. Conjunctiva secretes mucus.   2. Lacrimal glands secrete salt water and lysozyme.   3. Tarsal glands secrete oil. Circle the lacrimal gland secretion.

5. 1. L or refraction.   2. A or accommodation.   3. F or emmetropia.   4. H or hyperopia.   5. K or photopupillary reflex.   6. D or cataract.   7. I or myopia.   8. C or astigmatism.   9. G or glaucoma.   10. E or convergence.   11. B or accommodation pupillary reflex.   12. J or night blindness.

6. 1. Autonomic nervous system.

7. 1. Convex.   2. Real.   3. Behind.   4. Convex (converging).   5. In front of.   6. Concave (diverging).

8. 1. E or ciliary zonule.   2. A or aqueous humor.   3. L or sclera.   4. J or optic disk.   5. D or ciliary body.   6. C or choroid.   7. B or canal of Schlemm.   8. K or retina.   9. M or vitreous humor.   10. C or choroid.   11. and 12. D or ciliary body, H or iris.   13. G or fovea centralis.   14.–17. A or aqueous humor, F or cornea, I or lens, and M or vitreous humor.   18. F or cornea.   19. H or iris.

9. Figure 8–2:

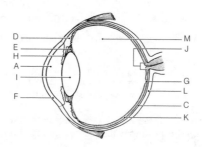

10. 1. In distant vision the ciliary muscle is relaxed, the lens convexity is decreased, and the degree of light refraction is decreased.   2. In close vision the ciliary muscle is contracted, the lens convexity is increased, and the degree of light refraction is increased.

11. Retina ⟶ Optic nerve ⟶ Optic chiasma ⟶ Optic tract ⟶ Synapse in thalamus ⟶ Optic radiation ⟶ Optic cortex.

12. 1. Three.   2.–4. Blue, green, and red.   5. At the same time.   6. Total color blindness.   7. Males.   8. Rods.

13. 1. Vitreous humor.   2. Superior rectus.   3. Far vision.   4. Proprioceptors.   5. Iris.   6. Iris.   7. Pigmented layer.

14. 1. Opsin.   2. Rhodopsin.   3. Bleaching of the pigment.   4. Yellow.   5. Colorless.   6. A.

The Ear: Hearing and Balance

15. 1.–3. E, I, and M.   4.–6. C, K, and N.   7.–9. A, F, and L.   10. and 11. K and N.   12. B.   13. M.   14. C.   15. B.   16. and 17. K and N.   18. G.   19. D.   20. H.

16. Figure 8–3: I, E, and M are yellow; A, F, and L are red; C is blue, and K (continuing to N) is green.

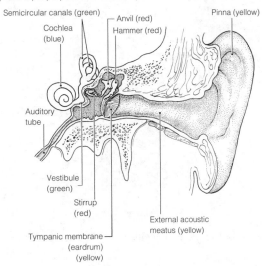

17. Eardrum → Hammer → Anvil → Stirrup → Oval window → Perilymph → Membrane → Endolymph → Hair cells.

18. Figure 8–4:

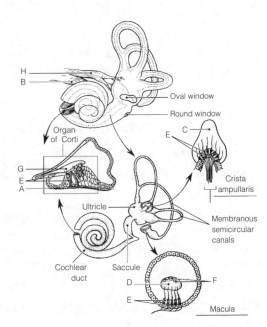

19. 1. C or dynamic.   2. I or semicircular canals.   3. A or angular/rotatory.   4. D or endolymph.   5. B or cupula. 6. J or static.   7. and 8. H or saccule, K or utricle.   9. E or gravity.   10. and 11. G or proprioception, L or vision.

20. 1. C.   2. S.   3. S.   4. C.   5. C, S.   6. C.   7. S.

21. Nausea, dizziness, and balance problems.

22. 1. Pinna.   2. Tectorial membrane.   3. Sound waves.   4. Auditory tube.   5. Optic nerve.

Chemical Senses: Smell and Taste

23. 1.–3. (in any order): VII-Facial, IX-Glossopharyngeal, X-Vagus.   4. I-Olfactory.   5. Mucosa of the "roof." 6. Sniffing.   7. Taste buds.   8. or 9. Fungiform, circumvallate.   10.–14. (in any order): sweet, salty, bitter, sour, umami.   15. Bitter.   16. Smell.   17. Dry.   18. Memories.

24. Figure 8–5:

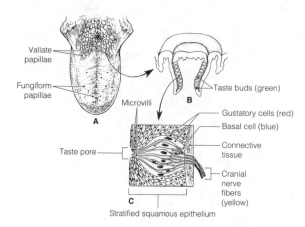

25. Figure 8–6: 1. Mucus "captures" airborne odor molecules.   2. Olfactory neurons are bipolar neurons.

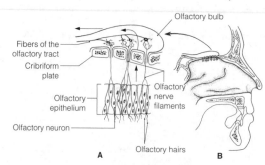

26. 1. Musky.   2. Epithelial cell.   3. Neuron.   4. Olfactory nerve.   5. Four receptor types.   6. Metal ions.

Developmental Aspects of the Special Senses

27. 1. Nervous system.   2. Measles (rubella).   3. Blindness.   4. Vision.   5. Hyperopic.   6. Elastic.   7. Presbyopia.   8. Cataract.   9. Presbycusis.

Incredible Journey

28. 1. Bony labyrinth.   2. Perilymph.   3. and 4. Saccule, utricle.   5. Gel (otolithic membrane).   6. Otoliths.   7. Macula.   8. Static.   9. Cochlear duct.   10. Organ of Corti.   11. Hearing.   12. Cochlear division of cranial nerve VIII.   13. Semicircular canals.   14. Cupula.   15. Crista ampullaris.   16. Dynamic.

At the Clinic

29. Patching the strong eye to force the weaker eye muscles to become stronger.

30. Cataract; UV radiation, smoking.

31. Vision is poor because, without pigment within the eye (in the choroid), light scatters before it can be properly focused.

32. Night blindness; vitamin A; rods.

33. The proximal end close to the oval window; sensorineural.

34. Otitis externa, most likely because of his exposure to pool bacteria. This diagnosis would be confirmed by presence of an inflamed external ear canal. If it is otitis media, the middle ear would be inflamed. Bulging of the eardrum would suggest that inserting ear tubes might be recommended.

35. Cranial nerve I, the olfactory nerve.

36. Abducens nerve, cranial nerve VI.

37. Glaucoma, inadequate drainage of aqueous humor; blindness due to compression of retina and optic disc.

38. Taste bud cells are subjected to friction and heat and hence are rapidly dividing cells that will be targeted by chemotherapeutic drugs. A chef must have a fine sense of taste to be successful.

The Finale: Multiple Choice

39. 1. D.   2. B, C.   3. B, D.   4. A, B, D.   5. C.   6. C.   7. A, B, C.   8. D.   9. B.   10. A, B, D.   11. B, C, D.   12. B.   13. A, D.   14. C.   15. A, C, D.   16. A, B, D.   17. A, C.   18. B.

Chapter 9 The Endocrine System

The Endocrine System and Hormone Function—An Overview

1. 1. F or slower and more prolonged.   2. E or nervous system.   3. B or hormones.   4. D or nerve impulses.
5. A or cardiovascular system.

2. 1. I or receptors.   2. N or target cell(s).   3. A or altering activity.   4. L or stimulating new or unusual activities.
5. K or steroid or amino acid–based.   6. G or neural.   7. C or hormonal.   8. D or humoral.   9. F or negative
feedback.   10. B or anterior pituitary.   11. J or releasing hormones.   12. E or hypothalamus.   13. H or
neuroendocrine.

3. Steroid hormones: B, C, D. Amino acid–based hormones: A, E.

The Major Endocrine Organs

4. Figure 9–1:

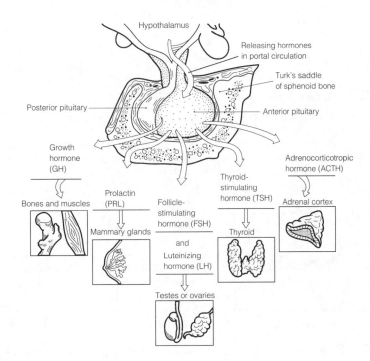

5. Figure 9–2: A. Pineal.   B. Posterior pituitary.   C. Anterior pituitary.   D. Thyroid.   E. Thymus.
F. Adrenal cortex.   G. Adrenal medulla.   H. Pancreas.   I. Ovary.   J. Testis.   K. Parathyroids.   L. Placenta.

6. 1. C.   2. B.   3.–4. F.   5. G.   6. I, L.   7. C.   8. H.   9. H.   10. C.   11. A.   12. B.   13. I, L.
14. C.   15. K.   16. C.   17. J.   18. E.   19. D.   20. C.

7. 1. Thyroxine.   2. Thymosin.   3. PTH   4. Cortisol (glucocorticoids).   5. Epinephrine.   6. Insulin.
7.–10. (in any order): TSH, FSH, LH, ACTH.   11. Glucagon.   12. ADH.   13. and 14. (in any order): FSH, LH.
15. and 16. (in any order): Estrogens, progesterone.   17. Aldosterone.   18. and 19. Prolactin, oxytocin.

8. 1. Estrogen/testosterone.   2. PTH.   3. ADH.   4. Thyroxine.   5. Thyroxine.   6. Insulin.
7. Growth hormone.   8. Estrogen/progesterone.   9. Thyroxine.

9. 1. Growth hormone.   2. Thyroxine.   3. PTH.   4. Glucocorticoids.   5. Growth hormone.
6. Androgens (testosterone).

10. 1. Polyuria—high sugar content in kidney filtrate causes large amount of water to be lost in the urine.
2. Polydipsia—thirst due to large volumes of urine excreted.   3. Polyphagia—hunger because blood sugar
cannot be used as a body fuel even though blood levels are high.

11. Figure 9–3:

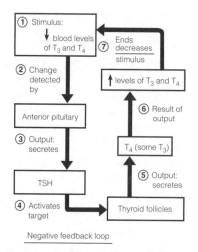

Negative feedback loop

12. 1. Anterior lobe.   2. Steroid hormone.   3. Cortisol.   4. Increases blood Ca^{2+}.   5. Growth hormone.   6. Parafollicular cells.

Other Hormone-Producing Tissues and Organs

13.

Hormone	Chemical makeup	Source	Effects
Gastrin	Peptide	Stomach	Stimulates stomach glands to secrete HCl
Secretin	Peptide	Duodenum	Stimulates the pancreas to secrete HCO_3^--rich juice and stimulates the liver to release more bile; inhibits stomach glands
Cholecystokinin	Peptide	Duodenum	Stimulates the pancreas to secrete enzyme-rich juice and the gallbladder to contract; relaxes sphincter of Oddi
Erythropoietin	Glycoprotein	Kidney in response to hypoxia	Stimulates production of red blood cells by bone marrow
Active vitamin D_3	Steroid	Skin; activated by kidneys	Enhances intestinal absorption of calcium
Atrial natriuretic peptide (ANP)	Peptide	Heart	Inhibits Na^+ reabsorption by kidneys; inhibits aldosterone release by kidneys
Human chorionic gonadotropin (hCG)	Protein	Placenta	Stimulates corpus luteum to continue producing estrogens and progesterone, preventing menses
Leptin	Peptide	Adipose tissue	Targets the brain; reduces appetite; increases energy expenditure

Developmental Aspects of the Endocrine System

14. 1. Neoplasm.   2. Hypersecretion.   3. Iodine.   4. Estrogens.   5. Menopause.   6. Bear children.   7. Insulin.

Incredible Journey

15. 1. Insulin.   2. Pancreas.   3. Posterior pituitary, or hypothalamus.   4. ADH.   5. Parathyroid.   6. Calcium.   7. Adrenal medulla.   8. Epinephrine.   9. Thyroxine.

At the Clinic

16. Pituitary dwarfs who secrete inadequate amounts of GH have fairly normal proportions; cretins (hypothyroid individuals) retain childlike body proportions.

17. Hypothyroidism; iodine deficiency (treated by dietary iodine supplements) or thyroid cell burnout (treated by thyroid hormone supplements).

18. Adrenal cortex.

19. For the giant, GH is being secreted in excess by the anterior pituitary, resulting in extraordinary height. For the dwarf, GH is deficient, resulting in very small stature but normal body proportions. For the fat man, T_3 and T_4 are not being produced adequately, resulting in depressed metabolism and obesity (myxedema). The bearded lady has a tumor of her adrenal cortex (androgen-secreting area) leading to excessive hairiness (hirsutism).

20. Stressor $\longrightarrow$ hypothalamus $\longrightarrow$ CRH (releasing hormone) released to the hypophysial portal system blood $\longrightarrow$ to anterior pituitary which releases ACTH $\longrightarrow$ acts on adrenal cortex to trigger release of glucocorticoids (cortisol, etc.).

21. Prolactin.

The Finale: Multiple Choice

22. 1. D.   2. A.   3. B.   4. A, B, C, D.   5. D.   6. B.   7. A, B, C.   8. C.   9. A, B, D.   10. A, B, C, D.
11. A, B, C, D.   12. A, C.   13. B.   14. A, C.   15. A.   16. C.   17. D.   18. C.   19. B.

Chapter 10   Blood

Composition and Functions of Blood

1. 1. Connective tissue.   2. Formed elements.   3. Plasma.   4. Clotting.   5. Erythrocytes.   6. Hematocrit.
7. Plasma.   8. Leukocytes.   9. Platelets.   10. One.   11. Oxygen.

2. 1. F or neutrophil.   2.–4. (in any order): C or eosinophil, D or basophil, F or neutrophil.   5. A or red blood.
6. and 7. (in any order): E or monocyte, F or neutrophil.   8. and 9. (in any order): E or monocyte, G or lymphocyte.
10. B or megakaryocyte.   11. H or formed elements.   12. C or eosinophil.   13. D or basophil.
14. G or lymphocyte.   15. A or red blood.   16. I or plasma.   17. E or monocyte.   18. D or basophil.
19.–23. (in any order): C or eosinophil, D or basophil, E or monocyte, F or neutrophil, G or lymphocyte.

3. Figure 10–1: 100–120 days.

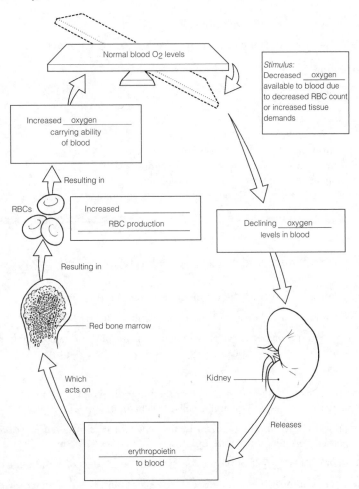

4. **Figure 10–2:** A is a neutrophil, B is a monocyte, C is an eosinophil, D is a lymphocyte.

5. 1. Diapedesis.   2. *T.*   3. Kidneys.   4. 7.35.   5. 5.5.   6. *T.*   7. *T.*   8. 4.5–5.5.   9. Hematocrit.
10. Less.   11. Monocytes.   12. Lymphocytes.

6. 1. Erythrocytes.   2. Monocytes.   3. Lymphocyte.   4. Platelets.   5. Aneurysm.   6. Hemoglobin.
7. Lymphocyte.

7. 1. 2.   2. 5.   3. 1.   4. 4.   5. 3.

8. Check 1, 2, 3.

Hemostasis

9. 1. A or break.   2. E or platelets.   3. I or serotonin.   4. K or tissue factor.   5. H or PF$_3$.   6. G or prothrombin activator.   7. F or prothrombin.   8. J or thrombin.   9. D or fibrinogen.   10. C or fibrin.   11. B or erythrocytes.

10. 1. 3–6 min.   2. Heparin.   3. *T.*

Blood Groups and Transfusions

11.

Blood type	Agglutinogens or antigens	Agglutinins or antibodies in plasma	Can donate blood to type	Can receive blood from type
1. Type A	A	anti-B	A, AB	A, O
2. Type B	B	anti-A	B, AB	B, O
3. Type AB	A, B	none	AB	A, B, AB, O
4. Type O	none	anti-A, anti-B	A, B, AB, O	O

12. Type O is the universal donor. AB is the universal recipient.

13. A reaction during which plasma antibodies attach to and lyse red blood cells different from your own.

Developmental Aspects of Blood

14. 1. F.   2. Jaundiced.   3. Sickle-cell.   4. Hemophilia.   5. Iron.   6. Pernicious.   7. B$_{12}$.   8. Thrombi.
9. Leukemia.

Incredible Journey

15. 1. Hematopoiesis.   2. Hemostasis.   3. Hemocytoblasts.   4. Neutrophil.   5. Phagocyte.   6. Erythropoietin.
7. Red blood cells.   8. Hemoglobin.   9. Oxygen.   10. Lymphocytes.   11. Antibodies.   12.–15. (in any order):
Basophils, eosinophils, monocytes, platelets.   16. Endothelium.   17. Platelets.   18. Serotonin.   19. Fibrin.
20. Clot.   21. Prothrombin activator.   22. Prothrombin.   23. Thrombin.   24. Fibrinogen.   25. Embolus.

At the Clinic

16. 1. Hemolytic disease of the newborn.

   2. Its RBCs have been destroyed by the mother's antibodies; therefore, the baby's blood is carrying insufficient oxygen.

   3. She must have received mismatched (Rh$^+$) blood previously in a transfusion.

   4. Give the mother RhoGAM to prevent her from becoming sensitized to the Rh$^+$ antigen.

   5. Fetal progress will be followed in expectation of hemolytic disease of the newborn; intrauterine transfusions will be given if necessary, as well as complete blood transfusion to the newborn.

17. No; A+.

18. The stem cells for hematopoiesis in red bone marrow are a rapidly dividing cell population. Hence, they would be targeted (along with other rapidly dividing cells) by chemotherapeutic drugs.

19. Virtually all bones contain red marrow and functional hematopoietic tissue in young children, but in adults only the sternum, ilium, and a very few long bone epiphyses contain red marrow.

20. Erythrocytes, which account for nearly half of blood volume, will be produced in the largest numbers.

21. Stomach cells are the source of intrinsic factor needed to absorb vitamin B$_{12}$. Apparently insufficient numbers of vitamin-producing cells remain after the stomach surgery. Vitamin B$_{12}$ cannot be absorbed orally, so it must be injected. If he refuses the shots, pernicious anemia will ensue.

The Finale: Multiple Choice

21. 1. A, B, D.   2. B.   3. A, B, C, D.   4. D.   5. A, B, D.   6. D.   7. C.   8. A.   9. C.   10. B, C, D.   11. C.
12. B, C, D.   13. A, D.   14. D.   15. B.   16. C.

Chapter 11   The Cardiovascular System

The Heart

1. 1. Thorax.   2. Diaphragm.   3. Second.   4. Aorta.   5. Right atrium.   6. Atria.   7. Ventricles.
8. Endocardium.   9. Epicardium.   10. Friction.   11. Cardiac muscle.

2. 1. Right ventricle.   2. Pulmonary semilunar.   3. Pulmonary arteries.   4. Lungs.   5. Right and left pulmonary veins.   6. Left atrium.   7. Mitral (bicuspid).   8. Left ventricle.   9. Aortic.   10. Aorta.   11. Capillary beds.
12. Superior vena cava.   13. Inferior vena cava.

In Figure 11–1, the white areas represent regions transporting O$_2$-rich blood. The gray vessels transport O$_2$-poor blood.

Figure 11–1:

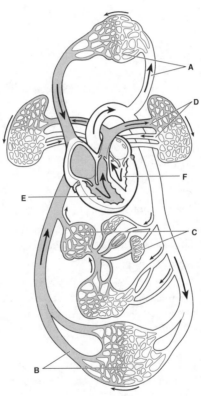

3. **Figure 11–2:** 1. Right atrium.   2. Left atrium.   3. Right ventricle.   4. Left ventricle.   5. Superior vena cava.
6. Inferior vena cava.   7. Aorta.   8. Pulmonary trunk.   9. Left pulmonary artery.   10. Right pulmonary artery.
11. Right pulmonary veins.   12. Left pulmonary veins.   13. Coronary circulation.   14. Apex of heart.
15. Ligamentum arteriosum.

4. **Figure 11–3:**

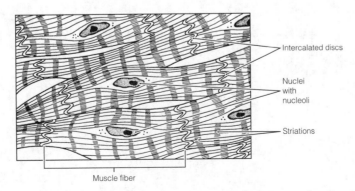

5. 1. Systole.   2. Diastole.   3. Lub-dup.   4. Atrioventricular.   5. Semilunar.   6. Ventricles.   7. Atria.
8. Atria.   9. Ventricles.   10. Murmurs.

6. **Figure 11–4:** Red arrows should be drawn from the left atrium to the left ventricle and out the aorta. Blue arrows should be drawn from the superior and inferior venae cavae into the right atrium, then into the right ventricle and out the pulmonary trunk. Green arrows should be drawn from #1 to #5 in numerical order.

1. SA node.   2. AV node.   3. AV bundle or bundle of His.   4. Bundle branches.   5. Purkinje's fibers.   6. Pulmonary valve.   7. Aortic valve.   8. Mitral (bicuspid) valve.   9. Tricuspid valve.

A. and B. (in any order): 6, 7.   C. and D. (in any order): 8, 9.   E. 9.   F. 8.   G. 1.   H. 2.

7. 1. C or electrocardiogram.   2. F or P wave.   3. H or T wave.   4. G or QRS wave.   5. B or bradycardia. 6. D or fibrillation.   7. I or tachycardia.   8. E or heart block.   9. A or angina pectoris.

8. **Figure 11–5:**

9. 1. Cardiac output.   2. Heart rate.   3. Stroke volume. 4. About 75 beats per minute.   5. 70 mL per beat. 6. 5250 mL per minute.   7. Minute.   8. Stroke volume.   9. Stretch.   10. Blood.

10. Check 1, 2, 4, 5, 6, 8, and 10.

11. 1. Fetal.   2. Rate of contraction.   3. Left.   4. *T.*   5. *T.*

12. 1. Left side of heart.   2. P wave.   3. AV valves opened. 4. Aortic semilunar valve.   5. Tricuspid valve.   6. Heart block.

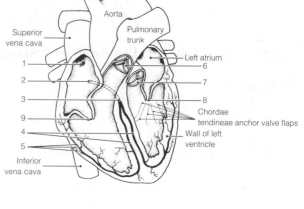

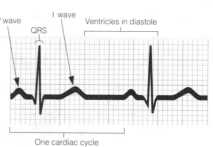

Blood Vessels

13. 1. Lumen.   2. Vasoconstriction.   3. Vasodilation.   4. Veins.   5. Arteries.   6. Arterioles.   7. Venules.

14. Arteries are high-pressure vessels. Veins are low-pressure vessels. Blood flows from high to low pressure. The venous valves help to prevent the backflow of blood that might otherwise occur in those low-pressure vessels.

15. Skeletal muscle activity and breathing (respiratory pump).

16. 1. A or tunica intima.   2. B or tunica media.   3. A or tunica intima.   4. A or tunica intima.   5. C or tunica externa. 6. B or tunica media.   7. C or tunica externa. **Figure 11–6:** A. Artery; thick media; small, round lumen. B. Vein; thin media; elongated, relatively collapsed lumen; a valve present. C. Capillary; single layer of endothelium. In A and B, the tunica intima is the innermost vessel layer, the tunica externa is the outermost layer, and the tunica media is the thick middle layer.

17. **Figure 11–7:**

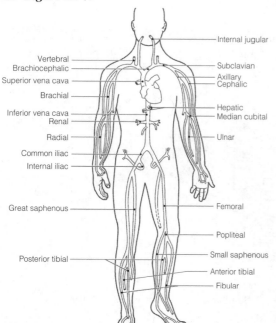

Figure 11–8:

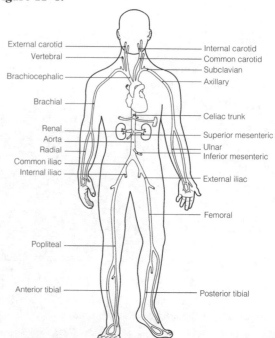

18. 1. and 2. S or radial and X or ulnar.   3. U or subclavian.   4. E or cardiac.   5. T or renal.   6. Q or internal jugular. 7. D or brachiocephalic.   8. and 9. A or anterior tibial and R or posterior tibial.   10. M or hepatic portal.   11. F or cephalic.   12. J or gonadal.   13. B or azygos.   14. O or inferior vena cava.   15. L or hepatic.   16.–18. I or gastric, N or inferior mesenteric, and V or superior mesenteric.   19. K or great saphenous.   20. G or common iliac. 21. H or femoral.

19. **Figure 11–9:** The right atrium and ventricle and all vessels with "pulmonary" in their name should be colored blue; the left atrium and ventricle and the aortic arch and lobar arteries should be colored red.

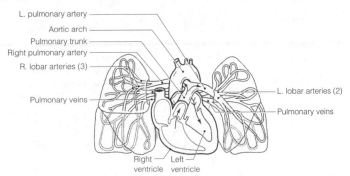

20. 1. F.   2. D.   3. B.   4. A.   5. D.   6. A.   7. E.

21. **Figure 11–10:**

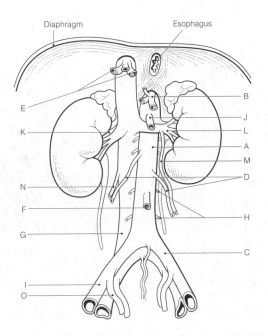

22. **Figure 11–11:**

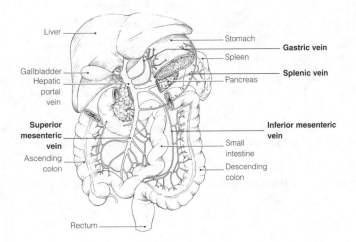

23. 1. and 2. F or common carotid and W or subclavian. 3. H or coronary. 4. and 5. P or internal carotid and Y or vertebral. 6. B or aorta. 7. J or dorsalis pedis. 8. I or deep artery of the thigh. 9. S or phrenic. 10. C or brachial. 11. C or brachial. 12. N or inferior mesenteric. 13. Q or internal iliac. 14. L or femoral. 15. C or brachial. 16. X or superior mesenteric. 17. G or common iliac. 18. E or celiac trunk. 19. K or external carotid. 20.–22. (in any order): A or anterior tibial, R or peroneal, T or posterior tibial. 23. U or radial. 24. B or aorta.

24. Figure 11–12:

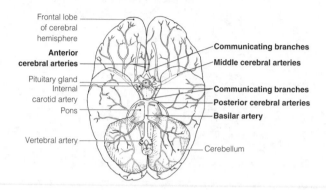

25. Figure 11–13:

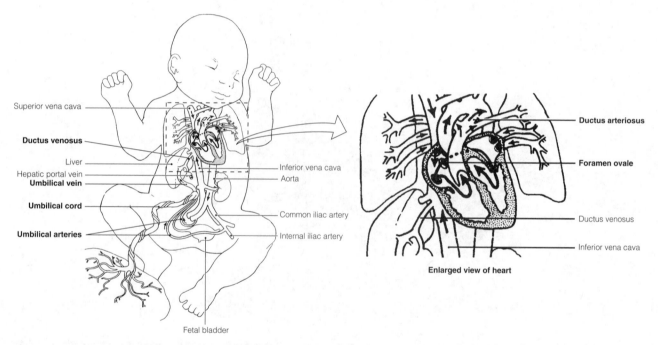

Enlarged view of heart

26. 1. C or circle of Willis. 2. J or umbilical vein. 3. E or ductus venosus. 4. A or anterior cerebral artery, G or middle cerebral artery. 5. B or basilar artery. 6. D or ductus arteriosus. 7. F or foramen ovale.

27. The fetal lungs are not functioning in gas exchanges, and they are collapsed. The placenta makes the gas exchanges with the fetal blood.

28. 1. Vein. 2. Carotid artery. 3. Vasodilation. 4. High blood pressure. 5. Vasodilation.

29. 1. H or pulse. 2. B or blood pressure. 3. and 4. C or cardiac output and F or peripheral resistance. 5. D or constriction of arterioles. 6. J or systolic blood pressure. 7. E or diastolic blood pressure. 8. A or over arteries. 9. G or pressure points. 10. I or sounds of Korotkoff.

30. 1. G or interstitial fluid. 2. C or diffusion. 3. E or fat soluble. 4.–6. (in any order): B or capillary clefts, D or fenestrations, I or vesicles. 7. D or fenestrations. 8.–9. B or capillary clefts; D or fenestrations.

31. 1. D. 2.–5. I. 6. D. 7. D. 8. I. 9.–11. D. 12. I. 13. I.

32. 1. Increase. 2. Orthostatic. 3. Brain. 4. Stethoscope. 5. Low. 6. *T*. 7. Vasoconstricting. 8. Hypertension.

Figure 11–14:

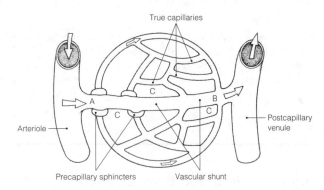

33. 1. Through the vascular shunt. 2. A. 3. Capillary blood. 4. Capillary hydrostatic (blood) pressure (Hp_C).
5. Blood pressure. 6. Capillary colloid osmotic pressure (Op_C). 7. Albumin. 8. At the arteriole end.
9. It is picked up by lymphatic vessels for return to the bloodstream.

34. 1. Femoral artery. 2. Brachial artery. 3. Popliteal artery. 4. Facial artery. 5. Radial artery.
6. Temporal artery.

Developmental Aspects of the Cardiovascular System

35. 1. Fourth. 2. Lungs. 3. Ductus venosus. 4. Umbilical vein. 5. Placenta. 6. Fetal liver. 7. Umbilical
arteries. 8. Occluded. 9. Deaths. 10. Atherosclerosis (and arteriosclerosis). 11. Menopause. 12. Aerobic
exercise. 13. Atherosclerosis. 14. Varicose veins. 15. and 16. Feet and legs.

Incredible Journey

36. 1. Left atrium. 2. Left ventricle. 3. Mitral. 4. Chordae tendineae. 5. Diastole. 6. Systole/contraction.
7. Aortic semilunar. 8. Aorta. 9. Superior mesenteric. 10. Endothelial. 11. Superior mesenteric.
12. Splenic. 13. Nutrients. 14. Phagocytic (Kupffer). 15. Hepatic. 16. Inferior vena cava. 17. Right atrium.
18. Pulmonary trunk. 19. Pulmonary. 20. Lungs. 21. Subclavian.

At the Clinic

37. Zero; myocardial infarction. The posterior interventricular artery supplies much of the left ventricle, the systemic
pump.

38. Bradycardia, which results from excessive vagal stimulation of the heart, can be determined by taking the pulse.

39. Peripheral congestion caused by right heart failure.

40. Thrombosis or atherosclerosis; an arterial anastomosis (circle of Willis), e.g., (1) Left internal carotid artery to left
anterior cerebral artery. Then through anterior communicating branch to right anterior cerebral artery and
(2) vertebral arteries to basilar artery to right posterior cerebral artery through the posterior communicating
branch to right middle cerebral artery.

41. High; polycythemia increases blood viscosity (thus peripheral resistance), which increases blood pressure.

42. The stiffened valve flaps would not close properly and the valve would become incompetent. A heart murmur
would be heard after the valve had (supposedly) closed and blood was flowing back through the valve.

43. Thrombophlebitis occurs when a thrombus (clot) forms in an inflamed blood vessel (a vein). The danger is that
the clot may detach, leading to a pulmonary embolism.

44. An ECG only reveals electrical problems. It is not useful for revealing valvular problems.

45. If anything, exercise extends life by making the cardiovascular and respiratory systems more efficient. Heart rate
drops and stroke volume increases.

46. When the environmental temperature is high, blood vessels serving the skin vasodilate, and much of the blood
supply will be found in dermal blood vessels. Then when you stand suddenly, there will initially be inadequate
blood volume in the larger, more central blood vessels to ensure that the brain receives a normal blood supply,
thus the dizziness.

47. A drug that blocks calcium channels will decrease the force of heart contraction. Because contractile force is directly
related to stroke volume, the SV will decrease.

48. Acetylcholine slows heart rate (this is the neurotransmitter released by the vagus nerves). Thus, with a longer filling time the heart's stroke volume will increase.

49. It reveals their elasticity. When the heart contracts and forces blood into the large arteries near the heart, they stretch to accommodate the greater blood volume (systolic pressure). Then, as the blood continues on in the circuit, their walls recoil, keeping pressure on the blood which keeps it moving (diastolic pressure).

The Finale: Multiple Choice

50. 1. A, D. 2. B. 3. A, D. 4. A, B, C, D. 5. D. 6. C. 7. A, C. 8. C. 9. A, B, C, D. 10. A, B, C. 11. A, B, C. 12. A, C. 13. B, D. 14. A, B, C. 15. C, D. 16. A. 17. A, B, C, D. 18. A, B, C. 19. B. 20. D. 21. A. 22. B. 23. D. 24. D.

Chapter 12   The Lymphatic System and Body Defenses

The Lymphatic System

1. 1. Pump. 2. Arteries. 3. Veins. 4. Valves. 5. Lymph. 6. 3 liters.

2. Figure 12–1:

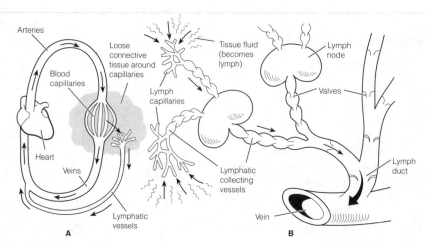

3. 1. Blood capillary. 2. Abundant supply of lymphatics. 3. High-pressure gradient. 4. Impermeable.

4. 1. C or spleen. 2. A or lymph nodes. 3. D or thymus. 4. B or Peyer's patches, E or tonsils. 5. B or Peyer's patches.

5. Figure 12–2: Shade in the right upper limb and right side of the thorax and head.

6. 1. B lymphocytes. 2. They produce and release antibodies. 3. T lymphocytes. 4. Macrophages, phagocytes. 5. This slows the flow of lymph through the node, allowing time for immune cells and macrophages to respond to foreign substances present in the lymph. 6. Valves in the afferent and efferent lymphatics. 7. Cervical, axillary, inguinal.

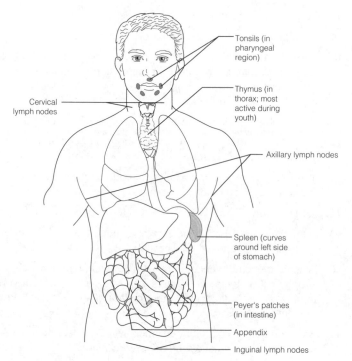

Figure 12–3:

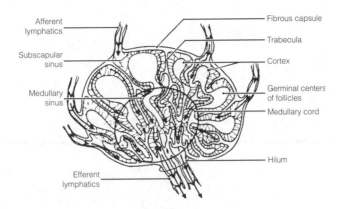

7. 1. C.   2. A.   3. D.   4. B.   5. C.

Body Defenses

8. 1. Surface membrane barriers, mucosae.   2. Natural killer cells.   3. Chemicals (inflammatory and antimicrobial).

9. 1. Tears and saliva.   2. Stomach and female reproductive tract.   3. Sebaceous (oil) glands; skin.   4. Digestive.

10. Figure 12–4:

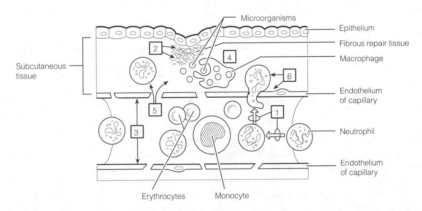

11. 1. Itching.   2. Natural killer cells.   3. Interferon.   4. Inflammation.   5. Antibacterial.

12. 1. B or lysozyme, F or sebum.   2. C or mucosae, G or skin.   3. A or acids, B or lysozyme, D or mucus, E or protein-digesting enzymes, F or sebum.   4. D or mucus.   5. A–G.

13. They propel mucus laden with trapped debris superiorly away from the lungs to the throat, where it can be swallowed or spat out.

14. Phagocytosis is ingestion and destruction of particulate material by certain cells. The rougher the particle, the more easily it is ingested.

15. Check 1, 3, 4.

16. 1. F or increased blood flow.   2. E or histamine.   3. G or inflammatory chemicals.   4. A or chemotaxis.   5. C or edema.   6. H or macrophages.   7. B or diapedesis.   8. I or neutrophils.   9. D or fibrin mesh.

17. 1. Proteins.   2. Activated.   3. Holes or lesions.   4. Water.   5. Lysis.   6. Opsonization.

18. Interferon is synthesized in response to viral infection of a cell. The cell produces and releases interferon proteins, which diffuse to nearby cells, where they prevent viruses from multiplying within those cells.

19. The adaptive immune system is antigen-specific, systemic, and has memory.

20. 1. Immune system.   2. Proteins.   3. Haptens.   4. Nonself.

21. 1. A or antigens.   2. E or humoral immunity.   3. D or cellular immunity.   4. and 5. B or B cells and I or T cells.   6. H or macrophages.   7. and 8. C or blood and F or lymph.   9. G or lymph nodes.

22. **Figure 12–5:**

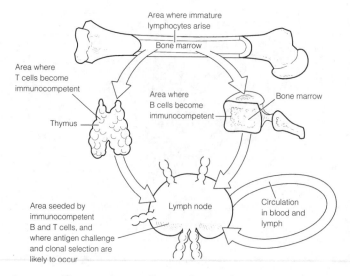

1. The appearance of antigen-specific receptors on the membrane of the lymphocyte.   2. Fetal life.   3. Its genes.
4. Binding to "its" antigen.   5. "Self."

23.

Characteristic	T cell	B cell
Originates in bone marrow from stem cells called hemocytoblasts	√	√
Progeny are plasma cells		√
Progeny include regulatory, helper, and cytotoxic cells	√	
Progeny include memory cells	√	√
Is responsible for directly attacking foreign cells or virus-infected cells	√	
Produces antibodies that are released to body fluids		√
Bears a cell-surface receptor capable of recognizing a specific antigen	√	√
Forms clones upon stimulation	√	√
Accounts for most of the lymphocytes in the circulation	√	

24. 1. Cytokines.   2. Hapten.   3. Liver.

25. **Figure 12–6:**

1. The V portion.
2. The C portion.

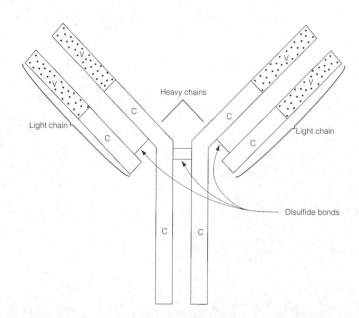

26. 1. B or IgD.   2. D or IgG.   3. E or IgM.   4. D or IgG, E or IgM.   5. E or IgM.   6. D or IgG.
7. C or IgE.   8. A or IgA.

27. 1. Antigen.   2. Complement activation and lysis.   3. Neutralization.   4. Agglutination.   5. IgM.
6. Precipitation.   7. Phagocytes.

28. 1. A.   2.–4. P.   5. A.   6. A.

29. 1. P.   2. P.   3. S.   4. P.   5. S.

30. 1. A or helper T cell.   2. A or helper T cell.   3. C or regulatory T cell.   4. B or cytotoxic T cell.   5. D or memory cell.

31. 1. G or interferon.   2. C or chemotaxis factors.   3. B or antibodies.   4. F or inflammation.
5. E or cytokines.   6. D or complement.   7. E or cytokines.

32. 1. Allografts; an unrelated person.   2. Cytotoxic (killer) T cells and macrophages.   3. To prevent rejection, the recipient's immune system must be suppressed. The patient is unprotected from foreign antigens, and bacterial or viral infection is a common cause of death.

33. Figure 12–7:

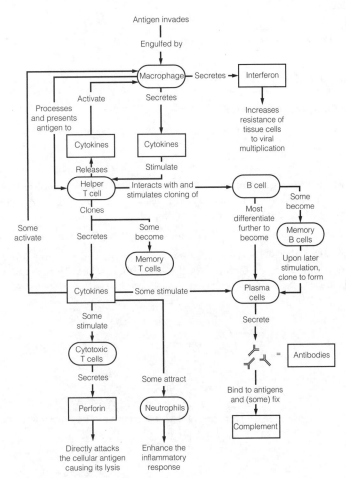

34. 1. C or immunodeficiency.   2. A or allergy.   3. A or allergy.   4. C or immunodeficiency.   5. B or autoimmune disease.   6. C or immunodeficiency.   7. B or autoimmune disease.   8. A or allergy.   9. A or allergy.

Developmental Aspects of the Lymphatic System and Body Defenses

35. 1. Veins.   2. Thymus.   3. Spleen.   4. Thymic.   5. Liver.   6. Lymphatic organs.   7. Birth (or shortly thereafter).   8. Declines.   9.–11. (in any order): Immunodeficiencies, autoimmune diseases, cancer.   12. IgA.

Incredible Journey

36. 1. Protein.   2. Lymph node.   3. B lymphocytes (B cells).   4. Plasma cell.   5. Antibodies.   6. Macrophage.
7. Antigens.   8. Antigen presenters.   9. T.   10. Clone.   11. Immunologic memory.

At the Clinic

37. Anaphylactic shock (histamine caused bodywide loss of fluid from the bloodstream); epinephrine injections.

38. Contact dermatitis (delayed hypersensitivity) probably caused by a reaction to the chemicals in the detergent used to launder the diapers.

39. James is suffering from AIDS.

40. She has the classic signs of hypothyroidism (probably due to neck trauma) and she appears to be exhibiting an autoimmune reaction to formerly "hidden antigens" in the thyroid gland colloid.

41. Hemorrhage; the spleen is a blood reservoir. No; the liver, bone marrow, and other tissues can take over the spleen's functions.

42. The acidity of the vaginal tract inhibits bacterial growth. Hence, anything that decreases vaginal acidity provides an opportunity for bacterial proliferation and vaginal inflammation.

43. Lymphedema or swelling caused by an accumulation of tissue fluid (lymph) in the area. No, the lymphatic vessels will eventually be replaced by budding from the veins in the area.

44. Most likely increased (or increasing) because it is the plasma cells that are the main source of antibodies.

45. Lipid-soluble because it enters the body through the skin cells.

The Finale: Multiple Choice

46. 1. C, D.   2. A, B, C, D.   3. A.   4. A, B, C.   5. A, B, D.   6. A, B, D.   7. B, D.   8. A, B, C.   9. A, C, D.
10. C.   11. B, C, D.   12. B, D.   13. B, C, D.   14. C.   15. C, D.   16. D.   17. A.   18. A.

Chapter 13   The Respiratory System

Functional Anatomy of the Respiratory System

1. 1. Nose, pharynx, larynx, trachea, bronchi and smaller branches.   2. To conduct air to the respiratory zone. 3. Alveoli.

2. 1. R.   2. L.   3. R.

3. 1. External nares or nostrils.   2. Nasal septum.   3.–5. (in any order): Warm, moisten, cleanse.   6. Paranasal sinuses.   7. Speech.   8. Pharynx.   9. Larynx.   10. Tonsils.   11. Cartilage.   12. Pressure.   13. Anteriorly. 14. Thyroid.   15. Vocal folds or true vocal cords.   16. Speak.

4. 1. Mandibular.   2. Alveolus.   3. Larynx.   4. Peritonitis.   5. Nasopharynx.   6. Main bronchus.

5. **Figure 13–1:** In color coding, the pharynx includes the nasopharynx, oropharynx, and laryngopharynx. The larynx runs from the laryngopharynx through the vocal folds to the trachea. The paranasal sinuses include the frontal and sphenoidal sinuses.

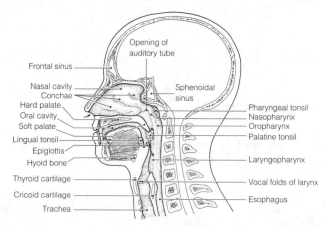

6. 1. B or bronchioles.   2. G or palate.   3. I or phrenic.   4. E or esophagus.   5. D or epiglottis.   6. K or trachea. 7. A or alveoli.   8. H or parietal pleura.   9. L or visceral pleura.   10. F or glottis.   11. C or conchae.   12. M or vocal cords.

7. 1. Elastic connective.   2. Gas exchange.   3. Surfactant.   4. Reduce the surface tension.

8. **Figure 13–2:** 1. Provides a patent airway; serves as a switching mechanism to route food into the posterior esophagus; acts in voice production (contains vocal folds).   2. Elastic.   3. Hyaline.   4. The epiglottis has to be flexible to be able to flap over the glottis during swallowing. The more rigid hyaline cartilages support the walls of the larynx.   5. Adam's apple.

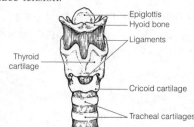

9. **Figure 13–3:**

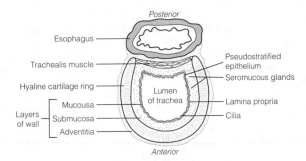

1. Prevents the airway from collapsing during the pressure changes that occur during breathing.
2. Allows eosphagus wall to bulge anteriorly when a large food bolus is being swallowed.
3. Contraction of the trachealis muscle reduces the diameter of the trachea, causng the air to rush superiorly and with greater force. Helps to clear mucus from the airway during coughing.

10. **Figure 13–4:**

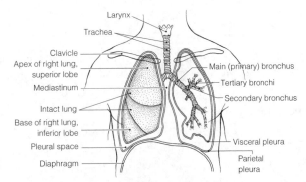

11. **Figure 13–5:** The intact alveoli are the saclike structures resembling grapes in part A; these should be colored yellow. The small vessels that appear to be spider webbing over their outer surface are the pulmonary capillaries. O_2 should be written inside the alveolar chamber and its arrow should move from the alveolus into the capillary. CO_2 should be written within the capillary and its arrow shown going from the capillary into the alveolar chamber.

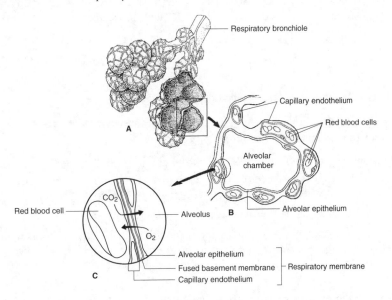

Respiratory Physiology

12. 1. C or intrapleural pressure.   2. A or atmospheric pressure.   3. and 4. B or intrapulmonary pressure.
    5. C or intrapleural pressure.   6. B or intrapulmonary pressure.   7. B or intrapulmonary pressure.

13. When the diaphragm contracts, the internal volume of the thorax increases, the internal pressure in the thorax decreases, the size of the lungs increases, and the direction of airflow is into the lungs. When the diaphragm relaxes, the internal volume of the thorax decreases, the internal pressure in the thorax increases, the size of the lungs decreases, and the direction of airflow is out of the lungs.

14. 1. C or inspiration.   2. D or internal respiration.   3. E or ventilation.   4. A or external respiration.   5. B or expiration.

15. 1. Transversus abdominis and external and internal obliques.   2. Internal intercostals and latissimus dorsi.

16. 1. Hiccup.   2. Cough.   3. Sneeze.   4. Yawn.

17. 1. E or tidal volume.   2. A or dead space volume.   3. F or vital capacity.   4. D or residual volume.
5. B or expiratory reserve volume.

18. Figure 13–6:

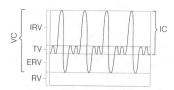

19. 1. F.   2. G.   3. H.   4. B.   5. E.   6. J.   7. D.   8. C.   9. I.

20. 1. Hemoglobin.   2. Bicarbonate ions.   3. Plasma.   4. Oxygen.

21. 1. Acidosis.   2. ↑pH.   3. Hyperventilation.   4. ↑Oxygen.   5. ↑CO_2 in blood.   6. ↑PCO_2.

Respiratory Disorders

22. 1. A or apnea.   2. F or eupnea.   3. D or dyspnea.   4. G or hypoxia.   5. E or emphysema.   6. C or chronic bronchitis.   7. B or asthma.   8. C or chronic bronchitis, E or emphysema.   9. H or lung cancer.
10. I or tuberculosis.

Developmental Aspects of the Respiratory System

23. 1. Infant respiratory distress syndrome.   2. Surfactant.   3. Lower the surface tension of the watery film in the alveolar sacs.   4. It keeps the lungs inflated so that gas exchange can continue.

24. 1. 40.   2. 12–18.   3. Asthma.   4. Chronic bronchitis.   5. Emphysema or TB.   6. Elasticity.
7. Vital capacity.   8. Respiratory infections, particularly pneumonia.

Incredible Journey

25. 1. Nasal conchae.   2. Pharyngeal tonsils.   3. Nasopharynx.   4. Mucus.   5. Vocal fold.   6. Larynx.
7. Digestive.   8. Epiglottis.   9. Trachea.   10. Cilia.   11. Throat (pharynx).   12. Main bronchi.   13. Left.
14. Bronchiole.   15. Alveolus.   16. Red blood cells.   17. Red.   18. Oxygen.   19. Carbon dioxide.
20. Cough.

At the Clinic

26. Pleurisy.

27. Michael most likely is suffering from carbon monoxide poisoning.

28. Sudden infant death syndrome (SIDS).

29. Chronic bronchitis; smoking inhibits ciliary action.

30. Atelectasis. The lungs are in separate pleural cavities, so only the left lung will collapse.

31. The mucus secreted by the respiratory mucosa will be abnormally thick and difficult to clear. As a result, respiratory passages will become blocked with mucus, which favors respiratory infections.

32. The pharyngeal tonsils, which lie at the dorsal aspect of the nasal cavity.

33. 1. The mucus increases the thickness of the respiratory membrane, impairing the efficiency of gas diffusion and exchange.   2. One gas in cigarette smoke is carbon monoxide, which competes with oxygen for binding sites on hemoglobin. Also smoking paralyzes the cilia, increasing the patient's risk of passageway obstruction by mucus and infection.

34. Shallow breaths flush air out of dead space (areas where the air does not participate in gas exchange). A deeper breath is more likely to contain air containing alcohol that is vaporizing from the blood into the alveoli.

35. Both sets of cilia move the mucus toward the esophagus where it can be swallowed. This prevents dust and germ-laden mucus from pooling in the lungs.

The Finale: Multiple Choice

36. 1. B, D.   2. B, C, D.   3. A.   4. B.   5. D.   6. A.   7. D.   8. B.   9. A.   10. B.   11. B, C, D.
12. B, C.   13. D.   14. B.   15. B, C, D.   16. C.

Chapter 14   The Digestive System and Body Metabolism

Anatomy of the Digestive System

1. 1. Oral cavity.   2. Digestion.   3. Blood.   4. Eliminated or excreted.   5. Feces.   6. Alimentary canal or GI tract.   7. Accessory.

2. **Figure 14–1:** The ascending, transverse, descending, and sigmoid colon are all part of the large intestine. The parotid, sublingual, and submandibular glands are salivary glands.

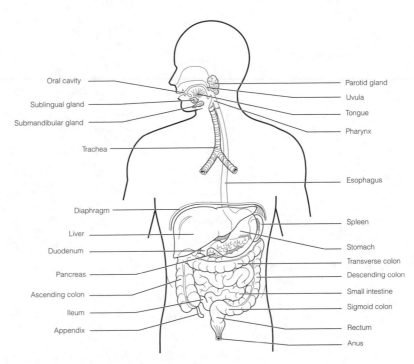

3. **Figure 14–2:** Color the frenulum red; the soft palate blue; the tonsils yellow; and the tongue pink.

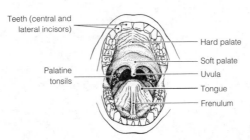

4. 1. B or intestinal glands.   2. E or salivary glands.   3. D or pancreas.   4. C or liver.   5. A or gastric glands.

5. 1. J or mesentery.   2. X or villi.   3. N or Peyer's patches.   4. P or plicae circulares.   5. L or oral cavity, U or stomach.   6. V or tongue.   7. O or pharynx.   8. E or greater omentum, I or lesser omentum, J or mesentery.   9. D or esophagus.   10. R or rugae.   11. G or haustra.   12. K or microvilli.   13. H or ileocecal valve.   14. S or small intestine.   15. C or colon.   16. W or vestibule.   17. B or appendix.   18. U or stomach.   19. I or lesser omentum.   20. S or small intestine.   21. Q or pyloric sphincter.   22. T or soft palate.   23. S or small intestine.   24. M or parietal peritoneum.   25. A or anal canal.   26. F or hard palate.   27. Y or visceral peritoneum.

6. 1. Esophagus.   2. Rugae.   3. Gallbladder.   4. Cecum.   5. Circular folds.   6. Frenulum.   7. Palatine.   8. Saliva.   9. Protein absorption.

7. **Figure 14–3:** On part B, the parietal cells should be colored red, the mucous neck cells yellow, and the chief cells blue.

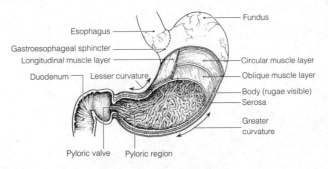

8. **Figure 14–4:**

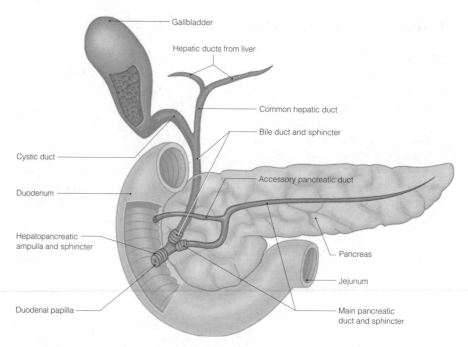

- Gallbladder
- Hepatic ducts from liver
- Common hepatic duct
- Bile duct and sphincter
- Accessory pancreatic duct
- Cystic duct
- Duodenum
- Pancreas
- Jejunum
- Hepatopancreatic ampulla and sphincter
- Duodenal papilla
- Main pancreatic duct and sphincter

9. **Figure 14–5:** 1. Mucosa.   2. Muscularis externa.   3. Submucosa.   4. Serosa.

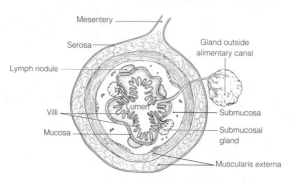

- Mesentery
- Serosa
- Lymph nodule
- Gland outside alimentary canal
- Villi
- Lumen
- Submucosa
- Mucosa
- Submucosal gland
- Muscularis externa

10. **Figure 14–6:**

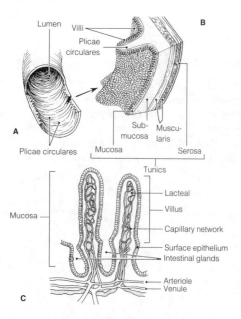

- Lumen
- Villi
- Plicae circulares
- B
- Plicae circulares
- A
- Sub-mucosa
- Muscu-laris
- Mucosa
- Serosa
- Tunics
- Lacteal
- Villus
- Capillary network
- Mucosa
- Surface epithelium
- Intestinal glands
- Arteriole
- Venule
- C

11. **Figure 14–7:**

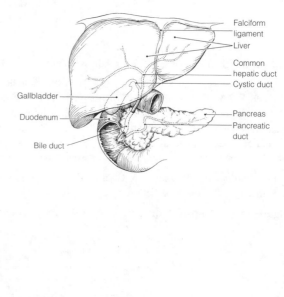

- Falciform ligament
- Liver
- Common hepatic duct
- Cystic duct
- Gallbladder
- Duodenum
- Pancreas
- Pancreatic duct
- Bile duct

12. 1. Deciduous.   2. 6 months.   3. 6 years.   4. Permanent.   5. 32.   6. 20.   7. Incisors.   8. Canine.
9. Premolars.   10. Molars.   11. Wisdom.

13. Figure 14–8: 1. A   2. B.   3. E.   4. C.

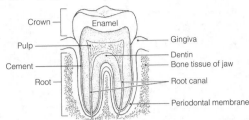

Physiology of the Digestive System

14. 1. D or eating.   2. G or swallowing, H or segmentation and peristalsis.   3. E or chewing, F or churning.
4. B or enzymatic breakdown.   5. A or transport of nutrients from lumen to blood.   6. C or elimination of feces.

15. 1. G or peritonitis.   2. E or heartburn.   3. F or jaundice.   4. H or ulcer.   5. C or diarrhea.   6. D or gallstones.
7. B or constipation.

16. 1. O or salivary amylase.   2. G or hormonal stimulus.   3. M or psychological stimulus.   4. I or mechanical
stimulus.   5. L or pepsin.   6. F or HCl.   7. K or mucus.   8. N or rennin.   9. E or churning.   10. C or brush
border enzymes.   11. A or bicarbonate-rich fluid.   12. H or lipases.   13. B or bile.

17. 1. A or cholecystokinin, C or secretin.   2. B or gastrin.   3. A or cholecystokinin.   4. C or secretin.

18. 1. C or fructose, D or galactose, E or glucose.   2. F or lactose, G or maltose, I or sucrose.   3. A or amino acids.
4. B or fatty acids.   5. E or glucose.

19. 1. P.   2. A.   3. A.   4. P.   5. A.   Circle fatty acids.

20. 1. Deglutition.   2. Buccal.   3. Pharyngeal-esophageal.   4. Tongue.   5. Uvula.   6. Larynx.   7. Epiglottis.
8. Peristalsis.   9. Cardioesophageal.   10. and 11. Peristalsis, segmental.   12. Segmental.   13. Mass movement.
14. Rectum.   15. Defecation.   16. Emetic.   17. Vomiting.

Nutrition and Metabolism

21. 1. B or carbohydrates.   2. C or fats.   3. A or amino acids.   4. C or fats.   5. C or fats.   6. A or amino acids.

22. 1. A or bread/pasta, D or fruits, H or vegetables.   2. B or cheese/cream.   3. G or starch.   4. C or cellulose.
5. B or cheese/cream, E or meat/fish.   6. I or vitamins.   7. F or minerals.

23. Figure 14–9:

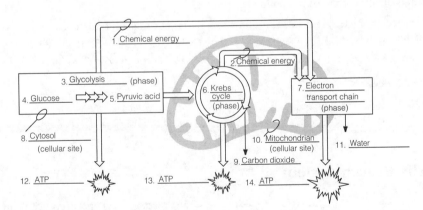

1. Glycolysis (#3) does not require oxygen.   2. Krebs cycle (#6) and the electron transport chains (#7) require oxygen.   3. In the form of hydrogen atoms bearing high-energy electrons.   4. and 5. The electron transport chain.

24. 1. K or glucose.   2. O or oxygen.   3. R or water.   4. H or carbon dioxide.   5. A or ATP.   6. N or monosaccharides.   7. and 8. (in any order): C or acetoacetic acid; D or acetone.   9. M or ketosis.   10. I or essential.
11. F or ammonia.   12. Q or urea.

25. 1. TMR.   2. ↓ Metabolic rate.   3. Child.   4. Fats.   5. Vasoconstriction.

26. 1. Albumin.   2. Clotting proteins.   3. Cholesterol.   4. Hyperglycemia.   5. Glycogen.   6. Hypoglycemia.
7. Glycogenolysis.   8. Gluconeogenesis.   9. Detoxification.   10. Phagocytic.   11. Lipoproteins.   12. Insoluble.
13. LDLs.   14. Membranes.   15. Steroid hormones.   16. Liver.   17. Bile salts.   18. Atherosclerosis.
19. A.   20. Iron.

27. 1. D or heat.   2. B or constriction of skin blood vessels, K or shivering.   3. A or blood.   4. F or hypothalamus.
5. J or pyrogens.   6. C or frostbite.   7. H or perspiration, I or radiation.   8. G or hypothermia.
9. E or hyperthermia.

Developmental Aspects of the Digestive System

28. 1. B or alimentary canal.   2. A or accessory organs.   3. D or cleft palate/lip.   4. N or tracheoesophageal fistula.
5. E or cystic fibrosis.   6. H or PKU.   7. K or rooting.   8. M or stomach.   9. C or appendicitis.
10. G or gastritis, O or ulcers.   11. I or periodontal disease.

Incredible Journey

29. 1. Mucosa.   2. Vestibule.   3. Tongue.   4. Salivary amylase.   5. Peristalsis.   6. Esophagus.   7. Larynx.
8. Epiglottis.   9. Stomach.   10. Mucus.   11. Pepsin.   12. Hydrochloric acid.   13. Pyloric.   14. Lipase.
15. Pancreas.   16. Villi.   17. Ileocecal.

At the Clinic

30. Many vegetables contain incomplete proteins. Unless complete proteins are ingested, the value of the dietary protein for anabolism is lost because the amino acids will be oxidized for energy. Beans and grains.

31. Heartburn due to a hiatal hernia; esophagitis and esophageal ulcers.

32. Heat exhaustion; they should drink a "sports drink" containing electrolytes or lemonade to replace lost fluids.

33. Bert has heat stroke. Heavy work in an environment that restricts heat loss results in a spiraling upward of body temperature and cessation of thermoregulation. Bert should be immersed in cool water immediately to bring his temperature down and avert brain damage.

34. Diverticula are small herniations of the mucosa through the colon walls, a condition called diverticulosis. They are believed to form when the diet lacks bulk and the volume of residue in the colon is small. The colon narrows, and contractions of its circular muscles become more powerful, increasing the pressure on its walls. Diverticulitis is a painful condition in which the diverticula become inflamed. This woman has diverticulitis caused by the inflammation of her diverticula.

35. Lack of lactase (lactose intolerance); add lactase drops to milk before drinking it.

36. Examination of the blood plasma would quickly reveal the presence of lipid breakdown products at above-fasting levels.

37. Yo-yo dieting causes dramatic drops in metabolic rate and causes the enzyme that unloads fats from the blood (to be stored in fat deposits) to become much more efficient. Furthermore, if the individual doesn't exercise, when he or she is not dieting, excess calories are stored as fat rather than being built into muscle, or being used to sustain the higher metabolic rate of muscle tissue.

38. Iron. She has hemorrhagic anemia compounded by iron loss.

39. Appendicitis is caused by bacterial infection. If untreated, bacterial proliferation may cause the appendix to rupture, resulting in contamination of the peritoneal cavity with feces and life-threatening peritonitis.

40. Fat-soluble vitamins (A, D, E, etc.) because these are absorbed as fat breakdown products are absorbed.

The Finale: Multiple Choice

41. 1. A, C, D.   2. B.   3. C.   4. D.   5. A, B, C, D.   6. C.   7. C.   8. D.   9. A, B.   10. A, C, D.   11. C.
12. D.   13. B.   14. B, D.   15. D.   16. A, B, C, D.   17. A, C.   18. A, B.   19. A, B, C, D.   20. B, D.
21. B, D.   22. A.   23. B, C, D.   24. A.   25. A, B, C.   26. C.   27. D.   28. D.

Chapter 15   The Urinary System

1. 1. Nitrogenous.   2. Water.   3. Acid-base.   4. Kidneys.   5. Ureters.   6. Peristalsis.   7. Urinary bladder.
8. Urethra.   9. 8.   10. $1\frac{1}{2}$.

Kidneys

2. Figure 15–1:

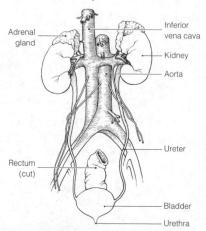

3. Figure 15–2: The fibrous membrane surrounding the kidney is the *fibrous capsule*; the basin-like *pelvis* is continuous with the ureter; a *calyx* is an extension of the pelvis; *renal columns* are extensions of cortical tissue into the medulla. The *cortex* contains the bulk of the nephron structures; the striped-appearing *medullary pyramids* are primarily formed by collecting ducts.

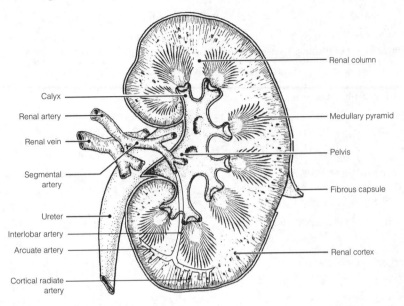

4. 1. Intraperitoneal.   2. Urethra.   3. Glomerulus.   4. Glomerulus.   5. Collecting duct.   6. Cortical nephrons.
7. Collecting duct.   8. Glomeruli.   9. Nephron loop.

5. Figure 15–3: 1. Glomerular capsule.   2. Afferent arteriole.   3. Efferent arteriole.   4. Cortical radiate artery.
5. Cortical radiate vein.   6. Arcuate artery.   7. Arcuate vein.   8. Interlobar artery.   9. Interlobar vein.
10. Nephron loop.   11. Collecting duct.   12. Distal convoluted tubule.   13. Proximal convoluted tubule.
14. Peritubular capillaries.   15. Glomerulus. Relative to the coloring instructions, #1 is green, #15 is red, #14 is blue, #11 is yellow, and #13 is orange.

6. Figure 15–4: 1. Black arrows: Site of filtrate formation is the glomerulus. Arrows leave the glomerulus and enter glomerular (Bowman's) capsule.   2. Red arrows: Major site of amino acid and glucose reabsorption. Shown going *from* the PCT interior and passing through the PCT walls to the capillary bed surrounding the PCT (the latter not shown). Nutrients *leave* the filtrate.   3. Green arrows: At site of ADH action. Arrows (indicating water movement) shown *leaving* the interior of the collecting duct and passing through the walls to enter the capillary bed surrounding that duct. Water *leaves* the filtrate.   4. Yellow arrows: Site of aldosterone action. Arrows (indicating Na^+ movement) *leaving* the collecting duct and the DCT and passing through their walls into the surrounding capillary bed. Na^+ *leaves* the filtrate.   5. Blue arrows: Site of tubular secretion. Arrows shown *entering* the PCT to enter the filtrate.

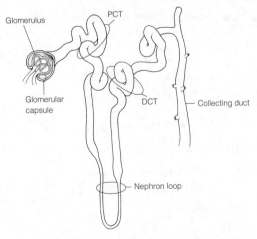

7. 1. Afferent.   2. Efferent.   3. Blood plasma.   4. and 5. Diffusion; active transport.   6. Microvilli.   7. Secretion.
8.–10. Diet, cellular metabolism, urine output.   11. 1–1.8.   12. Urochrome.   13.–15. Urea, uric acid, creatinine.
16. Lungs.   17. Evaporation of perspiration.   18. Decreases.   19. Dialysis.

8. 1. A.   2. B.   3. A.   4. A.   5. B.

9. 1. D.   2. D.   3. I.   4. D.   5. I.   6. I.

10. 1. L. 2. G. 3. G. 4. A. 5. L. 6. A. 7. G. 8. G. 9. G. 10. A. 11. G. 12. L.

11. 1. Hematuria; bleeding in urinary tract. 2. Ketonuria; diabetes mellitus, starvation. 3. Albuminuria; glomerulonephritis, pregnancy. 4. Pyuria; urinary tract infection. 5. Bilirubinuria; liver disease. 6. (No official terminology); kidney stones. 7. Glycosuria; diabetes mellitus.

12. 1. All reabsorbed by tubule cells. 2. Usually does not pass through the glomerular filter.

13. 1. Chemical buffering; response in less than 1 second. 2. Adjustment in respiratory rate and depth to regulate CO_2 levels; response in minutes. 3. Regulation by kidneys; response in hours to days.

14. 1. Male adult. 2. Lean adult. 3. Intracellular fluid. 4. Nonelectrolyte. 5. ↑ADH. 6. ↑K^+ reabsorption.

Ureters, Urinary Bladder, and Urethra

15. 1. Kidney. 2. Forms urine. 3. Continuous with renal pelvis. 4. Female

16. 1. B or urethra. 2. A or bladder. 3. A or bladder. 4. B or urethra. 5. B or urethra; C or ureter. 6. B or urethra. 7. C or ureter. 8. A or bladder; C or ureter. 9. B or urethra.

17. 1. Micturition. 2. Stretch receptors. 3. Contract. 4. Internal urethral. 5. External urethral. 6. Voluntarily. 7. About 600. 8. Incontinence. 9. Infants, toddlers. 10. and 11. Emotional/neural problems; Pressure (pregnancy). 12. Urinary retention. 13. Prostate.

18. 1. A or cystitis. 2. C or hydronephrosis. 3. F or uremia. 4. E or pyelonephritis. 5. B or diabetes insipidus. 6. D or ptosis.

Fluid, Electrolyte, and Acid-Base Balance

19. 1. N. 2. E. 3. E. 4. E. 5. N. 6. E.

20. 1. Aldosterone. 2. Secretion. 3. ↑K^+ reabsorption. 4. ↓BP. 5. ↓K^+ retention. 6. ↑HCO_3^- in urine. 7. Dilute urine. 8. ↓BP.

21. Figure 15–5: 1. *T*. 2. Hydrostatic pressure. 3. D. 4. Lymphatic vessels. 5. Tissue cell. 6. Plasma.

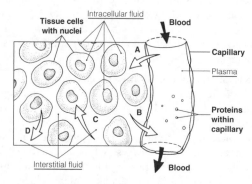

22. Most water (60%) comes from ingested fluids. Other sources are moist foods and cellular metabolism.

23. The greatest water loss (60%) is from excretion of urine. Other routes are as water vapor in air expired from lungs, through the skin in perspiration, and in feces. Insensible water loss is water loss of which we are unaware. This type continually occurs via evaporation from skin and in water vapor that is expired from the lungs. It is uncontrollable.

24. 1. E. 2. F. 3. C. 4. B. 5. A.

25. 1. B. 2. C. 3. E. 4. D.

26. 1. H^+ and HCO_3^- are ions. The others are molecules. 2. H_2CO_3 is a weak acid. HCO_2^- is a weak base. 3. Right.

Developmental Aspects of the Urinary System

27. 1. Placenta. 2. Polycystic disease. 3. Hypospadias. 4. Males. 5. Bladder. 6. 18–24. 7. Glomerulonephritis. 8. Antigen-antibody. 9. and 10. Proteins; Blood. 11. Arteriosclerosis. 12. Tubule. 13. and 14. Urgency; Frequency.

Incredible Journey

28. 1. Tubule. 2. Renal. 3. Afferent. 4. Glomerulus. 5. Glomerular capsule. 6. Plasma. 7. Proteins. 8. Nephron loop. 9. Microvilli. 10. Reabsorption. 11. and 12. Glucose, amino acids. 13. 7.4 (7.35–7.45). 14. Nitrogenous. 15. Sodium. 16. Potassium. 17. Urochrome. 18. Antidiuretic hormone. 19. Collecting duct. 20. Pelvis. 21. Peristalsis. 22. Urine. 23. Micturition. 24. Urethra.

At the Clinic

29. Anuria; renal dialysis.

30. Perhaps Eddie is a very heavy sleeper and is thus unresponsive to the "urge" to urinate.

31. High sodium content and copious urine volume (although the glucocorticoids can partially take over the role of aldosterone).

32. People who are under prolonged stress activate hypothalamic centers that regulate stress by controlling the release of ACTH by the anterior pituitary. Release of ACTH by the anterior pituitary in turn causes both catecholamines and corticosteroids to be released by the adrenal glands to counteract the stressor by raising blood pressure and blood sugar levels. The elevated blood pressure explains his headache.

33. The alcohol interferes with the action of ADH, which causes the kidneys to retain water. Hence, excessive body water is being lost in urine.

34. Mrs. Rodriques is in a diabetic coma due to lack of insulin. Her blood is acidic, and her respiratory system is attempting to compensate by blowing off carbon dioxide (hence, the elevated breathing rate). Her kidneys are reabsorbing bicarbonate.

35. The test will check for the presence of proteins in the person's urine, which is a symptom of kidney disease. More importantly, the urine test checks for the presence of drugs in the urine.

36. Hypertension would be the major symptom.

The Finale: Multiple Choice

37. 1. C, D.   2. A.   3. D.   4. B, C, D.   5. C, D.   6. C.   7. D.   8. B.   9. A, D.   10. D.   11. A.
12. A, C, D.   13. A, C.   14. A, B, D.   15. C, D.   16. A, B, C.   17. D.   18. C.   19. B, D.   20. B.
21. A, B, C, D.   22. A, B, C, D.   23. A, B, D.   24. C, D.

Chapter 16   The Reproductive System

Anatomy of the Male Reproductive System

1. Seminiferous tubule → Rete testis → Epididymis → Ductus deferens.

2. When body temperature (or external temperature) is high the scrotal muscles relax, allowing the testes to hang lower and farther away from the warmth of the body wall. This causes testicular temperature to drop. When the external temperature is cold, the scrotal muscles contract to draw the testes closer to the warmth of the body wall.

3. 1. E or penis.   2. K or testes.   3. C ductus deferens.   4. L or urethra.   5. A or bulbo-urethral glands, G or prostate, H or seminal vesicles, K or testes.   6. I or scrotum.   7. B or epididymis.   8. F or prepuce.
9. G or prostate.   10. H or seminal vesicles.   11. A or bulbo-urethral glands.   12. J or spermatic cord.

4. **Figure 16–1:** The spongy tissue is the erectile tissue in the penis; the duct that also serves the urinary system is the urethra; the structure providing ideal temperature conditions is the scrotum; the prepuce is removed at circumcision; the glands producing a secretion that contains sugar are the seminal vesicles; the ductus deferens is cut or cauterized during vasectomy.

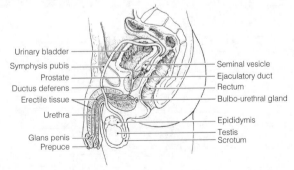

5. **Figure 16–2:** The site of spermatogenesis is the seminiferous tubule. Sperm mature in the epididymis. The fibrous coat is the tunica albuginea.

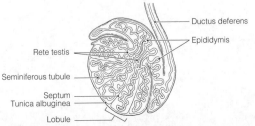

Male Reproductive Functions

6. 1. D or spermatogonium.   2. C or secondary spermatocyte, E or sperm, F or spermatid.   3. C or secondary spermatocyte.   4. F or spermatid.   5. E or sperm.   6. A or FSH, G or testosterone.

Figure 16–3:

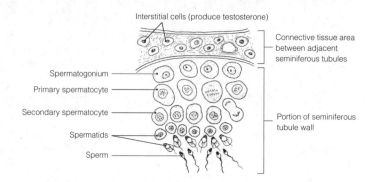

7. Figure 16–4:

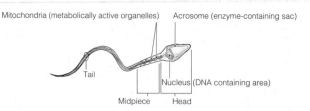

8. 1. A or mitosis.   2. B or meiosis.   3. C or both mitosis and meiosis.   4. A or mitosis.   5. B or meiosis.
6. A or mitosis.   7. A or mitosis.   8. B or meiosis.   9. C or both mitosis and meiosis.   10. B or meiosis.
11. B or meiosis.

9. Deepening voice; formation of a beard and increased hair growth all over body, particularly in axillary/genital regions; enlargement of skeletal muscles; increased density of skeleton.

Anatomy of the Female Reproductive System

10. 1. Uterus.   2. Vagina.   3. Uterine, or fallopian, tube.   4. Clitoris.   5. Uterine tube.   6. Hymen.
7. Ovary.   8. Fimbriae.

11. Figure 16–5: The egg travels along the uterine tube after it is released from the ovary. The round ligament helps to anchor the uterus. The ovary produces hormones and gametes. The homologue of the male scrotum is the labium majus.

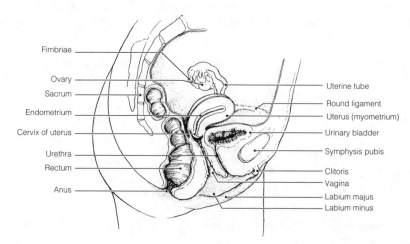

12. **Figure 16–6:** The clitoris should be colored blue; the hymen yellow; and the vaginal opening red.

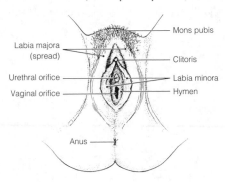

Female Reproductive Functions and Cycles

13. 1. B or primary oocyte.   2. C or secondary oocyte.   3. C or secondary oocyte.   4. D or ovum.

14. The follicle (granulosa) cells produce estrogen, the corpus luteum produces progesterone, and oocytes are the central cells in all follicles. Event A = ovulation.   1. No.   2. Peritoneal cavity.   3. After sperm penetration occurs. 4. Ruptured (ovulated) follicle.   5. One ovum; three polar bodies.   6. Males produce four spermatids → four sperm.   7. They deteriorate.   8. They lack nutrient-containing cytoplasm.   9. Menopause.

    Figure 16–7:

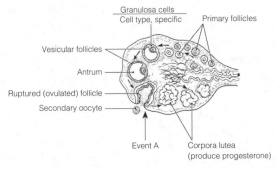

15. Because of this structural condition, many "eggs" (oocytes) are lost in the peritoneal cavity; therefore, they are unavailable for fertilization. The discontinuity also provides infectious microorganisms with access to the peritoneal cavity, possibly leading to PID.

16. 1. Follicle-stimulating hormone (FSH).   2. Luteinizing hormone (LH).   3. Estrogen and progesterone. 4. Estrogen.   5. LH.   6. LH.

17. Appearance of axillary/pubic hair, development of breasts, widening of pelvis, onset of menses.

18. 1. A or estrogens, B or progesterone.   2. B or progesterone.   3. A or estrogens.   4. B or progesterone. 5. and 6. A or estrogens.

19. **Figure 16–8:** From left to right on part C the structures are the primary follicle, the secondary (growing) follicle, the vesicular follicle, the ovulating follicle, the corpus luteum, and an atretic (deteriorating) corpus luteum. In part D, menses is from day 0 to day 4, the proliferative phase is from day 4 to day 14, and the secretory phase is from day 14 to day 28.

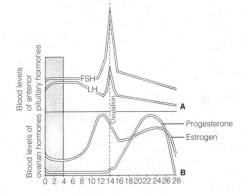

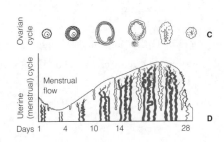

Mammary Glands

20. Figure 16–9: The alveolar glands should be colored blue, and the rest of the internal breast, excluding the duct system, should be colored yellow.

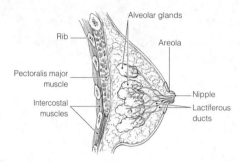

Survey of Pregnancy and Embryonic Development

21. 1. Just its head (the nucleus).   2. Digests away the cement holding the follicle cells together; allows sperm to reach the oocyte.

22. Figure 16–10: 1. Fertilization (sperm penetration).   2. Fertilized egg (zygote).   3. Cleavage.   4. Blastocyst (chorionic vesicle).   5. Implantation.   6. The polar body has virtually no cytoplasm. Without nutrients it would be unable to live until it reached the uterus.

23. 1. H or zygote.   2. F or placenta.   3. B or chorionic villi, C or endometrium.   4. A or amnion.   5. G or umbilical cord.   6. B or chorionic villi.   7. E or fetus.   8. F or placenta.   9. D or fertilization.

24. The blastocyst and then the placenta release hCG, which is like LH and sustains the function of the corpus luteum temporarily until the placenta can take over.

25. 1. B or mesoderm.   2. C or endoderm.   3. A or ectoderm.   4. B or mesoderm.   5. A or ectoderm.   6. B or mesoderm.   7. C or endoderm.   8. C or endoderm.

26. Oxytocin and prostaglandins.

27. 1. Prolactin.   2. Oxytocin.

28. Check 1, 3, 5, 9, 10, 11, 12.

29. False labor (irregular, ineffective uterine contractions). These occur because rising estrogen levels make the uterus more responsive to oxytocin and antagonize progesterone's quieting influence on the myometrium.

30. 1. Dilation stage: The period from the beginning of labor until full dilation (approx. 10-cm diameter) of the cervix; the longest phase.   2. Expulsion stage: The period from full dilation to the birth (delivery).   3. Placental stage: Delivery of the placenta, which follows delivery of the infant.

31. Figure 16–11:

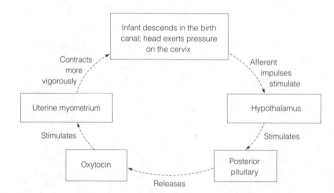

32. Each pass forces the baby farther into the birth passage. The cycle ends with the birth of the baby.

33. The response to the stimulus enhances the stimulus. For example, the more a baby descends into the pelvis and stretches the uterus, the more oxytocin is produced and the stronger the contractions become.

Developmental Aspects of the Reproductive System

34. 1. Y and X.   2. 2 Xs.   3. Male external genitalia and accessory structures.   4. Female external genitalia and duct system.   5. Cryptorchidism.   6.–8. (in any order) *Escherichia coli,* STDs or venereal disease, yeast infections. 9. PID (pelvic inflammatory disease).   10. Venereal disease microorganisms.   11. Breast.   12. Cervix of the uterus.   13. Pap smear.   14. Menopause.   15. Hot flashes.   16. Declines.   17. Rise.   18. Estrogen. 19. Vaginal.   20. Prostate.   21. and 22. Urinary; reproductive.

Incredible Journey

35. 1. Uterus.   2. Ovary.   3. Fimbriae.   4. Ovulation.   5. Secondary oocyte.   6. Follicle.   7. Peristalsis.   8. Cilia. 9. Sperm.   10. Acrosomes.   11. Meiotic.   12. Ovum.   13. Polar body.   14. Dead.   15. Fertilization. 16. Zygote (fertilized egg).   17. Cleavage.   18. Endometrium.   19. Implantation.   20. Vagina.

At the Clinic

36. Pitocin will act on the placenta, stimulating production and release of prostaglandins. Pitocin and prostaglandins are powerful uterine muscle stimulants. Oxytocin normally causes frequent and vigorous contractions of the uterine wall.

37. Megadoses of testosterone would inhibit anterior pituitary gonadotropin (FSH) release. Spermatogenesis is inhibited in the absence of FSH stimulation.

38. Her tubes were probably scarred by PID. Hormonal testing and the daily basal temperature recordings would have indicated her anovulatory condition.

39. His scrotal muscles had contracted to draw the testes closer to the warmth of the abdominal cavity.

40. Mary's fetus might have respiratory problems or even congenital defects caused by her smoking, because smoking causes vasoconstriction, which would hinder blood delivery to the placenta.

41. Cervical cancer.

42. There is little possibility that she is right. Body organs are laid down during the first trimester, and only growth and final differentiation occur after that.

43. Sexually transmitted diseases (STDs). It is important to inform his partner(s) that they might be infected also, particularly because some females do not exhibit any signs or symptoms of these particular infections but still need to be treated.

44. By the surgical procedure called a C-section (cesarian section).

45. These hormones exert negative feedback on the release of GnRH by the hypothalamus. This, in turn, would interfere with pituitary release of LH, thus interfering with ovulation.

46. Both procedures prevent the germ cells from reaching their normal destination during intercourse. Tubal ligation cuts through the fallopian tubes, whereas vasectomy interrupts the continuity of the vas deferens. Hence, these make the recipients sterile.

The Finale: Multiple Choice

47. 1. A, B.   2. B, C.   3. C.   4. D.   5. D.   6. B.   7. B.   8. B.   9. D.   10. C, D.   11. A, D.   12. B. 13. A, B.   14. A, D.   15. A, B, D.   16. A, C.   17. C.   18. C.   19. C.   20. A, B, C.   21. B.   22. A, C. 23. C.   24. A, C.   25. D.   26. B, D.   27. A, B, D.   28. B.   29. A, D.